Power System Fundamentals

Power System Fundamentals

Pedro Ponce • Arturo Molina • Omar Mata
Luis Ibarra • Brian MacCleery

CRC Press
Taylor & Francis Group
Boca Raton London New York

CRC Press is an imprint of the
Taylor & Francis Group, an **informa** business

CRC Press
Taylor & Francis Group
6000 Broken Sound Parkway NW, Suite 300
Boca Raton, FL 33487-2742

First issued in paperback 2021

Version Date: 20171114

ISBN 13: 978-1-03-224187-6 (pbk)
ISBN 13: 978-1-138-55443-6 (hbk)

DOI: 10.1201/9781315148991

Publisher's Note
The publisher has gone to great lengths to ensure the quality of this reprint but points out that some imperfections in the original copies may be apparent.

Visit the Taylor & Francis Web site at
http://www.taylorandfrancis.com

and the CRC Press Web site at
http://www.crcpress.com

To my mother Margarita, my children Pedro and Jamie, and my wife, Norma – ***Pedro***

To my lovely wife Silvia and children Julio and Monse – ***Arturo***

To my mother Fabiola, and my father Omar – ***Omar***

To my wife, Zyan Zavaleta – ***Luis***

To Eva Jane MacCleery – ***Brian***

Para Eva MacCleery, mi media naranja – ***Brian MacCleery***

Contents

Preface

Power systems are changing in the world because the conventional electric grid systems cannot face ecological and technological demands from the end-user of the electric grid. However, control systems, communication technologies, digital systems, and power electronics are opening new electric grids called smart grids. The Smart Grid has advanced engineering concepts that improve the efficiency inside the smart grid but at present there are new challenges that need to be solved; for instance, the real-time communication to meet customer electrical demand in the smart grid system.

This book shows the fundamentals of power systems that are the basic elements for understanding the smart grid technology, based on theoretical and experimental challenges in which the conventional and advanced electric grids have to be understood in undergraduate and graduate courses. Moreover, this book also describes the technological knowledge that has to be understood by the new generations of electrical engineers when they are dealing with power systems and smart grid technologies.

Although conventional electrical energy is a well-known engineering area, there are topics that have to be presented in a different manner because new technologies have arisen such as power electronics and digital systems. On the other hand, smart grid is a new concept for integrating several areas such as renewable energy, transmission lines, conventional electric sources, communications, SCADA systems, and so on. Smart grid will bring new possibilities to solve generation, distribution, and consumption problems of power systems. On the other hand, one of the main challenges is to generate engineers who understand power systems, so it is important to present in a clear and friendly manner the principles that drive modern power systems. As a result, this book illustrates step by step the topics that are needed to cover power systems, an introduction to smart grid courses.

The content of the book moves from electric circuits to renewable electric energy in order to connect the concepts that are needed; the book shows theoretical and experimental exercises so the reader can compare the simulation, co-simulation, experimental, and theoretical results. For additional material, visit https://www.crcpress.com/9781138554436

Preface

Acknowledgments

This book is a product of the Project 266632 "Laboratorio Binacional para la Gestión Inteligente de la Sustentabilidad Energética y la Formación Tecnológica" ["Bi-National Laboratory on Smart Sustainable Energy Management and Technology Training"], funded by the CONACYT SENER Fund for Energy Sustainability (Agreement: S0019-2014-01).

Also, we thank Tecnológico de Monterrey and National Instruments which always provide us with support in our research projects.

Finally, we thank the people who helped in the consummation of this project:

Antonio Rosales, Arturo Soriano, Eduardo Torres, Iván Villanueva, and David Balderas
Tecnologico de Monterrey
Mexico City, Mexico

The extraordinary editorial team of Taylor & Francis Group.

1

Linear Electric Circuits

1.1 Passive elements in electric circuits

An active element is defined as an element that is capable of furnishing an average power greater than zero to a particular device. However, a passive element cannot supply an average power that is greater than zero over an infinite time interval.

The resistor is the simplest passive element and is based on the statement of the fundamental relationship called Ohm's law. It states that the voltage across conducting materials is directly proportional to the current flowing through it,

$$v = Ri, \tag{1.1}$$

where the constant of proportionality R is called resistance. Its unit is called ohm, which is 1 $[V/A]$, and it is abbreviated by the Greek letter Ω.

Equation (1.1) is a linear equation; therefore it is considered to be the definition of a linear resistor. The linear resistor is an idealized circuit element; the voltage-current ratios of these physical devices are constant only within certain ranges of current, voltage, or power, and also depend on temperature and other external factors.

Resistors can be connected in series or parallel so the total resistance can be calculated according to the following expressions: series connection (R = R1 + R2 + R3 + ...), parallel connection (1 / R = 1 / R1 + 1 / R2 + 1 / R3 +...). Figure 1.1 shows a series connection.

Another passive element used in electric circuits is the capacitor. The capacitance C is defined as the following voltage-current relationship

$$i = C\frac{dv}{dt}, \tag{1.2}$$

where v and i are functions of time that satisfy the conventions for a passive element. Therefore the unit of capacitances is an ampere-second per volt, or coulomb per volt. So a *farad* $[F]$ is defined as one coulomb per volt and used as unit of capacitance.

A capacitor consists of two conducting surfaces separated by a thin insulating material with a very large resistance, in which electric charge may be stored.

Some of the characteristics of the capacitor are obtained from Equation (1.2). A constant voltage across a capacitor gives zero current passing through it; thus it can be seen as an "open circuit to DC voltage." Another characteristic comes to notice in the case of a sudden jump in the voltage; an infinite current is required, and since this is impossible, an instant change of voltage across a capacitor is prohibited.

The capacitor voltage may be expressed in terms of current by integrating the equation and solving for $v(t)$ (1.2)

$$\begin{aligned} dv &= \frac{1}{C} i(t) dt \\ v(t) &= \frac{1}{C} \int_{t_0}^{t} i(t)\, dt. \end{aligned} \tag{1.3}$$

In real problems, the initial voltage across the capacitor often cannot be discerned. In this case it is convenient to set $t_0 = -\infty$ and $v(-\infty) = 0$, so that

$$v(t) = \frac{1}{C} \int_{-\infty}^{t} i(c)\, dt'. \tag{1.4}$$

The last passive element is the inductor, and as the capacitor, its inductance L is defined by a voltage-current equation:

$$v = L \frac{di}{dt}. \tag{1.5}$$

The unit in which inductance is measured is the henry $[H]$. The operating principle of the inductor began with Michael Faraday, when he discovered that a changing magnetic field could induce a voltage, proving that this voltage was proportional to the time rate of change of current producing the magnetic field.

Equation (1.5) is the mathematical model of an ideal inductor. A physical inductor can be constructed by winding a length of wire into a coil. Some electrical characteristics can be obtained through this equation. First it is shown that the voltage across an inductor is proportional to the time rate of change of the current through it, therefore for a constant current there is no voltage across the inductor, so it can be viewed as a "short circuit to DC." Another fact is that a sudden or discontinuous change in the current must be associated with an infinite voltage across the inductor, but in reality an infinite voltage does not exists.

The inductor current may be expressed in terms of voltage, solving Equation (1.5) for $i(t)$:

$$\begin{aligned} di &= \frac{1}{L} v dt \\ i(t) &= \frac{1}{L} \int_{-\infty}^{t} v dt' + t(t_0). \end{aligned} \tag{1.6}$$

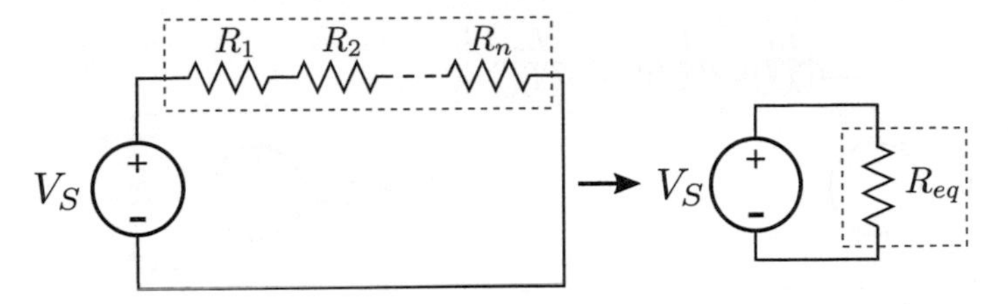

FIGURE 1.1
Series combination of N resistors

Solving a realistic problem, as was done for the capacitor, t_0 is considered $-\infty$ to ensure no current or energy is in the inductor:

$$i(t) = \frac{1}{L}\int_{-\infty}^{t} v dt'. \tag{1.7}$$

1.2 Type of connection of passive elements

When working with multiple elements of the same kind in a circuit, it is possible to replace complicated element combinations with a single equivalent element of the same kind. This technique is useful when the current, voltage, or power associated to each element is not needed; instead the voltage, current, or power of the whole circuit is of interest. There are two types of connections in which passive elements can be combined.

1.2.1 Series combination

The simplification or combination of multiple elements into an equivalent one can be applied only to one kind of passive element at the same time. Consider the series combination of N resistors as shown in Figure 1.1.

To simplify the circuit of N resistors in series, the equivalent resistor is obtained as the algebraic sum of all the resistors:

$$R_{eq} = R_1 + R_2 + \cdots + R_N. \tag{1.8}$$

All resistors share the same current, so the current in the equivalent resistor must be the same. On the other hand, the voltage of the source is divided in each of the resistors, so the voltage in the equivalent resistor must be the sum of all individual resistor voltages.

Inductors connected in series can also be combined into one equivalent inductor as shown in Figure 1.2.

In the same way an equivalent resistor is obtained, the equivalent inductor is equal to the sum of each inductor in the series connection:

$$L_{eq} = L_1 + L_2 + \cdots + L_N. \tag{1.9}$$

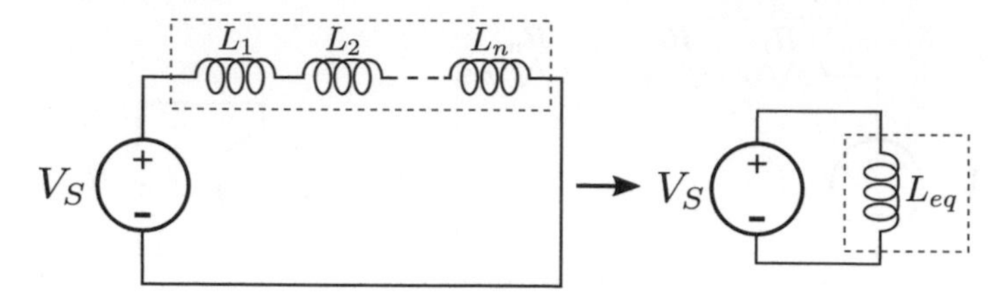

FIGURE 1.2
Series combination of N inductors

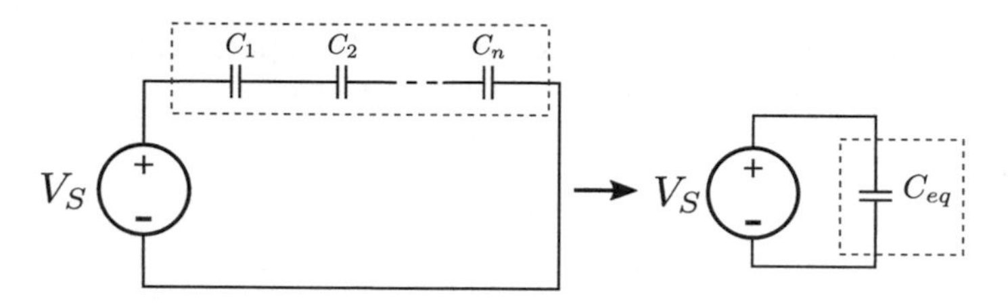

FIGURE 1.3
Series combination of N capacitors

In case the capacitors are connected in series (see Figure 1.3), the way to obtain its equivalent value is not the same as for the previous passive elements.

The equivalent capacitor for a series connection is obtained as the inverse sum of each capacitor inversed:

$$\frac{1}{C_{eq}} = \frac{1}{C_1} + \frac{1}{C_2} + \cdots + \frac{1}{C_N}. \tag{1.10}$$

1.2.2 Parallel combination

A parallel connection between the same kinds of passive elements is defined when two or more elements share two common nodes. Consider the parallel combination of N resistors as shown in Figure 1.4.

To simplify the circuit into one equivalent resistor, it must be noticed that the sum of all currents of each resistor must be equal to the current given by

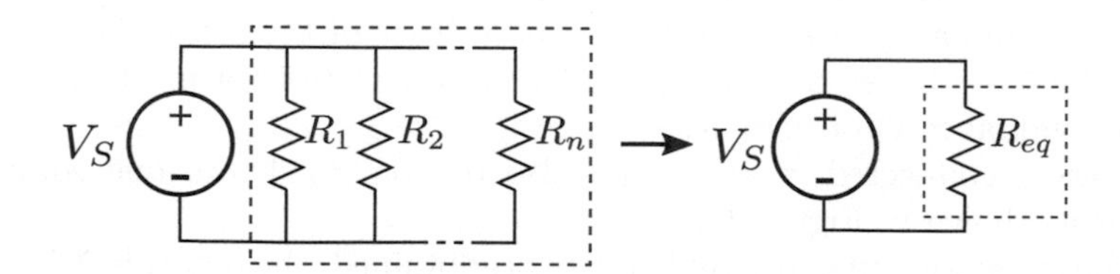

FIGURE 1.4
Parallel combination of N resistors

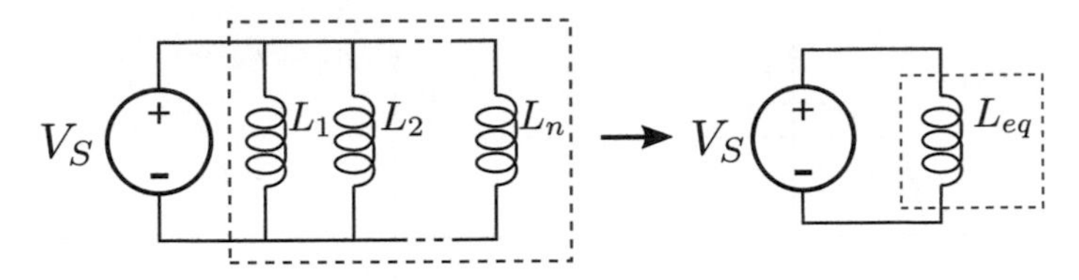

FIGURE 1.5
Parallel combination of N inductors

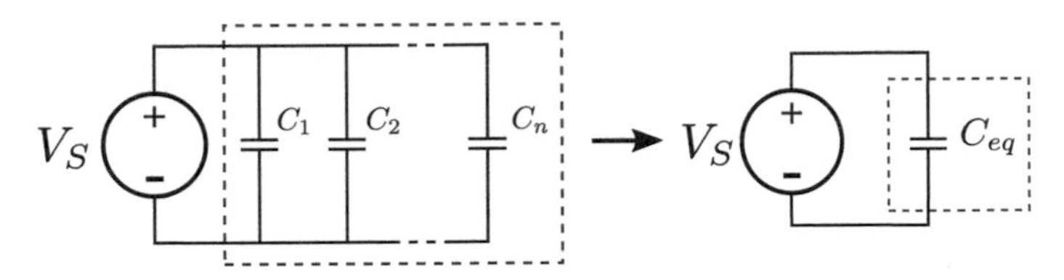

FIGURE 1.6
Parallel combination of N capacitors

the source, and substituting with ohm's law, this can be obtained:

$$\frac{1}{R_{eq}} = \frac{1}{R_1} + \frac{1}{R_2} + \cdots + \frac{1}{R_N}. \tag{1.11}$$

For inductor elements, the equivalent circuit is obtained the same way as for the resistors.

The equation of the equivalent inductor, represented in Figure 1.5, for a parallel connection is

$$\frac{1}{L_{eq}} = \frac{1}{L_1} + \frac{1}{L_2} + \cdots + \frac{1}{L_N}. \tag{1.12}$$

Finally, for N capacitors in a parallel combination as shown in Figure 1.6, the equation of the equivalent capacitor is not the same as the last two, but the algebraic sum of the capacitors in parallel.

The equation of the equivalent capacitor for a parallel connection is

$$C_{eq} = C_1 + C_2 + \cdots + C_N. \tag{1.13}$$

1.3 Frequency-domain analysis

1.3.1 Phasors

A sinusoidal signal like voltage or current at a given frequency is characterized by two parameters, amplitude and a phase angle. Consider the sinusoidal

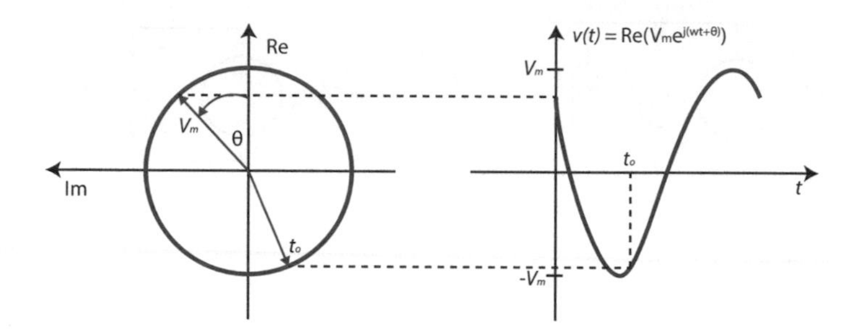

FIGURE 1.7
Frequency domain representation and time domain representation

voltage

$$v(t) = V_m \cos(\omega t + \theta), \tag{1.14}$$

where Vm is the amplitude of the sinusoid, ω the angular frequency in $[rad/s]$, and θ is the phase in $[rad]$.

Equation (1.14) can also be expressed as the real part of a complex quantity accordingly to Euler's identity

$$v(t) = Re\{V_m e^{j(\omega t+\theta)}\}. \tag{1.15}$$

Thus, a phasor is a complex number that represents the amplitude and phase of a sinusoid, the factor $e^{j\omega t}$ contains no useful information, and the voltage can be simplified to the expression

$$V = V_m e^{j\theta}. \tag{1.16}$$

Since a complex number can also be represented in a polar form, the phasor is generally represented as

$$V = V_m \angle \theta. \tag{1.17}$$

To differentiate a complex number from one that is not, it will be printed as boldface type.

This process is called a phasor transformation, in which a function is taken from a time-domain representation to a frequency-domain representation. It is noted that the frequency-domain expression of a voltage or a current does not implicitly include the frequency, as the time-domain expression does.

To obtain the sinusoid corresponding to a given phasor V, it has to be multiplied by the time factor $e^{j\omega t}$ and just the real part is taken; see Figure 1.7.

The importance of the phasor in electrical engineering lies in the fact that it is possible to define algebraic relationships between the voltage and current for each of the three passive elements instead of using differential equations [80]. The simplest case to begin with is the resistor, where

$$v(t) = Ri(t).$$

Applying complex voltage and current so that

$$V_m e^{j(\omega t+\theta)} = RI_m e^{j(\omega t+\theta)},$$

and dividing by $e^{j\omega t}$, the phasor relation for a resistor is found:

$$V_m e^{j\theta} = RI_m e^{j\theta} \rightarrow V = RI. \tag{1.18}$$

The second passive element is the inductor. Its defining equation in a time-domain is

$$v(t) = L\frac{di(t)}{dt}.$$

It can be shown that the derivate of the voltage or the current is transformed to the phasor domain as $j\omega X$, thus the complex expression for the inductor is

$$V_m e^{j(\omega t+\theta)} = j\omega LI_m e^{j(\omega t+\theta)},$$

and dividing by $e^{j\omega t}$, the phasor relation for an inductor is found:

$$V_m e^{j\theta} = j\omega LI_m e^{j\omega} \Rightarrow V = j\omega LI. \tag{1.19}$$

Finally, the time-domain current voltage relationship for the capacitor is

$$i(t) = C\frac{dv(t)}{dt},$$

and doing the same procedure as for the inductor, the phasor relation for a capacitor is

$$I = j\omega CV. \tag{1.20}$$

1.3.2 Impedances

An impedance can be defined as the ratio of the phasor voltage to the phasor current that represents the opposition that the circuit exhibits to the flow of sinusoidal current. Therefore an impedance can be treated, in a circuit analysis, as a resistance with the exception that the impedance is a complex quantity and its algebraic manipulation must be the one for a complex number [27].

Previously, the phasor relation for the three passive elements, Equations (1.5–1.7), were obtained,

$$V = RI,\ V = j\omega LI,\ V = \frac{I}{j\omega C}.$$

Writing Equations (1.5–1.7) in terms of the ratio of the phasor voltage to the phasor current, the impedance of each element is obtained:

$$\frac{V}{I} = Z_R = R,\ \frac{V}{I} = Z_L = j\omega L,\ \frac{V}{I} = Z_C = \frac{1}{j\omega C}.$$

The impedance is symbolized by the letter Z and has dimensions of ohms $[\Omega]$. It is important to note that the impedance is not a phasor, because it is not a sinusoidal varying quantity; hence it cannot be transformed to the time domain by multiplying by $e^{j\omega t}$ and taking the real part. Instead, these three passive elements can be seen as resistance, inductance, and capacitance in the time domain and in the frequency domain as their respective impedances R, $j\omega L$, $1/j\omega C$.

Since the impedance can be treated as a resistance, the combinations in serial or parallel of impedances have the same rules established for the resistances. The equivalent impedance of a series combination is

$$Z_{eq} = Z_1 + Z_2 + \cdots + Z_n, \tag{1.21}$$

and the equivalent impedance of a parallel combination is

$$\frac{1}{Z_{eq}} = \frac{1}{Z_1} + \frac{1}{Z_2} + \cdots + \frac{1}{Z_n}. \tag{1.22}$$

1.3.3 Phasor diagrams

Previously, the voltage and current phasors were introduced as a complex number that represents the magnitude and phase of the sinusoidal waveform for either the voltage or current. Hence, a phasor behaves as a vector whose graphical representation in the complex plane is known as a phasor diagram.

Let's take a complex voltage in its rectangular form, $V_1 = 4 + j3$ $[V]$. The desired voltage is located in the complex plane by an arrow drawn from the origin to a point marked in the x axis with the real part of the voltage and in the y axis with the imaginary part of the voltage. Another way to represent the same voltage is to transform the complex voltage into its polar form, $V_1 = 5\angle 36.87°$ $[V]$, so it can be represented as a vector of magnitude 5 and an angle of 36.87° counterclockwise with respect to the x axis, as shown in Figure 1.8.

Since a phasor behaves as a vector, it is compliant with all the linear algebra operations on vectors. Thus, phasors may be easily added or subtracted in a phasor diagram as shown in Figure 1.9, and as for the multiplication and division, may result in the addition and subtraction of angles and change in amplitude [80]

In addition, if both voltage and current are plotted in the same phasor diagram, it is easy to determine which waveform is leading and which is lagging. It is said that a signal is leading another if it is in the counterclockwise direction; however, if it is in the clockwise direction it is instead lagging; see Figure 1.10.

1.3.4 Phasor measurement unit

A phasor measurement unit (PMU) is defined by the IEEE as a device that produces synchronized phasor estimates from voltage and/or current signals at

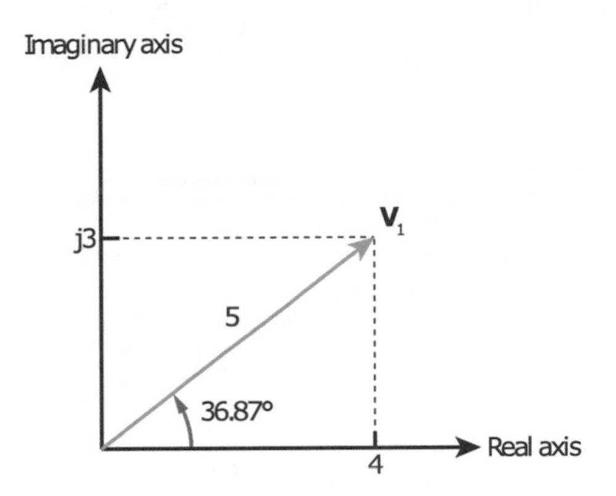

FIGURE 1.8
Phasor diagram for V_1

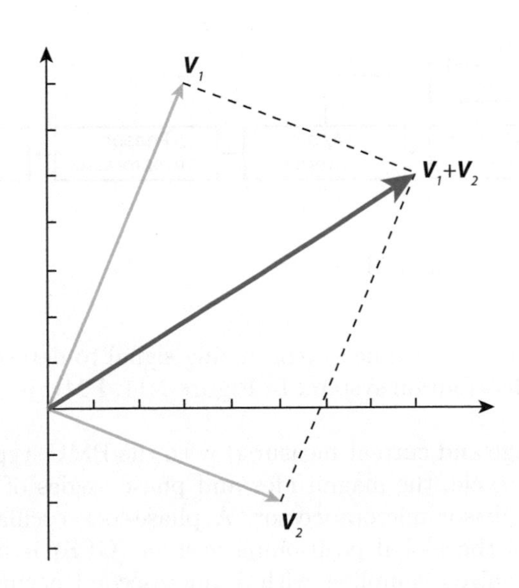

FIGURE 1.9
Phasorial addition of two phasors

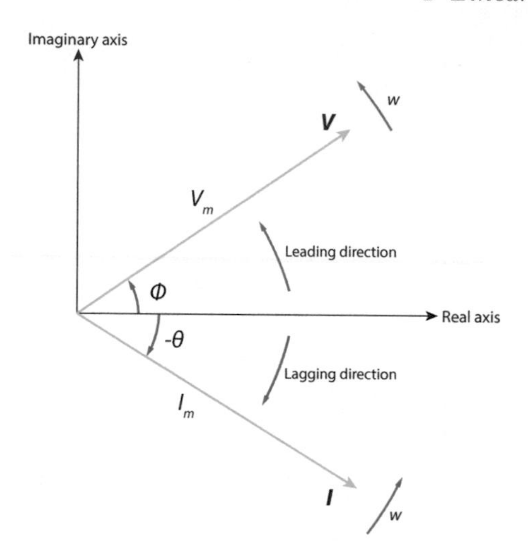

FIGURE 1.10
Leading or lagging direction of phasors

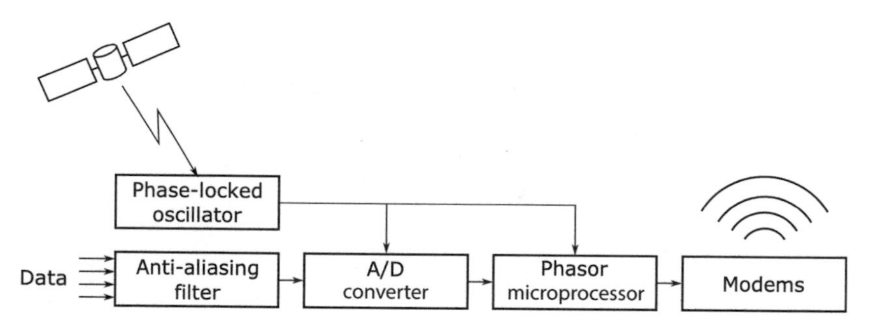

FIGURE 1.11
Phasor measurement unit (PMU)

a high sampling rate and a time synchronizing signal to determine the health of the electricity distribution system. In Figure 1.11, PMU is represented in a block diagram.

From the voltage and current measured with the PMU, typically at a rate of 48 samples per cycle, the magnitudes and phase angles of the signals are calculated in the phasor microprocessor. A phase-lock oscillator along with the clock signal of the global positioning system (GPS) is used to provide high-speed synchronized sampling with 1 microsecond accuracy for adding a timestamp to every measured point. The measured phasors are known as synchrophasors [44].

When data from multiple PMUs are combined together, the information provides a global behavioral view of the interconnection. This made possible some applications like the automation in power systems for smart grids, increasing the reliability of the power grid by detecting faults early, prevention of power outages, increasing power quality, etc.

1.3.5 Experimental exercises

Different types of loads can be found in household appliances. These loads may be resistive, inductive, capacitive, or a combination of them. An example of a resistive load can be any heater or toaster. An inductive load is a device that includes motors or fluorescent lights. While a purely capacitive load is difficult to find, they are useful to counteract inductive loads, as will be reviewed in the next sections.

Recommended equipment and connections:

- Three-phase supply unit
- Maximum demand meter
- Inductive load
- Resistive load
- Capacitive load
- LCR meter

With an LCR meter, measure three values of resistance **R**[**Ω**], capacitance **C**[**F**], and inductance **L**[**H**] from the respective load modules used, and find their impedance **Z** [Ω] (assume a 60 Hz frequency operation).

	R[Ω]	Z_R	C[F]	Z_c[Ω]	L[H]	Z_L[Ω]
1	1074.2	1074.2	2.465μ	-j1076.1	2.636	j993.75
2	761.2	761.2	4.056μ	-j654	1.847	j696.3
3	445.7	445.7	6.294μ	-j421.4	1.158	j436.5

Calculate the current I for the three measured impedances if the voltage is $226\angle 0°$ $[V_{rms}]$. ($V = IZ$).

Feed the three modules individually with 220 $[V_{rms}]$ from the three-phase supply unit and measure the current drawn with the maximum demand meter (see Diagram 1.1). Write down your results.

Draw a phasor diagram and write R, C, L next to the corresponding phasor current according to its nature and your previous calculations. Also write if it is lagging, leading, or in phase with the voltage phasor.

The current phasor of a capacitive load leads its corresponding voltage phasor while the current phasor of an inductive load lags the voltage phasor. In this exercise the 90° shift of the capacitive load and the −90° shift of the inductive load can be noticed.

When the resistive load phasors are in phase, it means that there is no shift between the current phasor and voltage phasor.

	Resistive		Capacitive		Inductive	
	Theory	Practical	Theory	Practical	Theory	Practical
$\mathbf{I}[A]$	$210m\Omega$	$210m\Omega$	$210m\Omega$ $\angle 90°$	$212m\Omega$	$227m\Omega$ $\angle -90°$	$158m\Omega$
$\mathbf{I}[A]$	$297m\Omega$	$298m\Omega$	$345m\Omega$ $\angle 90°$	$345m\Omega$	$324m\Omega$ $\angle -90°$	$226m\Omega$
$\mathbf{I}[A]$	$507m\Omega$	$506m\Omega$	$536m\Omega$ $\angle 90°$	$546m\Omega$	$517m\Omega$ $\angle -90°$	$382m\Omega$

1.4 AC power analysis

1.4.1 Instantaneous power

One of the most important parts of circuit analysis is to calculate either power delivered or power absorbed. First, a power analysis in time-domain will be made to obtain the instantaneous and average power. Then some other power concepts will be introduced to finally get to a frequency-domain analysis where all the relevant power information is gathered.

It is known that the instantaneous power $p(t)$ delivered to any element is the product of the instantaneous voltage $v(t)$ across the element and the instantaneous current $i(t)$ through it,

$$p(t) = v(t)i(t). \tag{1.23}$$

Since an AC power analysis is being made, let the voltage and current at the terminals of the circuit be

$$v(t) = V_m \cos(\omega t + \theta_v),$$
$$i(t) = I_m \cos(\omega t + \theta_i).$$

The instantaneous power absorbed by the circuit in the sinusoidal steady state is

$$p(t) = V_m I_m \cos(\omega t + \theta_v) \cos(\omega t + \theta_i).$$

Applying the trigonometric identity

$$\cos A \cos B = \frac{1}{2}[\cos(A - B) + \cos(A + B)]$$

to $p(t)$, an instantaneous power expression made of two terms is obtained:

$$p(t) = \frac{1}{2} V_m I_m \cos(\theta_v - \theta_i) + \frac{1}{2} V_m I_m \cos(2\omega t + \theta_v + \theta_i). \tag{1.24}$$

The first term of Equation (1.24) is not a function of time, this is constant

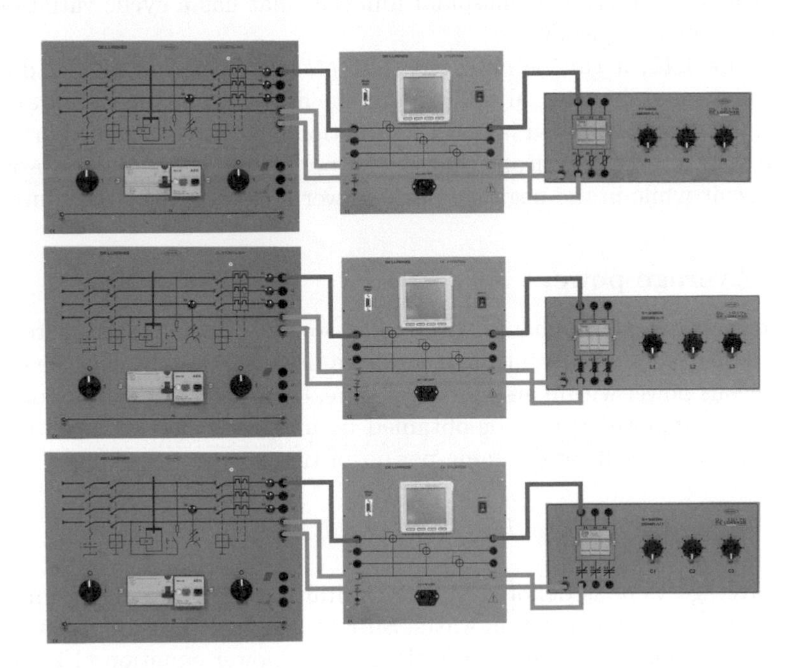

Diagram 1.1
Connections for different loads

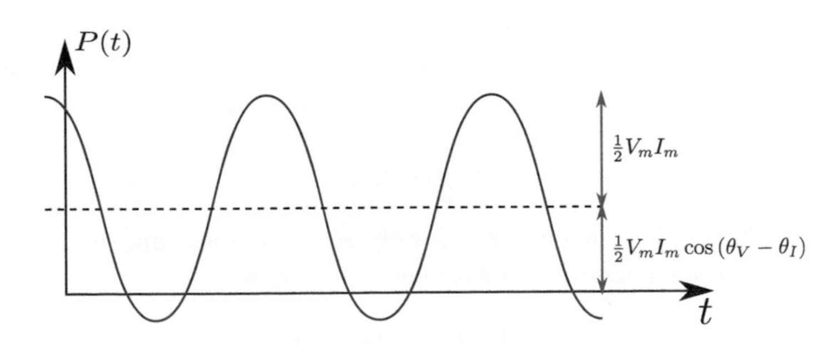

FIGURE 1.12
Instant power function

and depends on the phase difference between voltage and current. The second term of equation (1.24) is a sinusoidal function that has a cyclic variation at twice the applied frequency.

In Figure 1.12, it can be observed that $p(t)$ has a period of $T/2$ due to the second term and a vertical shift due to the first term of the instantaneous power expression. Also noted is that the function is positive in some parts and negative in others, which means that in the positive parts power is absorbed by the circuit while in the negative parts power is absorbed by the source.

1.4.2 Average power

The instantaneous power changes over time, so it is difficult to measure. The average power is more convenient to measure because it is the average of the instantaneous power within an interval of time, the period of the function [27].

The average power P may be obtained by integrating $p(t)$ over its period and dividing the result by the same period of time,

$$P = \frac{1}{T}\int_0^T p(t)dt. \tag{1.25}$$

The average value is denoted by the capital letter P; it is not a function of time and its dimensions are in watts. Substituting the instantaneous power $p(t)$ for a sinusoidal steady state into the average power equation (1.25) gives

$$P = \frac{1}{T}\int_0^T \frac{1}{2}V_m I_m \cos(\theta_v - \theta_i)dt + \frac{1}{T}\int_0^T \frac{1}{2}V_m I_m \cos(2\omega t + \theta_v + \theta_i)dt.$$

Since the first term of P is a constant, independent of t, the average value must be that constant itself. The second term of P is a cosine function whose average value over a period is zero. Thus,

$$P = \frac{1}{2}V_m I_m \cos(\theta_v \theta_i). \tag{1.26}$$

Consider two special cases. When $\theta_v = \theta_i$ the circuit is purely resistive, meaning that a resistive load absorbs power at all times:

$$P = \frac{1}{2}V_m I_m;$$

and when $\theta_v – \theta_i = \pm 90°$ the circuit is purely reactive, meaning that an inductive or capacitive load absorbs no average power at all:

$$P = \frac{1}{2}V_m I_m \cos 90° = 0.$$

1.4.3 Effective values or RMS values

In Mexico, most of the household power outlets available deliver a sinusoidal voltage with a frequency of 60 Hz and a voltage of 127 $[V]$. This voltage can be measured with a voltmeter, and it is certainly not the instantaneous value, because the voltage is not constant, nor is the average voltage, because it must be zero, and neither is the amplitude V_m that appeared in all of the equations mentioned before. So, what kind of value is this 127 $[V]$? It is the effective value of the sinusoidal voltage. An effective value identifies the effectiveness of a voltage or current source in delivering power to a resistive load [27]. For a periodic waveform, either voltage or current, the effective value is the direct current or voltage that delivers the same average power to the resistor as does the periodic waveform. The RMS value for any periodic function $x(t)$ is given by

$$X_{rms} = \sqrt{\frac{1}{2}\int_0^T x^2 dt}. \tag{1.27}$$

Let the sinusoidal voltage to be

$$v(t) = I_m \cos(\omega t + \theta_v),$$

so that the RMS value is

$$V_{rms} = \sqrt{\frac{1}{T}\int_0^T I^2{}_m \cos^2(\omega t + \theta)dt} = \frac{V_m}{\sqrt(2)}. \tag{1.28}$$

Similarly, for $i(t) = I_m \cos(\omega t + \theta_i)$, it can be proven that

$$I_{rms} = \frac{I_m}{\sqrt{2}}. \tag{1.29}$$

Therefore, it can be said that the effective value of a sinusoidal waveform is a real quantity numerically equal to $1/\sqrt{2}$ times the amplitude of the waveform and furthermore is independent of the phase angle. This is only true when the periodic function is sinusoidal.

1.4.4 Apparent power and power factor

Let's take the average power P and put it in terms of effective values:

$$P = \frac{1}{2}V_m I_m \cos(\theta_v - \theta i) = V_{rms} I_{rms} \cos(\theta_v - \theta_i).$$

This expression involves a new term called apparent power:

$$S = V_{rms} I_{rms}. \tag{1.30}$$

The apparent power is the product of the effective values of the voltage and current for sinusoidal waveforms. It is called apparent because of its resemblance to the average power if the voltage and current responses had been DC quantities [27]. It is measured in volt-amperes $[VA]$ to distinguish it from other kinds of power already seen.

The ratio of the average power to the apparent power is called the power factor (PF), so it is the cosine of the phase difference between voltage and current,

$$PF = \frac{P}{S} = \cos(\theta_v - \theta_i). \tag{1.31}$$

The PF may be seen as the factor that relates the apparent power to the average power. Its importance lies in the effectiveness in which power is transferred and therefore it affects the consumption cost of electricity. For a low power factor (≈ 0) a load draws more current than for a high power factor (≈ 1) with the same load, so the utility company may charge more to a customer with a low power factor.

1.4.5 Complex power

So far, all the power quantities have been expressed with real numbers. By making the power a complex quantity, the power relations are made simpler. For the power to become a complex quantity, an imaginary term is added, the reactive power. The complex power became important because it contains all the information pertaining to the power delivered to a given load.

For a voltage $v(t)$ and current $i(t)$ whose phasor representations are $V = V_m \angle \theta_v$ and $I = I_m \angle \theta_i$, the complex power S delivered to a load is

$$S = \frac{1}{2} V I^*. \tag{1.32}$$

In terms of effective values:

$$S = V_{rms} I^*_{rms}. \tag{1.33}$$

Since S is a complex quantity it can also be expressed in rectangular form,

$$S = P + jQ, \tag{1.34}$$

where P is the average power and Q is the new imaginary term, reactive power. The dimensions of Q have to be the same for P and S, but in order to avoid confusion, the unit of Q is defined as volt-ampere-reactive [VAr]. While P is the actual power dissipated by the load, Q is a measure of the energy exchange between the reactive components of the load and the source.

A physical interpretation can be made based on three cases for the reactive power Q [27]; see Figure 1.13.

1. $Q = 0$ represents resistive loads (unity PF)
2. $Q > 0$ represents capacitive loads (leading PF)
3. $Q < 0$ represents inductive loads (lagging PF)

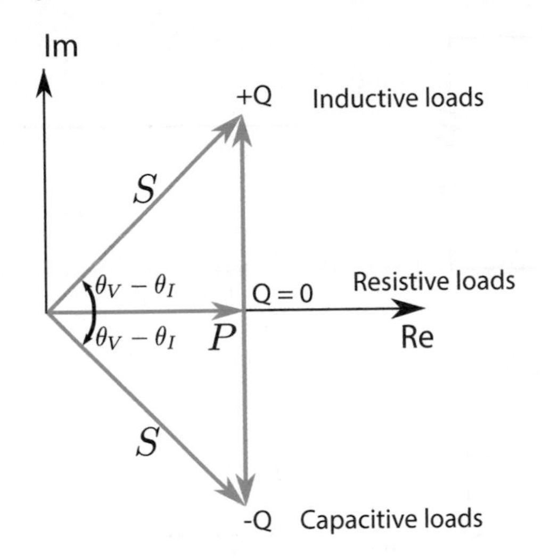

FIGURE 1.13
Power triangle with different reactive powers

It was said before that the complex power contains all the relevant power information in a given load, which is shown in the next recap:

$$\begin{aligned}
\text{Complex power} &= S = V_{rms} I^*_{rms} = P + jQ = V_{rms} I_{rms} \angle(\theta_v - \theta_i) \\
\text{Average power} &= P = Re(s) = S\cos(\theta_v - \theta_i) \\
\text{Reactive power} &= Q = Im(S) = S\sin(\theta_v - \theta_i) \\
\text{Apparent power} &= S = |S| = \sqrt{P^2 + Q^2} = V_{rms} I_{rms} \\
\text{Power factor} &= \frac{P}{S} = \cos(\theta_v - \theta_i)
\end{aligned}$$

1.4.6 Power factor correction

Most of the household appliances are inductive loads that operate at a low lagging power factor. As mentioned before, low power factors draw more current, therefore power companies have to build larger current-carrying capacity generators to provide this larger current to the consumers, making electric power more expensive. Although the inductive nature of the household loads cannot be changed, it is possible to increase the power factor of the system in which they are connected. The process of increasing the power factor without altering the voltage or current to the original load is known as power factor correction [27].

Let's say an inductive circuit is formed by a resistor and an inductor has a voltage V that draws a current I_L. Making a phasor diagram for the inductive load shows that the circuit has a power factor of $\cos(\theta_1)$, as shown in Figure 1.14.

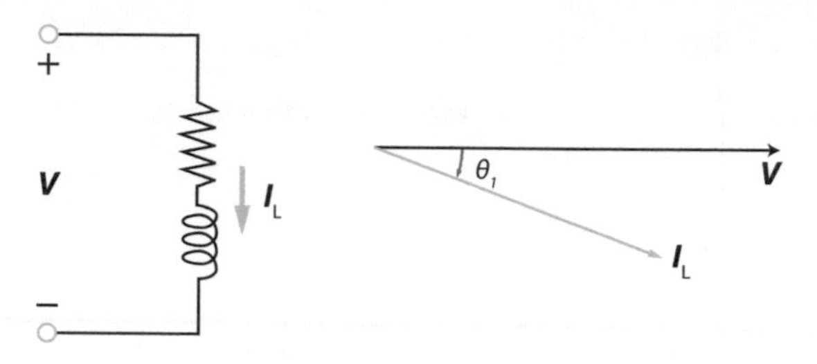

FIGURE 1.14
Circuit with an RL load

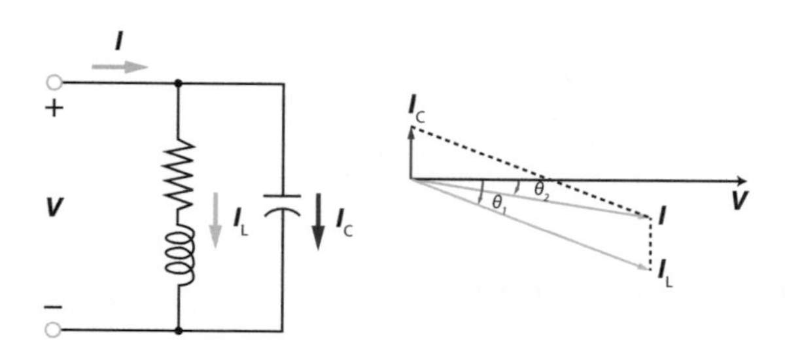

FIGURE 1.15
Circuit with an RL load in parallel with a capacitor

If a capacitor C is added in parallel with the load into the circuit, a new current I_C is generated so that the total current drawn from the generator becomes $I = I_C + I_L$. Again, making the phasor diagram, it can be seen that the power factor is increased by the reduction of the phase angle between the supplied voltage and the current from θ_1 to θ_2; see Figure 1.15.

Let us note that the magnitude of current I is less than the current I_L with the same supplied voltage. So, the reason to make a power factor correction is to reduce the current drawn from the generators, representing savings for both the power companies and the consumer.

To determine the value of the required capacitor for improving the PF of a circuit, it is necessary to take into consideration that the supplied real power has to remain the same.

From the power triangle for an inductive load with an apparent power S_1 (see Figure 1.16),

$$P = S_1 \cos(\theta_1), Q_1 = S_1 \sin(\theta_1) = P \tan(\theta_1).$$

Therefore, the modified reactive power has to be

$$Q_2 = P \tan(\theta_2).$$

So, the reduction in reactive power Q_C is due to a shunt capacitance C

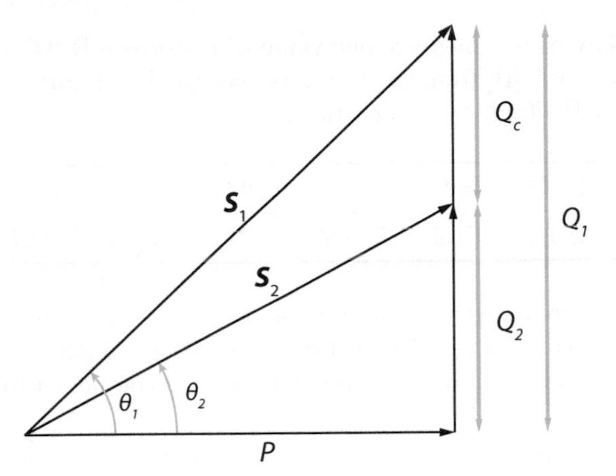

FIGURE 1.16
Power triangle with Q reduction

determined as

$$C = \frac{Q}{\omega V_{rms}^2} = \frac{P(\tan(\theta_1 - \theta_2))}{\omega V_{rms}^2}, \tag{1.35}$$

where P = supplied real power, $\tan(\theta_1)$ = present power factor, $\tan(\theta_2)$ = target power factor, ω = frequency of operation, V_{rms} = effective voltage.

1.4.7 Exercises

Most of the household appliances are a combination of inductive and resistive loads. In this exercise a comparison between an inductive-resistive load and a capacitive-resistive load will be made regarding their power consumption and power factor. Finally, a correction in power factor for the inductive-resistive load will be made to improve the power stats.

Recommended equipment and connections:

- Three-phase supply unit
- Maximum demand meter
- Inductive load
- Resistive load
- Capacitive load
- Switchable capacitor battery
- LCR meter

With an LCR meter, measure one value of resistance R [Ω], capacitance C [F], and inductance L [H] from the load modules used and find their impedance Z (assume a 60 Hz frequency operation).

	R [Ω]	Z_R[Ω]	C [F]	Z_c[Ω]	L [H]	Z_L[Ω]
1	761.2	761.2	6.294μ	-j421.4	2.636	j993.75

For a series combination of the previously selected resistor and capacitor, calculate the current I of the circuit if it has a voltage of $226\angle 0°$ $[V_{rms}]$. Then obtain the total complex power of the load and fill the next table.

Parameter	Number	Unit
Complex power (rectangular)	$51.36 - j28.43$	[VA]
Complex power (polar)	$58.7\angle -28.97°$	[VA]
Apparent power	58.7	[VA]
Average power	51.36	[W]
Reactive power	-28.43	[VAr]
Power factor	0.875	

Connect the previously selected resistive load and capacitive load in series, then feed them with 220 $[V_{rms}]$ from the three-phase supply unit. Measure all the power parameters of the load with the maximum demand meter and compare your results (see Diagram 1.2).

Parameter	Theoretical	Practical
Apparent power	58.7	59
Average power	51.36	51
Reactive power	-28.43	-20
Power factor	0.875	0.87

Now, for a series combination of the previously selected resistor and inductor, calculate the current I of the circuit if it has a voltage of $226\angle 0°$ $[V_{rms}]$. Then obtain the total complex power of the load and fill the next table:

Parameter	Number	Unit
Complex power (rectangular)	$24.81 + j32.39$	VA
Complex power (polar)	$40.8\angle 52.55°$	VA
Apparent power	40.8	VA
Average power	24.81	W
Reactive power	32.39	VAr
Power factor		

Connect the previously selected resistive load and inductive load in series,

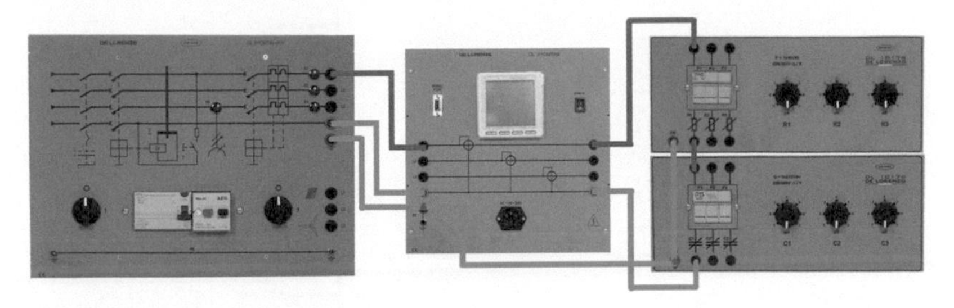

Diagram 1.2
RC load connection

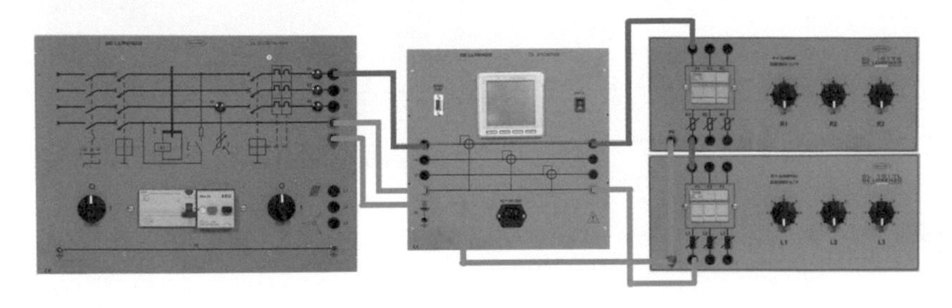

Diagram 1.3
RL load connection

then feed them with 220 [V_{rms}] from the three-phase supply unit. Measure all the power parameters of the load with the maximum demand meter and compare your results (see Diagram 1.3).

Parameter	Number	Unit
Complex power (rectangular)	$24.81 + j32.39$	[VA]
Complex power (polar)	$40.8\angle 52.55°$	[VA]
Apparent power	40.8	[VA]
Average power	24.81	[W]
Reactive power	32.39	[VAr]
Power factor		

For the RL load that is already connected, calculate the value of the capacitor that has to be added in parallel to improve the power factor to 0.97.

$$C = 1.36\ [\mu F].$$

Finally, connect the switchable capacitor battery module in parallel with the RL load (see Diagram 1.4). Activate the first switch from left to right; this will add a 2 μF capacitor to the circuit. Observe the new power factor

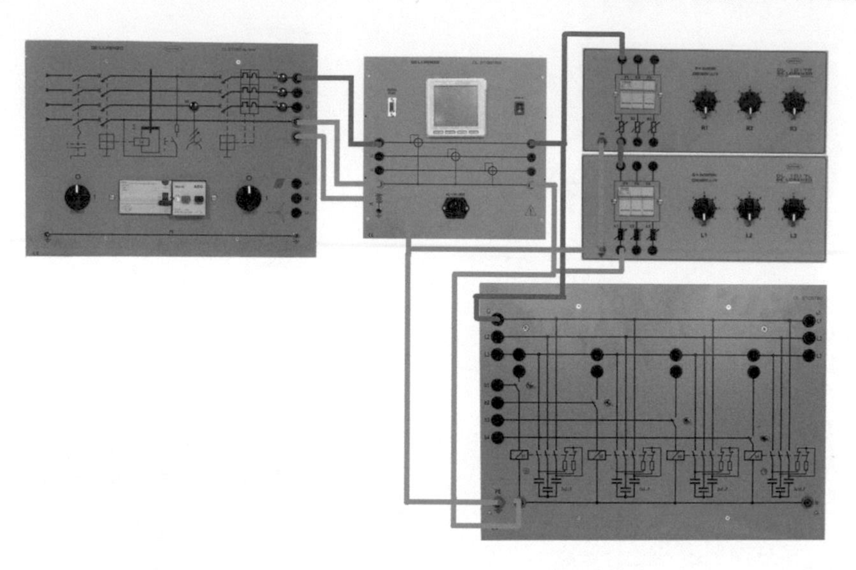

Diagram 1.4
RL load with power factor correction

in the maximum demand meter: is it improved? Activate different capacitors and combinations of them and observe their effect in the power factor.

As seen in Subsection 1.4.6, a capacitor is used to improve the power factor of an inductive load. In this exercise, when the 2 μF capacitor is added, an increase of power factor can be noticed, reaching almost a unitary factor. If the other available capacitors are also activated, the power factor will begin to decrease because the load is now more capacitive.

1.5 Polyphase circuits

So far, every AC analysis made consisted of a generator connected through a pair of wires to a load. A polyphase circuit or system instead refers to multiple AC sources operating at the same frequency but in different phases. The most common polyphase systems are a balanced three-phase system followed by the single-phase three-wire system (although it is called single-phase, actually it is a balanced two-phase system).

Before beginning with the description of the polyphase systems, it is important to know the double-subscript notation. This notation describes the path a current flows or the voltage existent between two points, by two letters in subscript notation. For example, the voltage of point a with respect to point b is V_{ab} while the current that flows from point a to point b is I_{ab} [80].

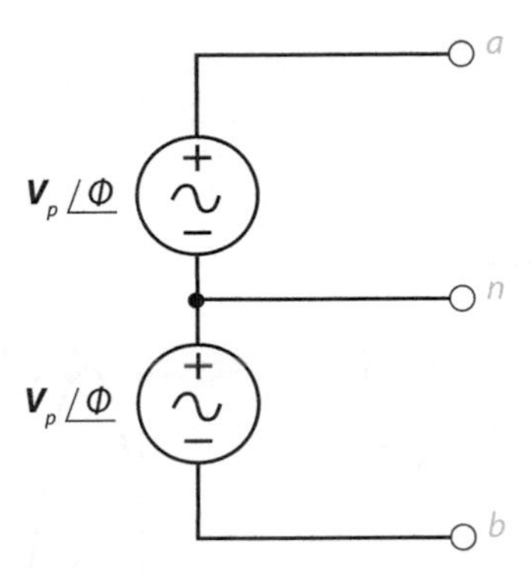

FIGURE 1.17
Single-phase three-wire system

1.5.1 Single-phase three-wire systems

In North America a normal household system is a single-phase three-wire system, permitting the operation of both 110 [V] and 220 [V] appliances at 60 [Hz]. This kind of system is defined as a source with three output terminals (a, n, and b) and two identical sources $V_p\angle\phi$ where V_p is the RMS magnitude of the source and ϕ is the phase, as shown in Figure 1.17.

As mentioned before, the voltages V_{an} and V_{nb} must have the same amplitude and phase, but if seen between the outer wires and the central wire they are exactly 180° out of phase. From this point of view it is strictly a balanced two-phase system; however, the two-phase system refers to two voltage sources in which one lags the other by 90°.

1.5.2 Balanced three-phase voltages

Three-phase voltages are produced with an AC generator that consists of three separate windings or coils placed 120° apart around the stator. As the rotor rotates, the magnetic field induces voltages equal in magnitude in each coil but out of phase by 120° because of their placement around the stator [27]. In Figure 1.18, a cross-sectional view from a generator is shown along with the generated voltages.

A three-phase system is equivalent to three single-phase circuits because each coil of the generator can be regarded as a single-phase generator. This means that a typical three-phase system consists of three voltage sources that can be either Y-connected or Δ-connected, as shown in Figure 1.19.

Let's consider only balanced three-phase sources, which implies that

$$V_{an} + V_{bn} + V_{cn} = 0, \tag{1.36}$$
$$|V_{an}| = |V_{bn}| = |V_{cn}|, \tag{1.37}$$

where V_{an}, V_{bn}, V_{cn} are respectively between lines a, b, and c, and the neu-

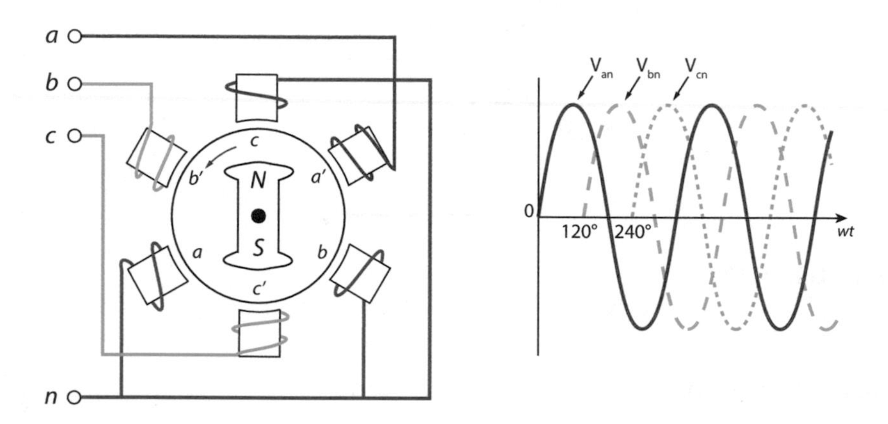

FIGURE 1.18
A three-phase generator with generated voltages

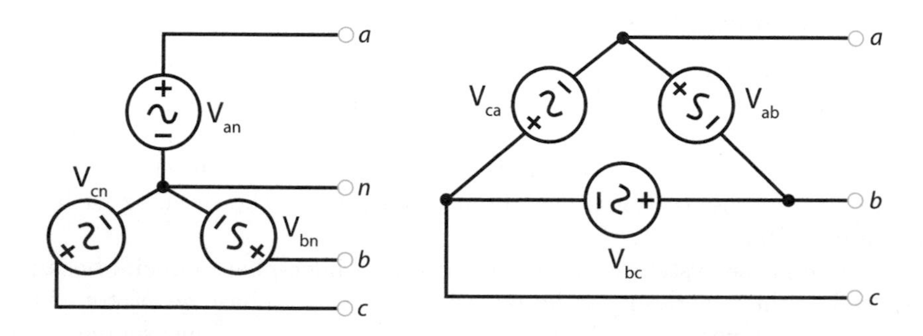

FIGURE 1.19
Y-connected and Δ-connected source voltages

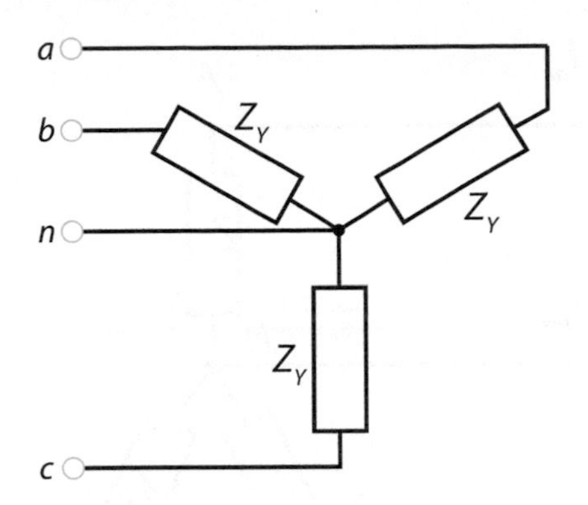

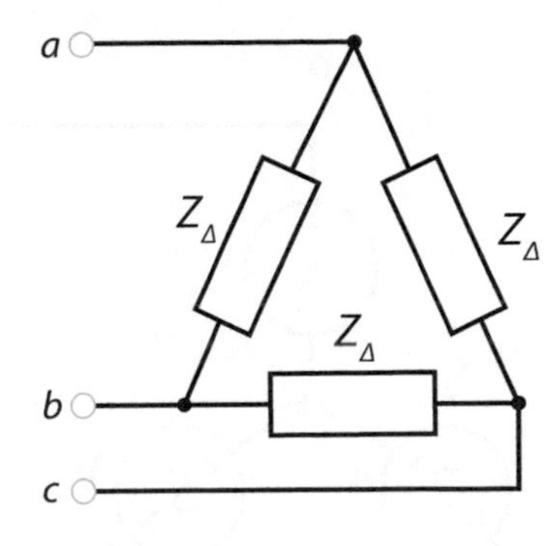

FIGURE 1.20
Y-connected and Δ-connected loads

tral line n known as phase voltages. Since the three voltages are 120° out of phase and arbitrarily V_{an} is chosen as the reference, there are two possible combinations of sequence:

1. $V_{an} = V_p\angle 0°$, $V_{bn} = V_p\angle - 120°$, $V_{cn} = V_p\angle - 240°$;
2. $V_{an} = V_p\angle 0°$, $V_{bn} = V_p\angle 120°$, $V_{cn} = V_p\angle 240°$.

The first one is called positive phase sequence and the second one is the negative phase sequence. The phase sequence of a physical three-phase source depends on the choice order of the tree terminals to be named a, b, and c in which the voltages pass through their respective maximum values [27, p. 507]. Working with a physical generator, it is common to choose the terminals for a positive phase sequence.

As the voltage sources, a three-phase load can be either Y-connected or Δ-connected. As shown in Figure 1.20, a neutral line in a Y-connected load may or may not appear, depending on whether the system is three- or four-wire. A load is balanced if the phase impedances are equal in magnitude and phase.

For a balanced Y-connected load, the impedance per phase is written Z_Y:

$$Z_a = Z_b = Z_c = Z_Y; \tag{1.38}$$

and similarly for a balanced Δ-connected load, the impedance per phase is written Z_Δ:

$$Z_A = Z_B = Z_C = Z_\Delta. \tag{1.39}$$

In this way any balanced load can be transformed from Y-connected to Δ-connected and vice versa using the next equations:

$$Z_Y = \frac{1}{3}Z_\Delta, \text{ or } Z_\Delta = 3Z_Y. \tag{1.40}$$

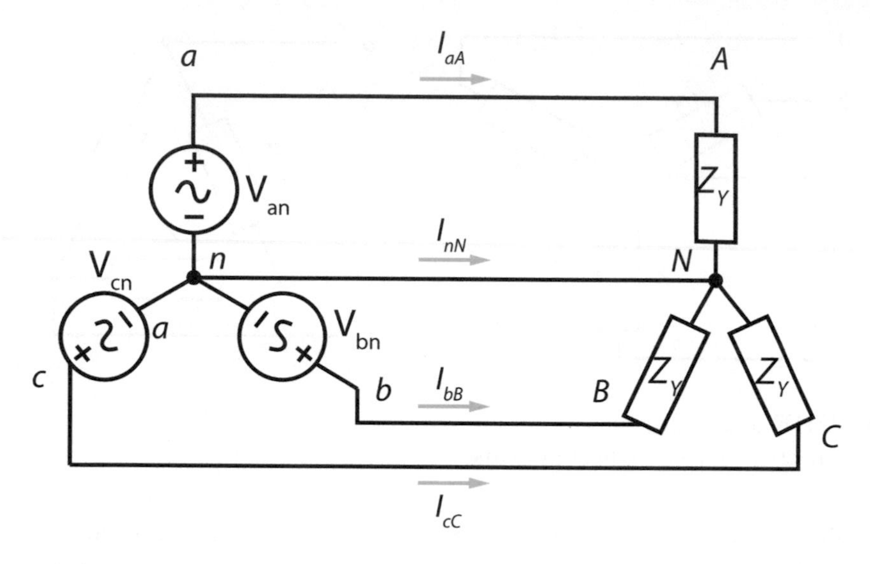

FIGURE 1.21
Balanced Y-Y circuit

Since the three-phase source and the three-phase load can be either Y-connected or Δ-connected, four different connections can be made [27, p. 508]:

1. Y-Y connection (Y-connected source and Y-connected load)
2. Y-Δ connection (Y-connected source and Δ-connected load)
3. Δ-Y connection (Δ-connected source and Y-connected load)
4. Δ-Δ connection (Δ-connected source and Δ-connected load).

1.5.3 Balanced Y-Y connection

This is the most important system because any balanced three-phase system can be reduced to an equivalent Y-Y system. Consider a balanced four-wire Y-Y system where the impedance Z_Y is the total load impedance per phase, which includes the line impedance Z_l and the load impedance Z_L in series; see Figure 1.21.

Assuming a positive sequence, the line voltages can be obtained with the relation of phase voltages,

$$V_{ab} = V_{an} - V_{bn} = V_p\angle 0^\circ - V_p\angle - 120^\circ = \sqrt{3}V_p\angle 0^\circ.$$

Similarly, for the other line voltages,

$$V_{bc} = \sqrt(3) - 90^\circ,$$
$$V_{ca} = \sqrt{3}V_p\angle - 210^\circ.$$

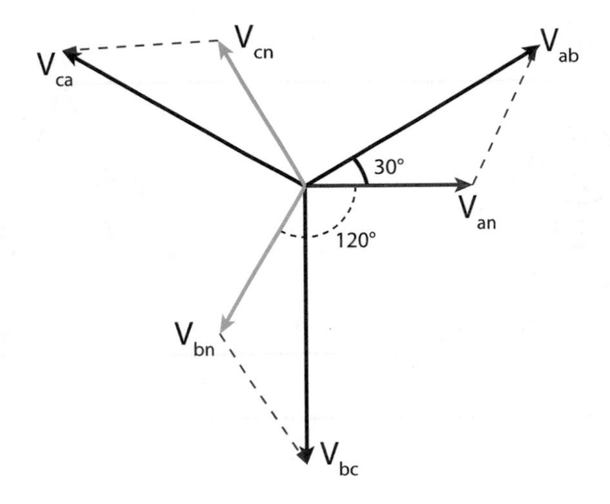

FIGURE 1.22
Phasor diagram for a Y-Y circuit.

Thus, it can be observed that the magnitude of the line voltages V_L is $\sqrt{3}$ the magnitude of V_p:

$$V_L = \sqrt{3}V_p. \tag{1.41}$$

It can also be noted that the line voltages are 120° out of phase between each other and 30° out of phase between their corresponding phase voltages. This is shown in the phasor diagram of Figure 1.22.

Now let's apply KVL to each phase to obtain the line currents I_{aA}, I_{bB}, and I_{cC}.

$$\begin{aligned} I_{aA} &= \frac{V_{an}}{Z_y}, \\ I_{bB} &= \frac{V_{bn}}{Z_y} = I_{aA}\angle -120°, \\ I_{cC} &= \frac{V_{cn}}{Z_y} = I_{aA}\angle -240°. \end{aligned}$$

Therefore,

$$I_{nN} = I_{aA} + I_{bB} + I_{cC} = 0. \tag{1.42}$$

Thus, no current flows through the neutral line if the source and the load are both balanced. This means that as long as the system is balanced, no matter the size of the impedance inserted in the neutral line, the neutral current will remain zero.

In this kind of system, the magnitude of the line voltage is different from the phase voltage by a factor of $\sqrt{3}$ while the line current is the same as the phase current.

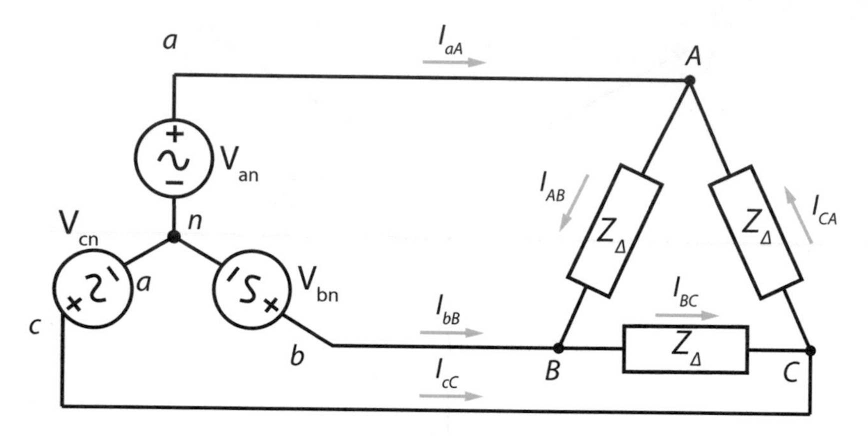

FIGURE 1.23
Balanced Y-Δ circuit

1.5.4 Balanced Y-Δ connection

This system consists of a balanced Y-connected source and a balanced Δ-connected load. This configuration is very common because of the ease of physical connection of the load impedances, and does not possess a neutral connection from source to load; see Figure 1.23.

Assuming a positive sequence, the phase voltages and line voltages are the same as for a Y-Y connection,

$$V_{an} = V_p\angle 0°, \qquad V_{bn} = V_p\angle{-120°}, \qquad V_{cn} = V_p\angle{-240°};$$
$$V_{ab} = \sqrt{3}V_p\angle 30°, \quad V_{bc} = \sqrt{3}V_p\angle{-90°}, \quad V_{ca} = \sqrt(3)V_p\angle{-210°}.$$

It is noted by inspection that the line voltages are equal to the voltages across the load impedances, making it possible to obtain the phase currents as

$$I_{AB} = \frac{V_{ab}}{Z_\Delta}, \; I_{BC} = \frac{V_{bc}}{Z_\Delta}, \; I_{CA} = \frac{V_{ca}}{Z_\Delta}. \tag{1.43}$$

Applying KCL at nodes A, B, and C, the line currents are obtained from the phase current previously calculated,

$$I_{aA} = I_{AB} - I_{CA}, \; I_{bB} = I_{BC} - I_{AB}, \; I_{cC} = I_{CA} - I_{BC}.$$

Since it is a balanced system, the three-phase currents have the same amplitude, so it can be shown that

$$I_L = \sqrt{3}I_p, \tag{1.44}$$

while the line current lags the corresponding phase currents by 30°, as shown in the phasor diagram of Figure 1.24.

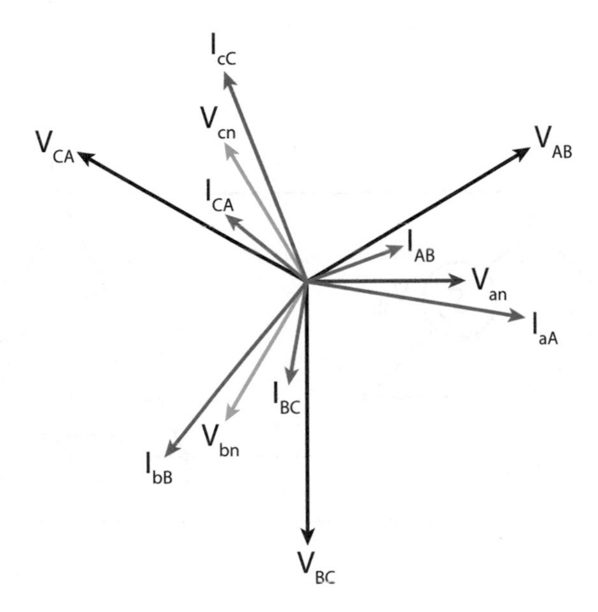

FIGURE 1.24
Phasor diagram for a Y-Δ circuit

1.5.5 Balanced Δ-Δ connection

This system consists of a balanced Δ-connected source and a balanced Δ-connected load, as shown in Figure 1.25.

Assuming a positive sequence, the phase voltages and line voltages are the same as long as there are no line impedances due to losses in the lines,

$$V_{ab} = V_{AB} = V_p\angle 0°,\ V_{bc} = V_{BC} = V_p\angle -120°,\ V_{ca} = V_{CA} = V_p\angle 240°.$$

Therefore, the phase currents can be obtained as for a Δ-Y connection system,

$$I_{AB} = \frac{V_{ab}}{Z_\Delta},\ I_{BC} = \frac{V_{bc}}{Z_\Delta},\ I_{CA} = \frac{V_{ca}}{Z_\Delta}. \tag{1.45}$$

Also, as for a Δ-Y connection system, each line current lags the corresponding phase current by 30° and is related by the factor $\sqrt(3)$,

$$I_L = \sqrt{3}I_p. \tag{1.46}$$

1.5.6 Balanced Δ-Y connection

The circuit Δ-Y consists of a balanced Δ-connected source feeding a balanced Y-connected load. In Figure 1.26, it can be noticed that the line voltages are the same as the phase voltages.

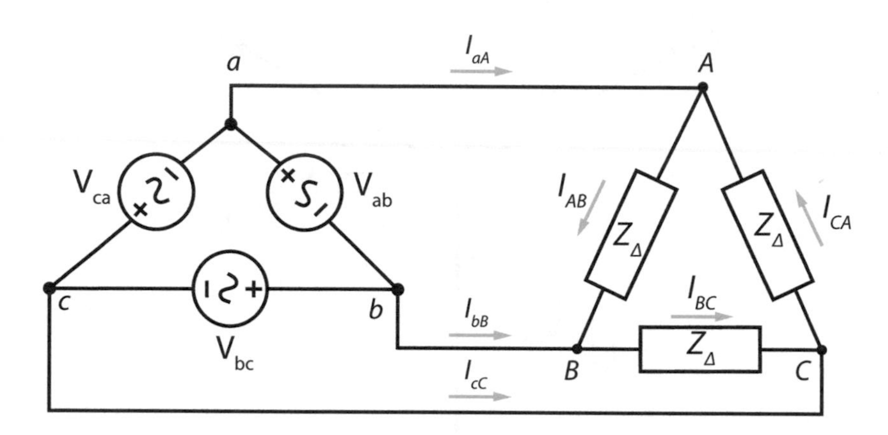

FIGURE 1.25
Balanced Δ-Δ circuit

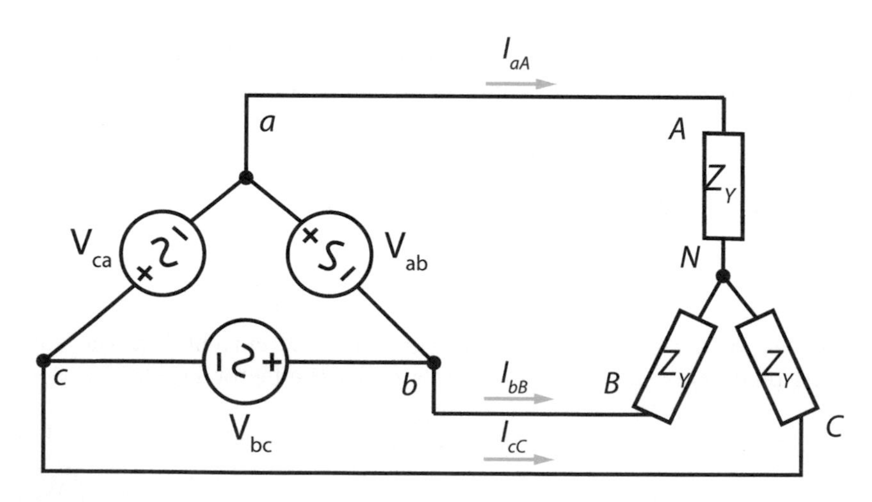

FIGURE 1.26
Balanced Δ-Y circuit

Assuming a positive sequence

$$V_{ab} = V_p\angle 0°,\ V_{bc} = V_p\angle -120°,\ V_{ca} = V_p\angle -240\circ,$$

one way to obtain the line currents is to apply KVL to the loop that involves I_{aA} and I_{bB}:

$$Z_Y I_{aA} - Z_Y I_{bB} = V_{ab}.$$

Thus,

$$I_{aA} - I_{bB} = \frac{V_p\angle 0°}{Z_y},$$

Since it is a balanced system in positive sequence, the line current I_{bB} must be the same as the current I_{aA} but 120° out of phase. Hence,

$$I_{aA} - I_{bB} = I_{aA}\sqrt{3}\angle 30°.$$

Substituting into the last equation, the line current is obtained:

$$I_{aA} = \frac{\frac{V_p}{\sqrt{3}}\angle -30°}{Z_y}. \quad (1.47)$$

Therefore, the other two line currents may be found by lagging 120° the current I_{aA}.

Another way to find the line currents is to either replace the Δ-connected source with an equivalent Y-connected source or to replace the Y-connected load with a Δ-connected load. Both cases were covered in the previous sections.

1.5.7 Unbalanced three-phase systems

Previously, four types of three-phase systems were analyzed, but all of them were balanced. It is said that a system is unbalanced if either one of two conditions is not fulfilled:

1. The source voltages must be equal in magnitude and their difference in phase must be symmetrical.

2. The load impedances must be the same for all phases.

Let us assume a balanced source voltage but an unbalanced Y-connected load, as shown in Figure 1.27. Since the load is unbalanced, each impedance is different, so the system is solved by direct application of mesh and nodal analysis.

The line currents are therefore determined by

$$I_{aA} = \frac{V_{AN}}{Z_A}, I_{bB} = \frac{V_{BN}}{Z_B}, I_{cC} = \frac{V_{CN}}{Z_C}. \quad (1.48)$$

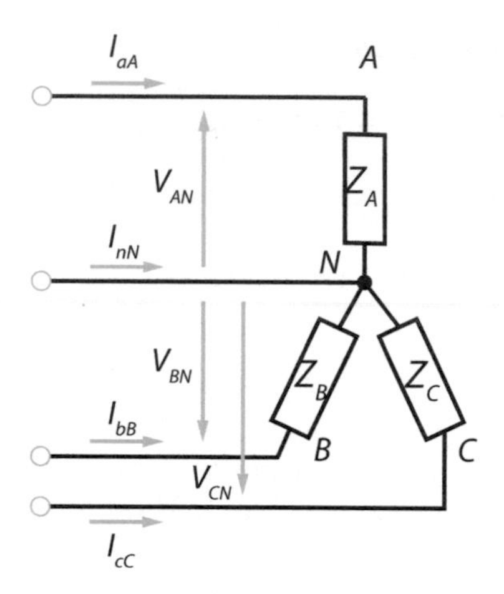

FIGURE 1.27
Unbalanced Y-connected loads

For a balanced system, it was said that the current in the neutral line must be zero, hence it could be removed. In this case each of the line currents produces a current though the neutral line, and applying KCL at node N, the neutral line current is obtained:

$$I_{nN} = -(I_{aA} + I_{bB} + I_{cC}). \tag{1.49}$$

To find the line currents for any of the other three-wire systems Δ-Y, Y-Δ, and Δ-Δ, a mesh analysis has to be made, taking into account that phase and line voltages and currents are no longer necessarily related by a $\sqrt{3}$ factor, nor a difference in phase of 120°.

1.5.8 Power in three-phase systems

As seen in previous sections, the complex power for an AC circuit is

$$S = V_{rms} I^*_{rms}.$$

To obtain the total complex power on a three-phase system, each load's power has to be analyzed independently per phase, whether for a Y-connected load or for a Δ-connected load, to later add each of them to get the total power.

The simplest case to begin with is the balanced three-phase system. Let a Y-connected load with $I_L = I_p$ and $V_L = \sqrt{3}V_p$, or a Δ-connected load with $I_L = \sqrt{3}I_p$ and $V_L = V_p$; see Figure 1.28. Each of the loads has to have the same power as the others, so the total complex power is easily calculated as three times the power of any of the loads:

$$S_{total} = 3S_p = 3V_p I^*_p = 3I_p^2 Z_p = \frac{3V_p^2}{Z^*_p}. \tag{1.50}$$

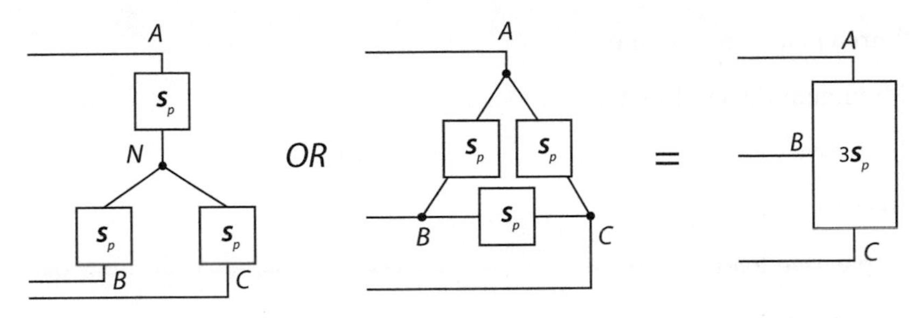

FIGURE 1.28
Power calculation for a balanced three-phase system

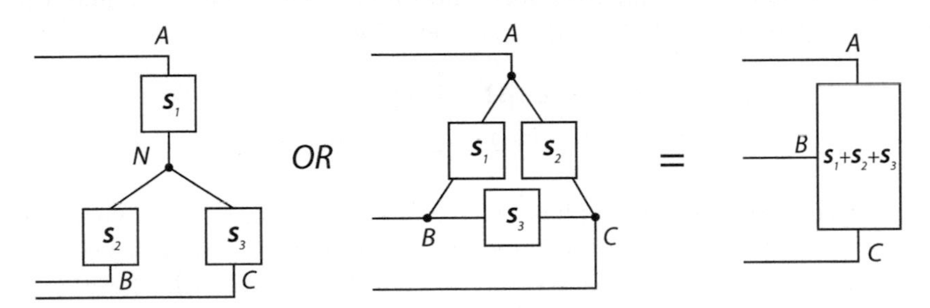

FIGURE 1.29
Power calculation for an unbalanced three-phase system

In the case for an unbalanced system, the total complex power cannot be obtained as three times the power of any of the loads because they are no longer equal; rather, the total complex power is the sum of the three loads (see Figure 1.29),

$$S_{total} = S_1 + S_2 + S_3. \tag{1.51}$$

It is important to notice that the relation of phase voltages to line voltages and phase currents to line currents by the factor $\sqrt{3}$ is no longer valid, therefore a mesh analysis is required to obtain the power of each phase.

An advantage of a three-phase system for power distributions is that uses a lesser amount of wire than the single-phase system for the same line voltage and the same absorbed power [27, p. 521].

1.5.9 Exercises

In some apartment buildings or condos, a three-phase motor is used to pump water from the underground deposit to a water tank in the roof. This motor can be connected in either a Y or Δ connection. In this exercise a comparison between these two connections can be made by measuring the power absorbed by the three-phase load circuit.

Recommended equipment and connections:

- Three-phase supply unit
- Maximum demand meter
- Inductive load
- Resistive load
- Capacitive load
- LCR meter

With an LCR meter, measure one value of resistance and inductance from the resistive load and inductive load modules, respectively, and find their impedance (assume a 60 Hz frequency operation).

	R[Ω]	$Z_R[\Omega]$	L[H]	$\mathbf{Z_L}$
1	761.2	761.2	2.636	j993.75

Connect a balanced three-phase Y-Y circuit (see Figure 1.30) where each source's phase is taken from the three phase supply unit and each load is an RL combination of the previous values chosen. Note that each phase has to have the same loads (see Diagram 1.5).

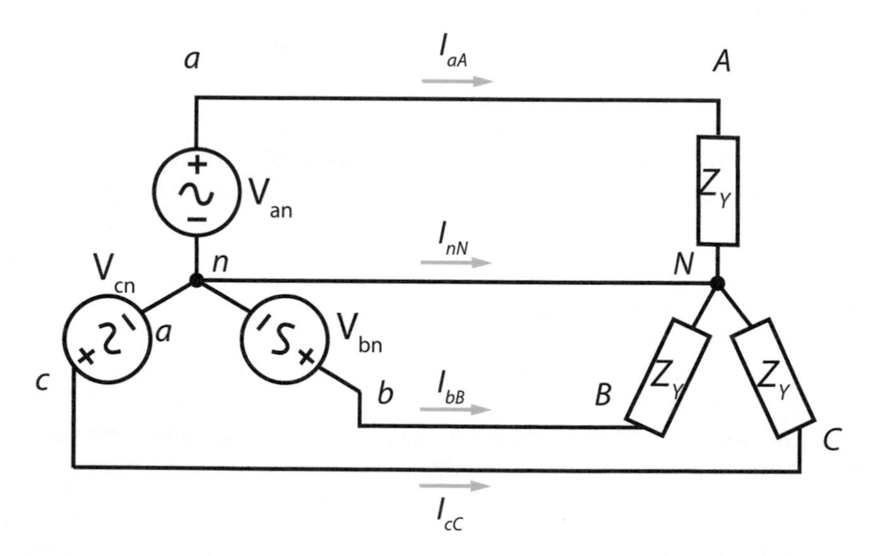

FIGURE 1.30
Three-phase Y-Y circuit electric diagram

Calculate the current I_{aA} of the circuit if it has a voltage $V_{an} = 226\angle 0°$ $[V_{rms}]$. Then obtain the power of the load Z_Y $(S = VI^*)$:

$$SY = 24.81 + j32.39[VA].$$

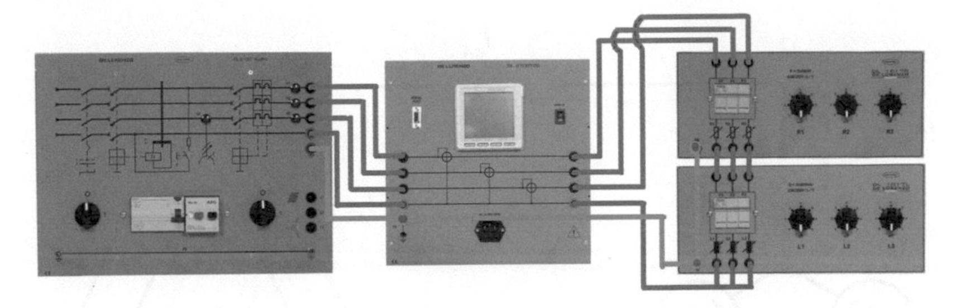

Diagram 1.5
Three-phase Y-Y circuit with RL load

This power corresponds to the absorbed power by the load of one of the three phases. Since it is a balanced three-phase circuit, the total absorbed power is obtained by multiplying the previous power by 3:

$$S_{TOTAL} = 74.43 + j97.17[VA].$$

With the maximum demand meter, measure the real power absorbed for each phase and the total power, and then compare your results.

	S_Y	S_{TOTAL}
Power (S)	16+j27	50+j80

Finally, modify the values of the loads for the second and third phase, now the circuit is unbalanced. With the maximum demand meter, measure the power being absorbed for each phase and the total.

	S_Y	S_{Y2}	S_{Y3}	S_{TOTAL}
Power (S)	16+j26	26+j31	12+j32	55+j89

Note that the total power absorbed now is obtained by the sum of each phase power.

Connect a balanced three-phase Y-Δ circuit (see Figure 1.31) where each source's phase is taken from the three-phase supply unit and each load is an RL combination of the previous values chosen. Note that each phase has to have the same loads (see Diagram 1.6).

With the first maximum demand meter (the one on the left side in Diagram 1.6), measure each of the line currents.

	I_{aA}	I_{bB}	I_{cC}
Current (I)	0.412	0.412	0.412

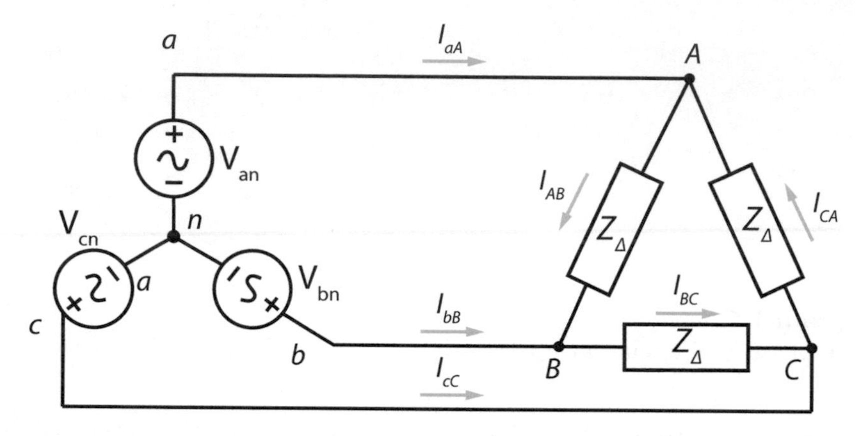

FIGURE 1.31
Three-phase Y-Δ circuit electric diagram

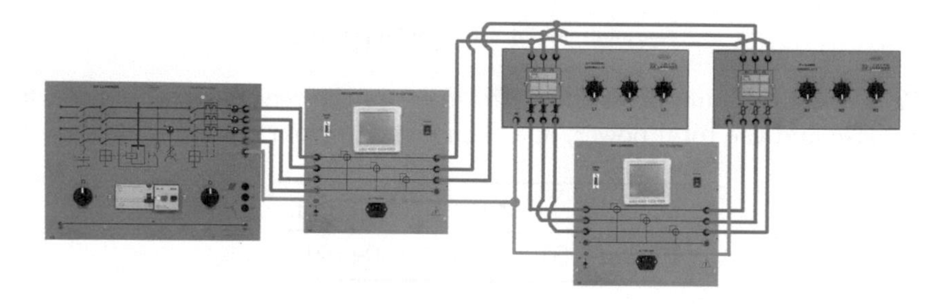

Diagram 1.6
Three-phase Y-Δ circuit with RL load

Now, with the second module maximum demand meter (the one on the right side of Diagram 1.6), measure each of the phase currents.

	I_{AB}	I_{BC}	I_{CA}
Current (I)	0.238	0.238	0.238

Note that in a balanced Y-Δ system, the phase current is related to its corresponding line current by a $\sqrt{3}$ factor.

Finally, modify the values of the loads for the second and third phase for the circuit to be unbalanced. Again, with the maximum demand meters, measure the phase and line currents.

Note that in an unbalanced Y-Δ system, the phase currents are no longer related to their corresponding line current by a $\sqrt{3}$ factor.

	$\mathbf{I_{aA}}$	$\mathbf{I_{bB}}$	$\mathbf{I_{cC}}$
Current (I)	0.238	0.238	0.238
	$\mathbf{I_{AB}}$	$\mathbf{I_{BC}}$	$\mathbf{I_{CA}}$
Current (I)	0.238	0.304	0.260

1.6 Harmonics

So far it has been assumed that the voltage supplied to the loads is a perfectly sinusoidal signal, but this is not necessarily true in reality. Mainly because of the increasing use of nonlinear loads, the voltage and current waveforms suffer from distortions, often expressed as harmonic distortion [50, p. 1].

Harmonics in AC power systems are components of a periodic current or voltage waveform which have frequencies that are integer multiples of the fundamental frequency of the system. A harmonic waveform is a perfectly sinusoidal signal that is added to another sinusoidal waveform, which results in a distorted waveform, as shown in Figure 1.32.

Figure 1.32 (left) shows the main waveform and its 3th, 5th, and 7th harmonics, which can be expressed as

$$v_1 = V_{m1} \sin(\omega t) \tag{1.52}$$
$$v_3 = V_{m3} \sin(3\omega t + \delta_3) \tag{1.53}$$
$$v_5 = V_{m5} \sin(5\omega t + \delta_5) \tag{1.54}$$
$$v_7 = V_{m7} \sin(7\omega t + \delta_7) \tag{1.55}$$

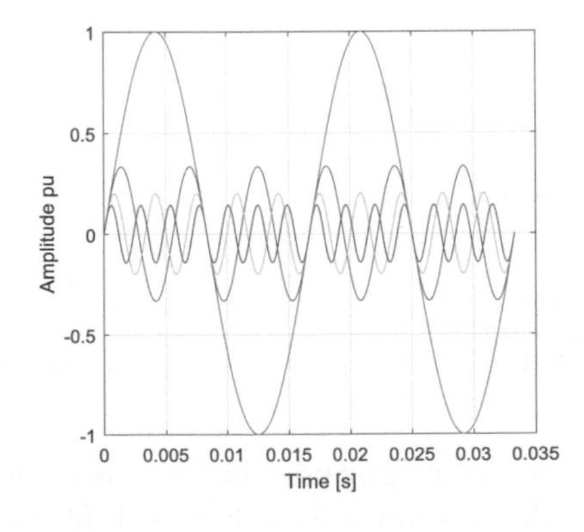

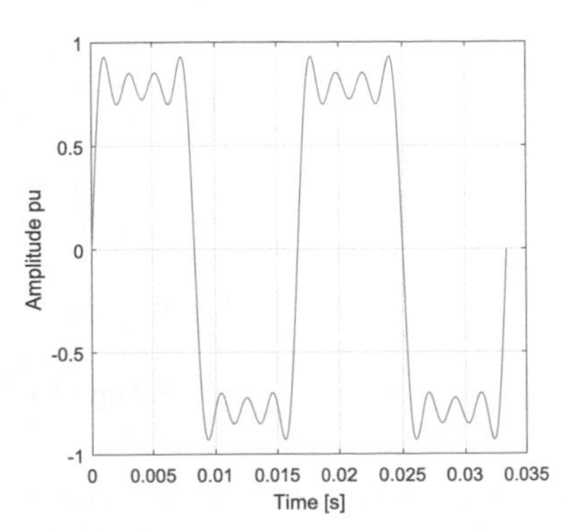

FIGURE 1.32
60 Hz signal and its 3th, 5th and 7th harmonics

Figure 1.32 (right) shows the resultant distorted waveform, which can be expressed as

$$v_{total} = V_{m1}\sin(\omega t) + V_{m3}\sin(3\omega t + \delta_3) + V_{m5}\sin(5\omega t + \delta_5) + V_{m7}\sin(7\omega t + \delta_7). \quad (1.56)$$

The frequency of any harmonic can be easily found with the expression

$$f_n = n(\text{fundamental frequency}), \quad (1.57)$$

where $f_n =$ is the n-th harmonic frequency, and n is an integer that represents the number of the harmonic.

The Fourier series represents an effective way to study and analyze harmonic distortion because it describes periodic functions and their contribution of sinusoidal functions of different frequencies. In this way, it allows inspecting the various constituents of a distorted waveform through decomposition [79, p. 7].

Any periodic waveform can be expanded by a Fourier series as

$$\begin{aligned} f(t) &= A_0 + \sum_{h=1}^{\infty} [A_h \cos(h\omega_0 t) + B_h \sin(h\omega_0 t)] \\ &= A_0 + \sum_{h=1}^{\infty} C_h \sin(h\omega_0 t + \psi_h), \end{aligned} \quad (1.58)$$

where $C_1 \sin(\omega_0 t + \psi_1)$ represents the fundamental component, $C_h \sin(h\omega_0 t + \psi_h)$ represents the hth armonic, and the coefficients are given by,

$$A_0 = \frac{1}{T}\int_0^T f(t)dt, \quad (1.59)$$

$$A_h = \frac{2}{T}\int_0^T f(t)\cos(h\omega_0 t)dt, \quad (1.60)$$

$$B_h = \frac{2}{T}\int_0^T f(t)\sin(h\omega_0 t)dt, \quad (1.61)$$

$$C_h = \sqrt{A_h^2 + B_h^2}, \quad (1.62)$$

$$psi_h = \tan^{-1}\left(\frac{A_h}{B_h}\right). \quad (1.63)$$

The harmonic spectrum is a bar plot of the magnitudes of harmonics described in the Fourier series. Let us take the waveform from Figure 1.32, and by applying an FFT we find the harmonic spectrum presented in Figure 1.33.

The total harmonic distortion (THD) is one of the power quality indices that exist; another one was previously introduced as the power factor. The THD is used to describe power quality issues in transmission and distribution systems [60, p. 28]. It measures the harmonic distortion present in a signal

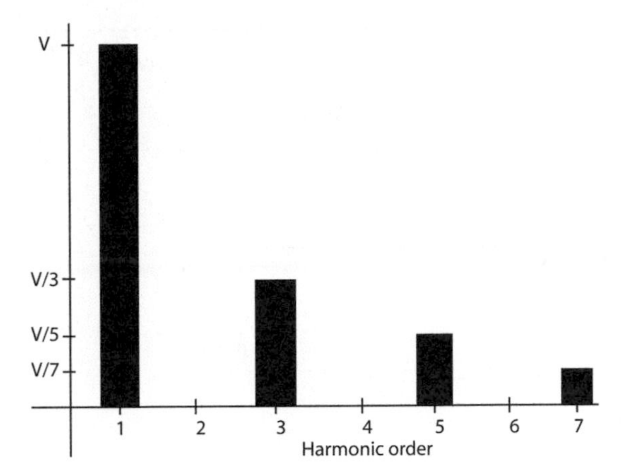

FIGURE 1.33
Harmonic spectrum

and is defined as the ratio between RMS values of signals, including harmonics and signals considering only the fundamental frequency:

$$\text{THD}_F = \frac{1}{F_1}\sqrt{\sum_{h=2}^{\infty} F_h^2}, \tag{1.64}$$

where F represents the function either voltage or current.

1.6.1 Power factor with nonlinear loads

As explained before, nonlinear loads distort the current waveform instead of typically just shifting it (lagging or leading), so the method revised for power factor correction is no longer always valid.

For nonlinear loads, the power triangle explained before becomes a three-dimensional relation with Q and P producing the apparent power and H as the distortion reactive power, as shown in Figure 1.34.

The true power factor (TPF) then becomes the combination of displacement power factor (dPF), which is still equal to $\cos\phi$ where ϕ is the angle between the fundamental current and voltage, and distortion power factor (hPF). Distortion PF is the true power factor divided by displacement PF. This is shown in (1.65).

$$pf = \frac{P}{S} = \frac{[\text{kW}]}{[\text{kVA}]} \neq \cos\phi \tag{1.65}$$

$$S = \sqrt{P^2 + Q^2 + H^2}$$

$$[kVA] = \sqrt{[\text{kW}]^2 + [\text{kVAr}]^2 + [\text{kVAr}]_H^2}$$

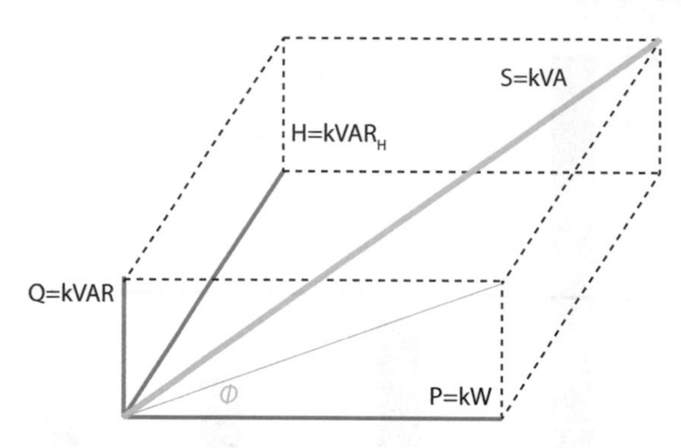

FIGURE 1.34
Power vector relation

Unlike displacement PF, distortion PF is neither leading nor lagging. Therefore, to improve a poor power factor with nonlinear loads, the harmonics have to be removed, so adding a capacitor is no longer the solution as they can resonate with the power system inductance.

Having a poor power factor not only implies a waste of energy or a damage to the electrical installation, but electrical utilities commonly charge a penalty for low PF to their customers within the demand charge. Since there is not a standard for charging this kind of penalty, each electrical utility has its own rules. In Mexico, CFE, an electrical utility, penalizes companies with PF lower than 90% with up to 120% of charges calculated, as shown in (1.66), and rewards companies with PF higher than 90% with a discount calculated as in (1.67).

$$\text{Penalty} = \frac{3}{5}\left(\frac{90}{\text{F.P.}} - 1\right) \text{x } 100 \tag{1.66}$$

$$\text{Reward} = \frac{1}{4}\left(1 - \frac{90}{\text{F.P.}}\right) \text{x } 100 \tag{1.67}$$

1.6.2 Exercises

In a "healthy" electric installation, the AC current has a sinusoidal waveform, but due to nonlinear loads, this waveform can be distorted by some harmonics introduced in the signal. In this exercise, a sinusoidal waveform is going to be observed in an oscilloscope with and without harmonics to appreciate the distortion in the signal.

Recommended equipment and connections:

- Lamps for photovoltaic trainer
- Photovoltaic inclinable module

- Inverter grid
- Electrical power digital measuring unit
- Maximum demand meter
- Three-phase supply unit
- Resistive load
- Oscilloscope with current clamp

Connect the lamps and head the light towards the photovoltaic panel (90°); the lamps act as the light source. The photovoltaic panel is then connected to the inverter grid, and this module is then connected in series to the thermo-magnetic differential switch to activate or deactivate the load. The electric power digital meter is connected to the switch in order to measure the output power coming from the inverter.

Finally, connect the three-phase supply unit, which will act as the "ideal" power generator, to the maximum demand meter and then to the load (see Diagram 1.7)

1. Set the resistive load to the 2nd value
2. Set the lamps to 100% of light intensity
3. Put the solar panel in front of the light at 90° degrees
4. Deactivate the switch of the circuit breaker so the load only receives energy from the generator. With the oscilloscope, observe the waveform of the current delivered to the load (see Figure 1.35). Is the waveform almost a perfect sine? Does it have any distortion? Why?
5. Activate the switch of the circuit breaker so the load now also receives energy from the solar panel. With the oscilloscope, observe the waveform of the current delivered to the load (see Figure 1.36). What difference can be noticed from the previous waveform? Does this new waveform have harmonics?

1.7 Theoretical problems

Problem 1.1. *Find the current I_x of the circuit of Figure 1.37 (assume all values in RMS).*

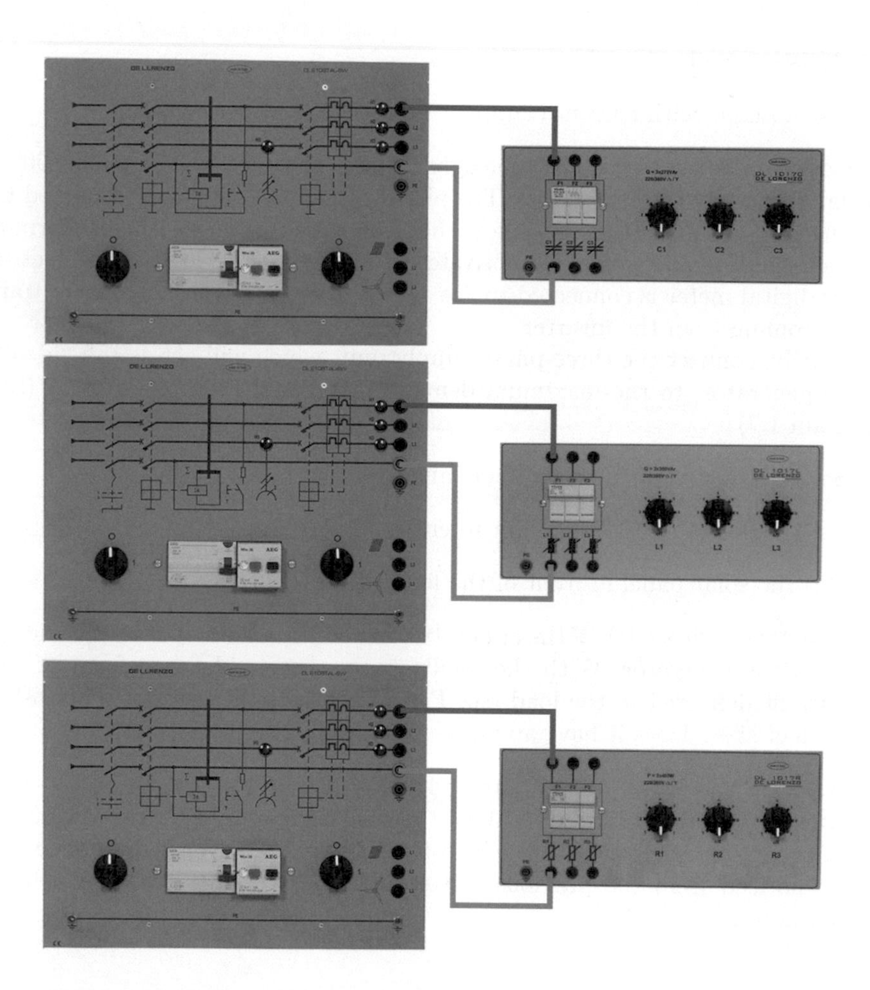

Diagram 1.7
Connection for harmonic distortion exercise

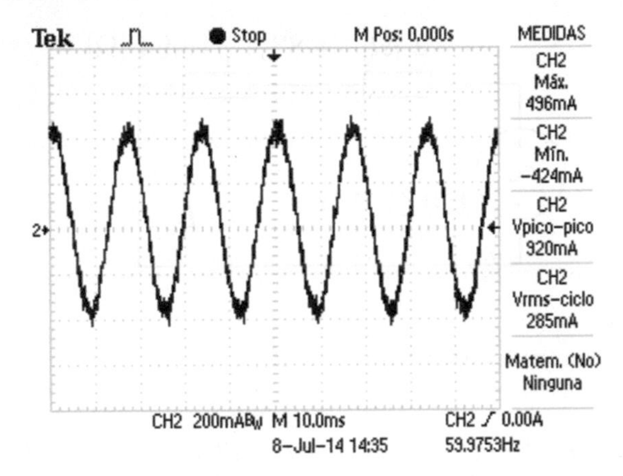

FIGURE 1.35
Current waveform in the oscilloscope

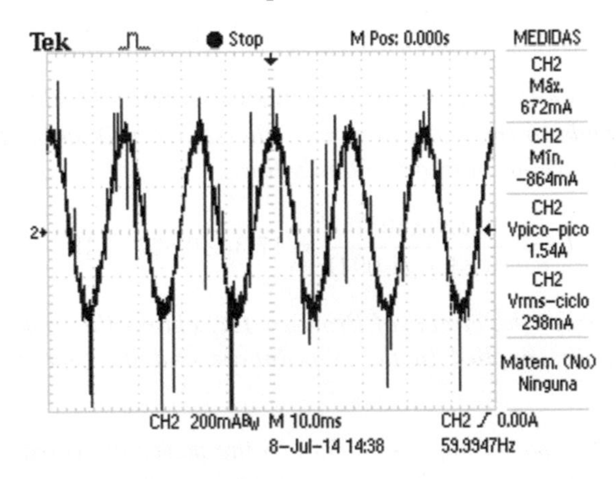

FIGURE 1.36
Current waveform with harmonics

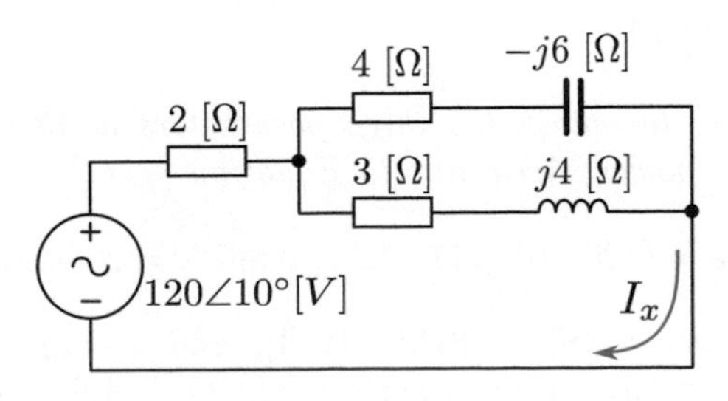

FIGURE 1.37
Theoretical problem circuit

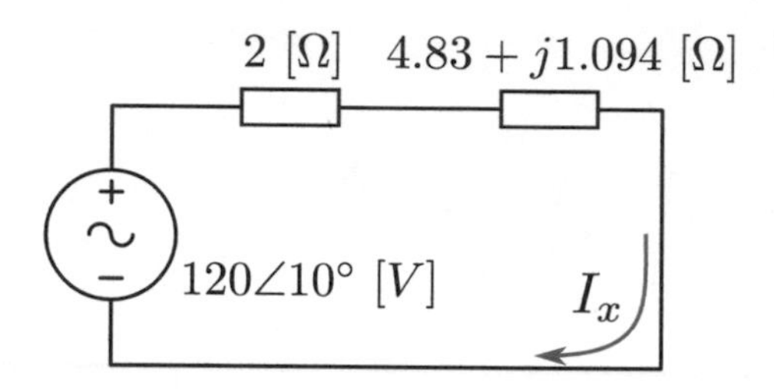

FIGURE 1.38
Theoretical problem reduced circuit

Solution 1.1. *Let's begin reducing the circuit to one equivalent impedance with a voltage source (see Figure 1.38). First note that the impedances of* 4 [Ω] *and* $-j6$ [Ω] *are in parallel with the impedances of* 3 [Ω] *and* $j4$ [Ω]*:*

$$Z_1 = \frac{(4-j6)(3+j4)}{(4-j6)+(3+j4)} = 4.83 + j1.094\Omega.$$

So, the equivalent impedance is $z_{eq} = 6.83 + j1.094$*. To obtain the current* I_x *that corresponds to the source current, the source voltage is divided by the equivalent impedance according to Ohm's law:*

$$I_x = \frac{(120\angle 10^\circ)}{(6.83 + j1.094)} = 17.34\angle 0.897^\circ \ [A].$$

Problem 1.2. *From the circuit of Problem 1.1, obtain the apparent, average, and reactive power supplied to the impedances and the power factor of the source.*

Solution 1.2. *The power supplied refers to the power delivered by the source. To obtain the average power, Equation (1.26) is used:*

$$P = V_{RMS} I_{RMS} \cos(\theta_v - \phi_i) = (120)(17.34)\cos(10 - 0.8987)$$
$$P = 2054.9 \ [W].$$

But the easiest way to obtain the three parameters at the same time is by getting the complex power. According to Equation (1.33):

$$S = V_{RMS} I^*_{RMS} = (120\angle 10^\circ)(17.34\angle -0.897^\circ) = 2080.8\angle 9.103^\circ \ [VA],$$

where apparent power is $|S| = 2080.8$ $[VA]$*, and average power is* $ReS =$ 2054.59 $[W]$*, and reactive power is* $ImS = 329.20$ $[VAr]$*.*

Now, the power factor can be easily obtained by the cosine of the angle of the complex power or by Equation (1.31):

$$PF = \frac{P}{V_{RMS} I_{RMS}} = \cos(\theta_v - \phi_i) = 0.9874.$$

Problem 1.3. *An AC motor can be seen as a load of* $Z_L = 4.2 + j3.6\ [\Omega]$ *and it is fed with* $220\ [V_{RMS}]$ $60\ [Hz]$. *Find its PF and the capacitor needed to achieve a* $PF = 1$.

Solution 1.3. *To calculate the capacitor needed for the power factor correction, Equation (1.35) is used:*

$$C = \frac{P(\tan(\theta_1 - \theta_2))}{\omega V_{RMS}^2}.$$

Note that the average power and the power factors of the source are required. First, let's obtain the complex power:

$$S = 220\left(\frac{220}{4.2 + j3.6}\right)^* = 6643.13 + j5694.12 = 8749.52\angle 40.6°.$$

Hence, the power factor is $PF = cos(40.6) = 0.7593$.

The required capacitor connected in parallel to increase the power factor to 1 *is:*

$$C = \frac{6643.13(\tan(40.6° - \tan(0°)))}{377(220^2)} = 312\mu[F].$$

Problem 1.4. *A 4 wire three-phase Y-Y unbalanced system with negative sequence has the following characteristics:* $V_{an} = 220\angle 0°\ [V]$, $Z_1 = 10{+}j8\ [\Omega]$, $Z_2 = 8 + j6\ [\Omega]$, $Z_3 = 5 - j5\ [\Omega]$. *Obtain the three-phase current and the total average power absorbed by the loads.*

Solution 1.4. *Since the sequence of the voltages is negative:*

$$\begin{aligned} V_{an} &= 220\angle 0°, \\ V_{bn} &= 220\angle 120°, \\ V_{cn} &= 220\angle 240°. \end{aligned}$$

Despite the system being unbalanced, it is easy to obtain the current of each phase because of the fourth wire (the neutral one). The system can be analyzed with three independent loops, each one having one of the source voltages and its corresponding phase load:

$$\begin{aligned} I_{aA} &= \frac{220\angle 0°}{10 + 8j} \\ I_{aA} &= 17.18\angle -38.66°\ [A]; \\ I_{bB} &= \frac{220\angle 120°}{8 + 6j} \\ I_{bB} &= 22\angle 83.13°\ [A]; \\ I_{cC} &= \frac{220\angle 240°}{5 + 5j} \\ I_{cC} &= 31.11\angle -75°\ [A]. \end{aligned}$$

To obtain the total average power, let's get the complex power for all three phases and then add just the real part of them:

$$
\begin{aligned}
S_1 &= V_{an} I_{aA}^* \\
S_1 &= 2951.36 + j2361.10 \ [VA]; \\
S_2 &= V_{bn} I_{bB}^* \\
S_2 &= 3872 + j2904 \ [VA]; \\
S_3 &= V_{cn} I_{cC}^* \\
S_3 &= 4839.58 - j4839.58 \ [VA].
\end{aligned}
$$

The total average power is

$$P_T = 11,662.94 \ [W].$$

1.8 Homework problems

Problem 1.5. *Consider a given system to consume* $i(t) = 2\cos(377t + 0.75)$ *fed with* $v(t) = 380\cos(377t)$. *Based on the average power equation, derive the active, reactive, and apparent power consumed by the system. Compare the first term of the instantaneous power equation and its graphical representation with the average power equation and the active power.*

Solution 1.5. *The offset term of the instantaneous power equation must coincide with the average power equation, as they both resemble the same characteristic of the resulting power function. If such an offset term is computed, it leads to*

$$\frac{1}{2} V_m I_m \cos(\theta_v - \theta_i) = \frac{1}{2}(380)(2)\cos(-0.75) = 278.04 \ [W].$$

This also matches the active power calculation:

$$
\begin{aligned}
V_{rms} = \frac{380}{\sqrt{2}} \ [V] \ and \ I_{rms} = \frac{2}{\sqrt{(2)}} \ [A], \ so \ S = V_{rms} I_{rms} = 380 \ [VA], \\
PF = \cos(\theta_v - \theta_i) = \cos(-0.75) = \cos(0.75), \\
P = S(PF) = 380\cos(0.75) = 278.04 \ [W].
\end{aligned}
$$

On the other hand, the instantaneous power equation leads to

$$
\begin{aligned}
p(t) &= \frac{1}{2} V_m I_m \cos(\theta_v - \theta_i) + \frac{1}{2} V_m I_m \cos(2\omega t + \theta_v + \theta_i) \\
&= 278.04 + 380\cos(754t + 0.75).
\end{aligned}
$$

Its average leads to

$$\frac{1}{T}\int_0^T 278.04 + 380\cos(754t + 0.75)dt.$$

The first part of the function is known, so only the second part needs to be computed. It can be written as

$$380\frac{377}{2\pi}\int_0^{\frac{2\pi}{377}} \cos(754t + 0.75)dt,$$

having

$$\int \cos(\omega t + \phi) = \frac{1}{\omega}\left(\cos(\omega t)\sin(\phi) + \cos(\phi)\sin(\omega t)\right), \ \textit{then,}$$
$$\int \cos(\omega t + \phi)|_{t=0} = \frac{\sin(\phi)}{\omega} \ \textit{and} \ \int \cos(\omega t + \phi)|_{t=T} = \frac{\sin(\phi)}{\omega},$$

and implying

$$380\frac{377}{2\pi}\int_0^{\frac{2\pi}{377}} \cos(754t + 0.75)dt = 0, \ \textit{so,}$$
$$\frac{1}{T}\int_0^T p(t)dt = \frac{1}{2}V_m I_m \cos(\theta_v - \theta_i) = 278.04 \ [W].$$

Up to this point, we have already calculated the apparent power and active power. The reactive power can then be obtained by any known method:

$$Q = S\sin(0.75) = 259 \ [VAr], \ \textit{or}$$
$$Q = \sqrt{S^2 - P^2} = 259 \ [VAr].$$

Problem 1.6. *Consider an inductive load connected to a variable frequency voltage source. If the inductance magnitude is L and it is fed by $V_m \sin(\omega t)$, find the equation that describes $||i(\omega)||$.*

Solution 1.6. *The inductance can be described by*

$$\frac{V}{i} = Z_L = j\omega L.$$

So, easily, the current can be derived from the above equation as

$$\frac{V(\omega)}{j\omega L} = i(\omega),$$

and accordingly,

$$\left\|\frac{V(\omega)}{v\omega L}\right\| = \|i(\omega)\|.$$

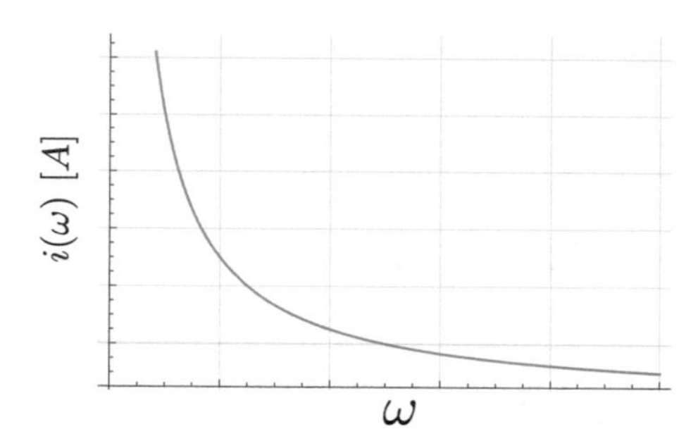

FIGURE 1.39
Current plot

Simplifying:

$$i_{rms}(\omega) = \frac{V_{rms}}{\omega L},$$

which can be plotted as shown in Figure 1.39.

As $\omega \to 0$, the inductance behaves as a short-circuit (its DC behavior).

Problem 1.7. *Consider a capacitive load connected to a variable frequency voltage source. If the capacitance magnitude is C and it is fed by $V_m \sin(\omega t)$, find the equation that describes $\|i(\omega)\|$.*

Solution 1.7. *The capacitance can be described by*

$$\frac{V}{i} = Z_C = \frac{1}{j\omega C}.$$

So, easily, the current can be derived from the above equation as

$$V(\omega)j\omega C = i(\omega),$$

and, accordingly,

$$\|V(\omega)j\omega C\| = \|i(\omega)\|.$$

Simplifying:

$$i_{rms}(\omega) = V_{rms}\omega C,$$

which can be plotted as shown in Figure 1.40. As $\omega \to 0$, the capacitance behaves as an open-circuit (its DC behavior).

Problem 1.8. *Power factor correction implies a current balancing between the inductive part of the load and the added capacitor. If a load of $Z_L = 4.2 + j3.6$ [Ω] is fed with 220 [V_{rms}] at 60 [Hz], find the required capacitor to obtain a PF= 1 without using Equation (1.22). Hint: see Figure 1.9.*

Solution 1.8. *We first need to find the current due to the inductive part of the load:*

$$\frac{220\ [V_{rms}]}{4.2 + j3.6[\Omega]} = 30.20 - j25.88\ [A_{rms}].$$

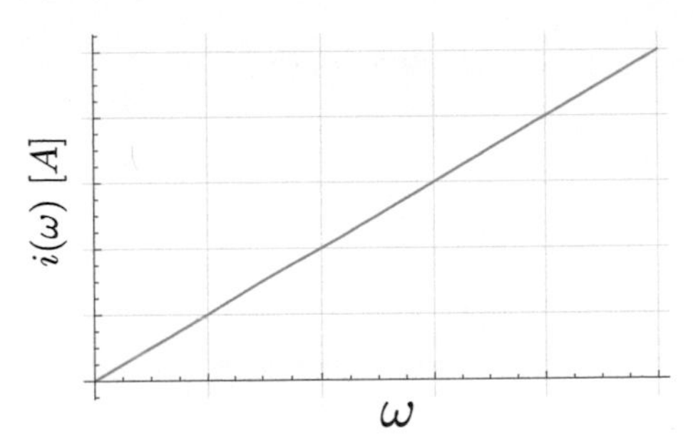

FIGURE 1.40
Current plot

A parallel capacitor is needed to compensate such current, so:

$$\|i_C\| = V_{rms}\omega C,$$

leading to $25.88/220(377) = 312[\mu F]$.

It can also be solved as follows. Adding a parallel capacitor implies that the resulting impedance will be of:

$$Z = \frac{(4.2 + j3.6)\frac{1}{j\omega C}}{(4.2 + j3.6) + \frac{1}{j\omega C}} = \frac{(4.2 + j3.6)}{1 - 1.36C + j1.58C} = \frac{5.53\angle 40.6^\circ}{\sqrt{1 - 2.72C + 4.35C^2}\angle \arctan\left(\frac{1.58C}{1-1.36C}\right)}.$$

A PF = 1 implies the current's phase angle is zero, so the angle of the impedance must be zero to achieve it. The magnitude can be disregarded, so:

$$\frac{\angle 40.6^\circ}{\arctan\left(\frac{1.58C}{1-1.36C}\right)} = 0^\circ,$$

implies that (in radians):

$$0.709 = \arctan\left(\frac{1.58C}{1 - 1.36C}\right).$$

Having

$$0.858 \times 10^{-3} = \frac{1.58C}{1 - 1.36C}$$

results in:

$$C = \frac{0.858 \times 10^{-3}}{2.75} = 312\ [\mu F].$$

It can also be graphically solved as follows (see above homework problems 1.6 and 1.7):

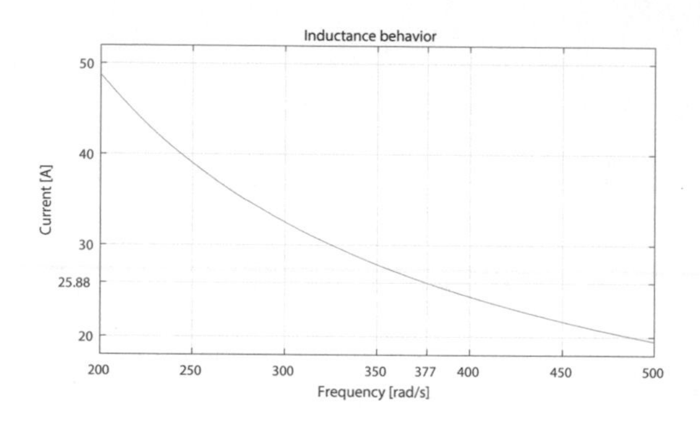

FIGURE 1.41
Inductance behavior

The inductance is of $9.55\ mH$*, so around* $377\ [rad/s]$ *its behavior is* $i_L(\omega)$ *(notice that the resistive part has been subtracted from the voltage as* $V_R = i_R R$*, and* i_R *is in phase with* V_{in}*):*

$$\|\imath_L(\omega)\| = \left\|\frac{V_L}{j\omega L}\right\| = \left\|\frac{220 - (30.2 \times .2)}{377 \times 9.55 \times 10^{-3})}\right\|,$$

as shown in Figure 1.41.

A matching parallel capacitor's current must then intersect inductor's current line at $377 rad/s$*.*

$$\|i_C(\omega)\| = \|V_C j\omega C\| = 220\omega C.$$

Sweeping capacitance values, $C = 312\ \mu F$ *can be found (see Figure 1.42).*

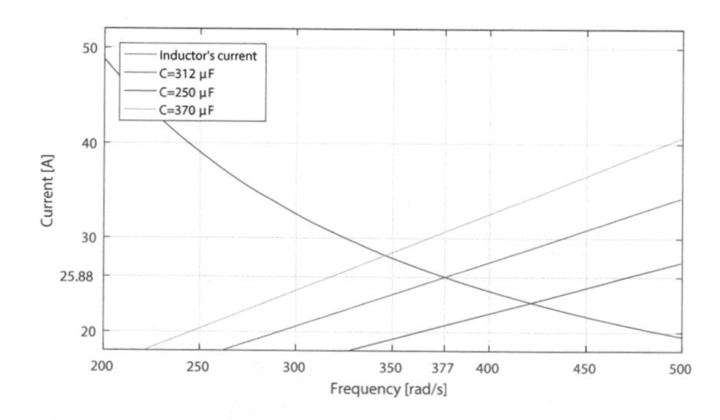

FIGURE 1.42
Sweeping capacitance values

Problem 1.9. *There are many buildings that take power from a three-phase* *installation and later perform a single-phase distribution throughout the different loads. To preserve balanced operating conditions, each set of single-phase loads must be balanced regarding the other phases; not doing so results in voltage variations and cables overheating. Suppose that, for the sake of simplicity, some building installation is distributed as follows:*

Phase 1: *Illumination (*$30\ [kVA]$*)*

Phase 2: *Computers (*$10\ [kVA]$*)*

Phase 3: *Heavy electrical equipment (*$140\ [kVA]$*)*

Find the voltage at each phase, considering the input electrical connection to be taken from a standard Y-225 kVA transformer, assuming $208\ V_{LL}$*, a series resistor of* $1.6\ [m\Omega]$*, and resistive load conditions.*

Solution 1.9. *First, the electric input is to be represented as a simplified voltage source, which can effectively account for the previously made assumptions (see Figure 1.43). On the other hand, each load is assumed to only exhibit resistive behavior, so they can be "transformed" to:*

Illumination: $480\ [m\Omega]$

Computers: $1.44\ [\Omega]$

Heavy electrical equipment: $102.9\ [m\Omega]$.

Each circuit can be solved separately, deriving: $i_a = 246.71\angle 0^\circ\ [A]$, $i_b = 82.96\angle 120^\circ\ [A]$, *and* $i_c = 1097.90\angle 240^\circ\ [A]$. *Thus, the neutral current is of* $i_n = 943.78\angle -111.36^\circ\ [A]$.

This unbalancing condition has several implications that must be addressed:

- *The load at phase A will "see" a voltage of* $480\ [m\Omega] \times i_a = 118.42\ [V]$
- *The load at phase B will "see" a voltage of* $1.44[\Omega] \times i_b = 119.46\angle 120^\circ\ [V]$
- *The load at phase C will "see" a voltage of* $102.9[m\Omega] \times i_c = 112.97\angle 240^\circ\ [V]$
- *There will be an unwanted voltage at the load neutral point of* $4.8\ [m\Omega] \times i_n = 4.53\angle -111.36^\circ\ [V]$*, accountable for the above unbalancing.*

The power at the neutral line is of $4.28\ kW$*, which is absolutely transformed into heat.*

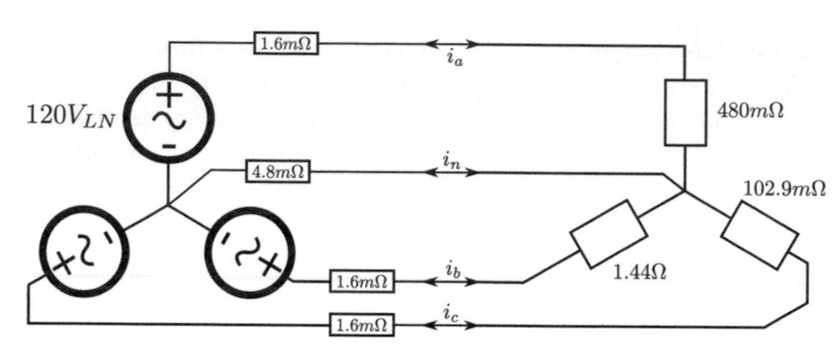

FIGURE 1.43
Electric circuit

1.9 Simulation

It is possible to analyze the effects of balanced or unbalanced loads in terms of voltage and current (see Figure 1.44). Besides the fact that unbalanced loads and harmonic effects will be a detriment to a grid's performance, it is difficult to find voltage variations due to the source's stiffness. However, current is directly affected, so it is more effective to measure current disturbances if some faulty source or load is being evaluated.

Furthermore, an unbalanced load will account for a differentiated power factor per phase. As the power factor is commonly measured as an average value, it will not represent the grid's reality for unbalanced networks. It is important to acknowledge the possible variations regarding unbalanced loads and harmonic distortion, and link their effects to common electric measurements.

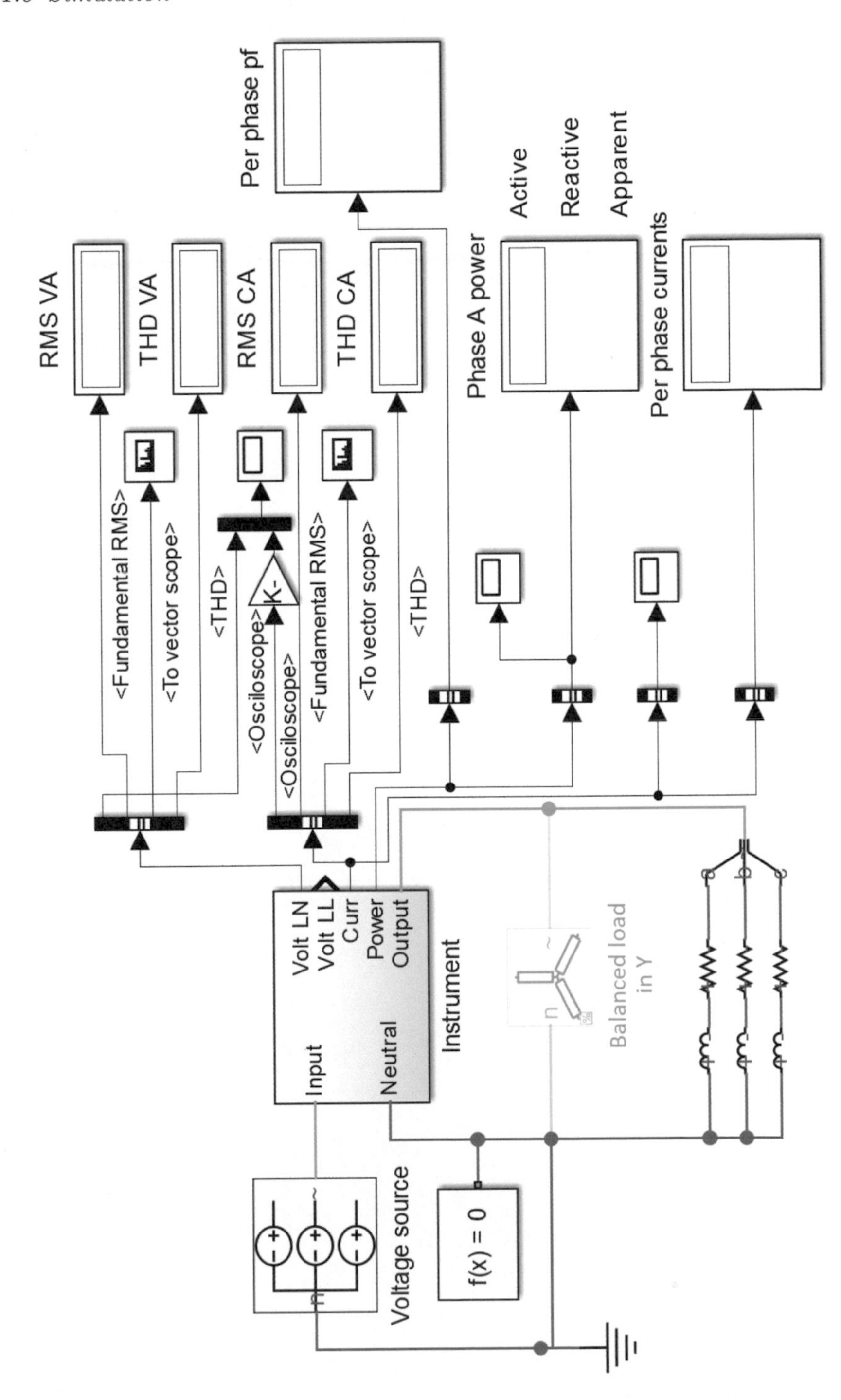

FIGURE 1.44
Load unbalancing simulation

2

Power Flow and Electric Machinery Basics

As a first step into the electric grid, the fundamental topics about power generation, transmission, distribution, and consumption must be established. Each phase must then be first described and appropriately placed among the others so the power flow is thoroughly understood. This chapter aims to introduce the electric grid by presenting how the power flow occurs and by offering the modeling basics around its major electromagnetic components.

An electric machine is a device that can convert either mechanical energy to electrical energy or electrical energy to mechanical energy. When such a device is used to convert mechanical energy to electrical energy, it is called a generator. When it converts electrical energy to mechanical energy, it is called a motor. Since any given electric machine can convert power in either direction, any machine can be used as either a generator or a motor. Almost all practical motors and generators convert energy through the interaction of a magnetic field, and only machines using magnetic fields to perform such conversions are considered in this book.

Many concepts must be established before performing any analysis of electric machines. The principle of electromechanical energy conversion is the most important law of machine analysis. This theory allows us to establish an expression of electromagnetic torque in terms of machine variables, like the currents and the displacement of the mechanical system. In this chapter, basic principles that will be mentioned are the equivalent circuit representations of magnetically coupled circuits, the concept of a sinusoidally distributed winding, and the winding inductances, among others.

2.1 A glance into power flow

2.1.1 Generators

The generators produce the electrical energy distributed by a power system. Almost all of the generators in use today produce electrical energy by converting mechanical energy to electrical energy through the action of a magnetic field. The mechanical energy comes from a prime mover, which is the device that spins the generator. Prime movers are usually some form of steam or water turbines, but diesel engines are sometimes used in remote locations. Modern generators generate electrical power at voltages of $13.8 - 24$ $[kV]$.

To construct an electric generator, the principle of "electromagnetic induction" is used, as one of the fundamental elements of matter is precisely the electromagnetic charge composed of a magnetic field and an associated particle-motion electric field. A bosons generator uses a magnetic field to energize electrons kinetically and cause an interaction with other electrons, which results in the generation of electric current and a voltage. Manipulating an electromagnetic force may induce the displacement or movement of electrons, and as a result, an electric current will flow.

As the rotor rotates at high speeds thanks to external mechanical energy from the turbine, current flows occur in the stator copper wires (1 $[A]$ is equal to 6.25×10^{18} electrons moving through a wire per second). All power plants have turbines and generators. Some turbines are powered by wind, water, and steam from the earth or from the combustion of biomass, fossil fuels, and other forms of energy. The electricity produced by a generator is transmitted through the transmission lines, taking the power from plants to homes, schools, and industry. Large-scale power generation requires huge power plants, composed of many turbines.

A specific view of modern generation technologies is portrayed in some chapters throughout this book: Hydroelectric generation is covered in Chapter 3, Chapter 4 introduces wind power, and Chapter 5 deals with photovoltaic solar power generation. On the other hand, the basic tools to perform electric analyses of transformers and electric machinery are provided in the remainder of this chapter.

2.1.2 Transformers

Transformers convert AC electrical energy at one voltage magnitude into AC electrical energy at another voltage magnitude. They are essential for the operation of a modern power system, since transformers allow power to be transmitted with minimal losses over long distances. Transformers are the "connection" that holds the entire system together by increasing line voltages and reducing currents for transmission over long distances, and then reducing the voltages to the levels required by the end-users. They make modern power systems possible, and furthermore, they do it with very high efficiency.

The invention of the transformer and the current development of AC power sources forever eliminated range and power-level restrictions of power systems. A transformer ideally changes one AC voltage level to another without affecting the actual power supplied. If the transformer steps up the voltage level of a circuit, it decreases the current due to energy conservation. Therefore, AC electric power can be generated at one central location and stepped up for transmission over long distances (exhibiting very low losses), to finally be stepped down for its final use. Since the transmission losses in the lines of a power system are proportional to the square of the current in the lines, raising the transmission voltage and reducing the resulting transmission currents by a factor of 10 reduces power transmission losses by a factor of 100. Without the transformer, it would not be possible to use electric power in many of the ways it is used today.

In modern power systems, transformers are literally found everywhere. Electric power is generated at voltages of $12 - 25$ $[kV]$. Transformers at the output of the generators step up the voltage to $110 - 1000$ $[kV]$ for transmission. Substation transformers then step down the voltage to the $13 - 34.5$ $[kV]$ range, and distribution transformers finally permit the power to be used safely in homes, offices, and factories at voltages as low as 120 $[V]$.

2.1.3 Power lines

Power lines connect generators to loads, transmitting electrical power from one to the other with minimal losses. Power lines are usually divided into two categories: transmission lines and distribution lines. Transmission lines are designed to transmit electrical power efficiently over long distances. They run at very high voltages to reduce the resistive (I^2R) losses in the lines. Once the power reaches the vicinity of the user, its voltage is stepped down, and the power is supplied through distribution lines to the final customers. Distribution lines carry much less power than transmission lines, and for shorter distances, so they can operate at lower power voltages without prohibitive losses.

Generator and load are connected together through transmission lines, which transport electric power from the place where it is generated to the place where it is used. Efficient transmission lines are very important to modern power systems, because power generation is usually done at large electric generating stations located a long way from the cities and industries where the power is consumed. Most electric power consumers use small amounts of power, and they are scattered over wide areas. Transmission lines must take the bulk power from the generators, transmit it to the locality where it will be used, and then distribute it to the individual homes and factories.

As a rule of thumb, the power-handling capability of a transmission line is proportional to the square of the voltage on the line. Therefore, very high voltage transmission lines are used to transmit electric power over long distances to the areas where it will be used. Once the power reaches the area where it will be used, its voltage is stepped down into distribution substations, and then delivered to the individual customers on lower-voltage transmission lines called distribution lines.

There are two types of power transmission lines: overhead lines and buried cables. An overhead transmission line usually consists of three conductors or bundles of conductors containing the three phases of the power system. The conductors are usually aluminum cable steel reinforced (ACSR), with a steel core and aluminum conductors wrapped around the core. The steel core provides strength, while the aluminum conductors have a low resistance to minimize the losses in the transmission line. In an overhead transmission line, the conductors are suspended from a pole or tower via insulators. In addition to the phase conductors, a transmission line usually includes one or two steel wires known as ground or shield wires. These wires are electrically connected to the tower and the ground, and are thus at ground potential. In large transmission lines, these wires are located above the phase conductors, shielding them from lightning strikes, etc.

Cable lines are designed to be placed underground or under the water. In cables, the conductors are insulated from one another and surrounded by a protective sheath. Cable lines tend to be more expensive than overhead transmission lines, and they are harder to maintain and repair. Furthermore, cable lines have capacitance problems that prevent them from being used over very long distances. Nevertheless, they are increasingly popular in new urban areas where overhead transmission lines are considered an eyesore.

Transmission lines are characterized by a series resistance and inductance per unit length, and by a shunt capacitance per unit length. These values control the power-carrying capacity of each transmission line, and the voltage drop in the transmission line at full load. The inductance and capacitance calculations for three-phase transmission lines with varying geometries and conductor bundle sizes will be covered in Chapter 6.

2.1.4 Protective devices

There is a wide variety of devices designed to protect the system, including current, voltage, power sensors, relays, fuses, and circuit breakers. There are two common types of failures in a power system: overloads and faults. Overloads are conditions in which some or all components in the power system are supplying more power than they can safely handle. Overloads can happen because the total demand on the power system simply exceeds the ability of the system to supply power. However, it is more common for overloads to occur in localized parts of the power system because of changes elsewhere within the system.

For example, two parallel transmission lines may be sharing the task of providing power to a city. If one of them is disconnected for some reason, the remaining line will supply the total power needed by the city. This may cause the line to be overloaded. If an overload occurs on a power system, it should be corrected, but power systems are robust enough for operators to have several minutes to correct the problem before severe damage occurs.

Faults are conditions in which one or more of the phases in a power system are shorted to ground or to each other. Faults also occur if a phase is open circuited. When a short circuit occurs, very large currents flow and these currents can damage the power system unless they are stopped quickly. Unlike overloads, faults must be cleared immediately, so relays are designed to automatically open circuit breakers and isolate faults as soon as they are detected.

Chapter 7 is devoted to providing a deeper view regarding fault conditions and protection systems.

2.1.5 Sensing transformers

A potential transformer (PT) is an especially sound transformer with a high-voltage primary and a low-voltage secondary. It has a very low power rating, and its sole purpose is to provide a sample of the power system's voltage to the instruments monitoring it. Since the principal purpose of the transformer is voltage sampling, it must be very accurate so as not to distort the true voltage values. Potential transformers of several accuracy classes may be purchased, depending on how accurate the readings must be for a given application.

On the other hand, current transformers (CT) measure the current in a line and reduce it to a safe and measurable level. The current transformer consists of a secondary winding wrapped around a ferromagnetic ring, with the single primary line running through the center of the ring. The ferromagnetic ring holds and concentrates a small sample of the flux from the primary line. That flux then induces a voltage and current in the secondary winding.

A current transformer differs from the other transformers described in this chapter as its windings are loosely coupled. Unlike all the other transformers, the mutual flux ΦM is smaller than the leakage flux ΦL. Because of the loose coupling, the voltage and current ratios do not hold. Nevertheless, the secondary current in a current transformer is directly proportional to the much larger primary current, and the device can provide an accurate sample of a line's current for measurement purposes. Current transformer ratings are given as ratios of primary to secondary current. A typical current transformer ratio might be $600 : 5$, $800 : 5$ or $1000 : 5$. A 5 $[A]$ rating is standard on the secondary of a current transformer.

It is important to keep a current transformer short-circuited at all times, since extremely high voltages can appear across its open secondary terminals. In fact, most relays and other devices using the current from a current transformer have a shorting interlock, which must be shut before the relay can be removed for inspection or adjustment. Without this interlock, very dangerous high voltages will appear at the secondary terminals as the relay is removed from its socket.

2.2 Modeling basics

2.2.1 Mechanical perspective

Almost all electric machines rotate about an axis, called the shaft of the machine. In general, a three-dimensional vector is required to completely describe the rotation of an object in space. However, machines normally turn on a fixed shaft, so their rotation is restricted to one angular dimension. Relative to a given end of the machine's shaft, the direction of rotation can be described as either clockwise (CW) or counterclockwise (CCW). A CCW angle rotation is assumed to be positive and a CW is assumed to be negative.

The angular position θ of an object is the angle at which it is oriented if measured from some arbitrary reference point. Angular position is usually measured in $[rad]$ or $[^\circ]$. It corresponds to the linear concept of distance along a line.

Similarly, angular velocity is the rate of change in angular position with respect to time. It is assumed positive if the rotation is in a CCW direction. Angular velocity is the rotational analog of the concept of velocity on a line. So, it is defined as the rate of change of the angular displacement with respect to time:

$$\omega = \frac{d\theta}{dt}. \tag{2.1}$$

Angular acceleration is the rate of change in angular velocity with respect to time. It is assumed positive if the angular velocity is increasing in an algebraic sense. Angular acceleration is the rotational analog of the concept of acceleration on a line. It is defined as

$$\alpha = \frac{d\omega}{dt}. \tag{2.2}$$

In linear mechanics, a force applied to an object causes its velocity to change. In the absence of a net force on the object, its velocity is constant. The greater the force applied to the object, the more rapidly its velocity changes. There exists a similar concept for rotation. When an object is rotating, its angular velocity is constant unless a torque is present on it. The greater the torque on the object, the faster the angular velocity of the object changes. The torque depends on the magnitude of the applied force and the distance between the axis of rotation and the line of action of the force.

For objects moving along a straight line, the following equation describes the relationship between the force applied toward its resulting acceleration:

$$F = ma, \tag{2.3}$$

where F is the net force applied to an object, m represents the mass of the object, and a is the acceleration. A similar equation describes the relationship between the torque applied to an object and its resulting angular acceleration. This relationship is given by

$$\tau = J\alpha, \tag{2.4}$$

where τ is the net applied torque, α is the resulting angular acceleration, and the term J serves the same purpose as an object's mass in linear motion, and is called the moment of inertia of the object.

For linear motion, work W is defined as the application of a force through a distance:

$$W = \int F dr. \tag{2.5}$$

It is assumed that the force is collinear with the direction of motion. For the special case of a constant force applied collinearly with the direction of motion, this equation becomes

$$W = F dr. \tag{2.6}$$

For rotational motion, work is the application of a torque through an angle:

$$W = \int \tau d\theta, \tag{2.7}$$

and if the torque is constant, we will get

$$W = \tau d\theta. \tag{2.8}$$

Power is the change of work in time,

$$P = \frac{dW}{dt}, \tag{2.9}$$

and it is usually measured in joules per second (Watts $[J/s] = [W]$), but also it can be measured in foot-pounds per second or in horsepower $[hP]$. By this definition and assuming that force is constant and collinear with the direction of motion, power is given by

$$P = \frac{dW}{dt} = \frac{d}{dt}(Fr) = F\left(\frac{dr}{dt}\right) = Fv. \tag{2.10}$$

Similarly, assuming constant torque, power in rotational motion is given by

$$P = \frac{dW}{dt} = \frac{d}{dt}(\tau\theta) = \tau\left(\frac{d\theta}{dt}\right) = \tau\omega. \tag{2.11}$$

Equation (2.11) is very important in the study of electric machinery because it can describe the mechanical power on the shaft of a motor or generator. This equation is the correct relationship among power, torque, and speed if power is measured in $[W]$, torque in $[Nm]$, and speed in $[rad/s]$.

2.2.2 Magnetics perspective

Electric machines are made of either magnetic materials or materials that have magnetic properties, so it is important to review the main magnetic principles, which are used in electric machines. The energy density of the magnetic core affects the size of the electric machine. Thus, ferromagnetic materials are used in electric machines because higher permeability gives a higher energy density. Usually the materials can be divided into ferromagnetic and nonmagnetic according to the magnetization curve, which is the relationship between the magnetic field intensity H (A/m) and the magnetic flux density B. This relationship is described by a curve that is well-known as the magnetization B-H curve. If this curve is a linear representation, the curve represents a nonmagnetic material. If this curve is a nonlinear description, it represents a ferromagnetic material [24].

Magnetic fields are the fundamental mechanism by which energy is converted from one form to another in motors, generators, and transformers. Four basic principles describe how magnetic fields are used in these devices:

- A current carrying wire produces a magnetic field in the area around it
- A time changing magnetic field induces a voltage in a coil of wire if it passes through that coil, which is the basis of transformer action
- A current carrying wire in the presence of a magnetic field has a force induced on it, which is the basis of motor action
- A moving wire in the presence of a magnetic field has a voltage induced in it, which is the basis of the generator action

One of the most important mathematical expressions that describe the magnetic field is the Gauss's law for magnetic fields. This expression describes the magnetic flux in a closed surface. It is important to remember that a magnetic field cannot be isolated in the south or north pole (monopoles); always, they are together in pairs. Thus, the integral form of Gauss can be defined as (2.12) and the result is zero (the dot product indicates how to determine the magnetic field parallel to $\hat{n}$ that is perpendicular to the surface).

$$\oint_S \overrightarrow{B} \cdot \hat{n} \, da = 0, \tag{2.12}$$

where $\hat{B}$ is the magnetic field in Teslas [T], $\hat{n}$ is the unit vector normal to the surface, and da is a change in the area in [m^2].

There are three cases for calculating the magnetic field. The first one is when the magnetic field vector is uniform and perpendicular to S. Thus, the magnetic flux is calculated by (2.13).

$$\phi_B = \overrightarrow{B} \times A, \tag{2.13}$$

where A is the area of the surface.

The second case is the magnetic field, uniform and at an angle to S ().

$$\phi_B = \overrightarrow{B} \cdot \hat{n} \times A. \tag{2.14}$$

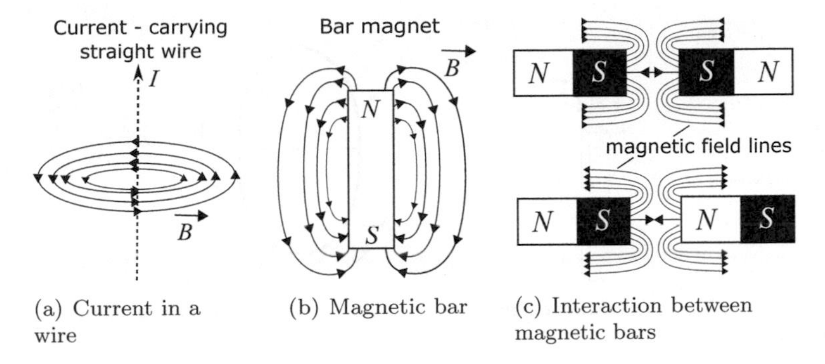

(a) Current in a wire

(b) Magnetic bar

(c) Interaction between magnetic bars

FIGURE 2.1
Magnetic fields

The third case is the magnetic field that is non-uniform and at a variable angle to S:

$$\oint_S \overrightarrow{B} \cdot \hat{n}\, da = 0 \tag{2.15}$$

The basic considerations for sketching magnetic fields are [61]:

- They form close loops because they do not begin and end on specific charges
- The magnetic field lines begin on the north pole and terminate on the south pole, and they are continuous loops
- The magnetic field can be defined as a vector so it can be added when two or more sources overlap at the same point. Thus, a total magnetic field at this point is generated.

Examples of magnetic fields are presented in Figure 2.1. It shows that the magnitude of the current is proportional to the magnitude of B but the magnitude of B is inversely proportional to the distance from the conductor.

In addition, Faraday's law deals with magnetic flux when it is changing. Furthermore, this equation offers a link between the change of magnetic field and the induced EMF. This law is expressed in (2.16). It is important to observe that the change of magnetic field can be generated under the following assumptions [61]:

- The strength of the magnetic field decreases or increases
- The angle between $\overrightarrow{B}$ and the surface normal changes
- The area of the surface S changes

Also, a combination of those assumptions can pop up. As a result, the right hand principle of Faraday's law will not be zero.

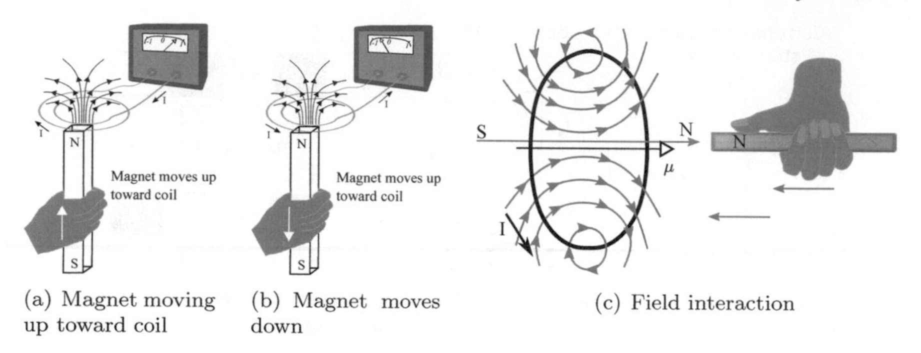

(a) Magnet moving up toward coil (b) Magnet moves down (c) Field interaction

FIGURE 2.2
Lenz law defines the direction of the current

$$\oint_C \overrightarrow{E} \cdot \overrightarrow{dl} = -\frac{d}{dt} \int_S \overrightarrow{B} \cdot \hat{n} \, da, \tag{2.16}$$

where the electric field is expressed by $\overrightarrow{E}$ that is a vector, and the increment segment of the path C is $\overrightarrow{dl}$.

The sign of the right-hand side of Faraday's law is defined by the Lenz law according to Figure 2.2. The EMF is produced when the magnetic flux is changing, so a current is generated. This current generates a magnetic field opposing the flux change that produced it. The Lenz law can be defined using the energy conservation principle because the direction of the current opposes the change. When the change appears, the current appears to limit it. Thus, the current avoids an increment of the change that induces a bigger current that increases the change.

An electrical charge in motion experiences a force when an electromagnetic field is an area of influence of this electromagnetic field. The Lorentz equation is defined by (2.17) in which the force direction and magnitude are described by the cross-product [61].

$$F = qu \times B, \tag{2.17}$$

where F is the force [N], Q is the value of the electric charge [C], u is the velocity of the charge [m/s], and B is the magnetic flux density [T].

If it is supposed that a current of magnitude $\mathbf{I}$ flows in a conductor of length l within a uniform magnetic field, it is possible to rewrite Equation (2.17) and get Equation (1.7)

$$F = l\mathbf{I} \times B. \tag{2.18}$$

A wire experiences a force when a current is flowing through it and an external magnetic field is acting across that current flow. The force is perpendicular to the magnetic field and the current as it is shown in Figure 2.3,

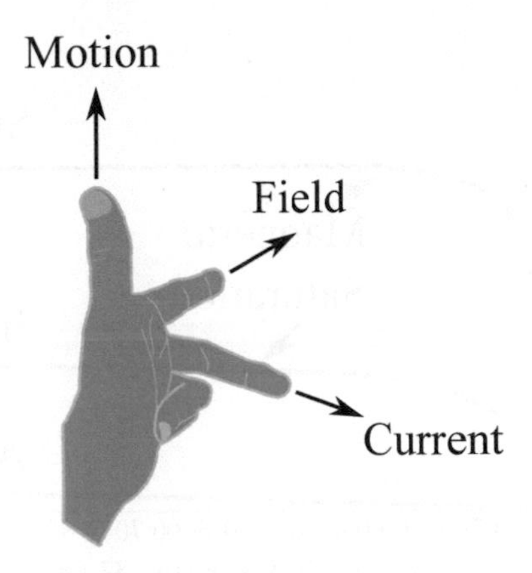

FIGURE 2.3
Fleming's left-hand rule

which depicts Fleming's left-hand rule [61] for determining the direction of the force (motion).

Another basic law governing the generation of a magnetic field by some current is Ampere's law:

$$\oint H \cdot dl = I_{net}, \tag{2.19}$$

where H is the magnetic field intensity produced by the current I_{net} , and dl is a differential element of length along the path of integration.

The magnetic field intensity H is, in a sense, a measure of the effort that current is putting into the establishment of the magnetic field. The strength of the magnetic field flux produced in the core also depends on the material of the core. The relationship between the magnetic field intensity H and the resulting magnetic flux density B produced within a material is given by

$$B = \mu H, \tag{2.20}$$

where μ is the magnetic permeability of the material. The actual magnetic flux density produced in a piece of material is thus given by a product of two terms: H, representing the effort exerted by the current to establish a magnetic field, as well as μ, representing the relative ease of establishing a magnetic field in a given material.

The permeability of any other material compared to the permeability of free space is called its relative permeability and it is defined as

$$\mu_r = \frac{\mu}{\mu_0}. \tag{2.21}$$

Relative permeability is a convenient way to compare the "magnetizability" of materials. For example, the steels used in modern machines have relative permeability of 2000 to 6000 (or even more). This means that, for a given

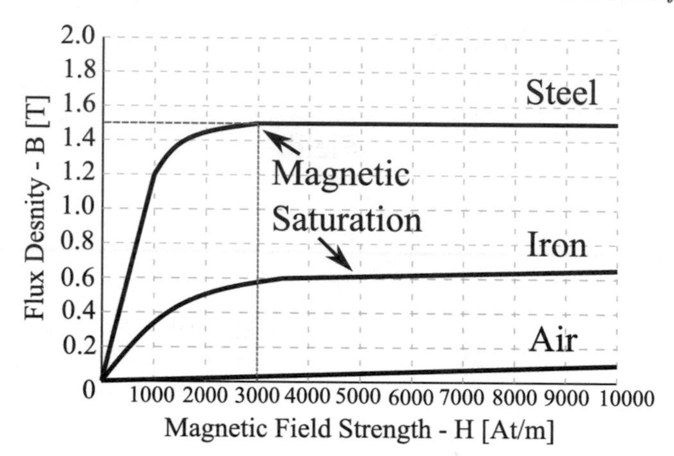

FIGURE 2.4
Magnetizing B-H curve

amount of current, 2000 to 6000 times more flux is established in a piece of steel than in a corresponding area of air. The permeability of air is essentially the same as the permeability of free space ($\mu_0 = 4\pi \times 10^{-7}$ $[H/m]$). Usually, the magnetizing B-H curve shows a linear region and saturation region. The transition between the saturation and the linear region is called the knee (see Figure 2.4).

Obviously, the metal core in a transformer or electric machine plays an extremely important part in increasing and concentrating the magnetic flux in the device. Also, because the permeability of iron is so much higher than that of air, the great majority of the flux in an iron core, as in Figure 2.8, remains inside the core instead of traveling through the surrounding air, which has much lower permeability. The small flux that does leave the iron core is very important in determining the flux linkages between coils and the self-inductances of coils in transformers and motors. In a core such as the one shown in Figure 2.8, the magnitude of the flux density is given by

$$B = \mu H = \frac{\mu N i}{l_c}. \tag{2.22}$$

Now, the total flux in a given area is given by

$$\phi = \int_A B \cdot dA, \tag{2.23}$$

where dA is the differential unit of area. If the flux density vector is perpendicular to a plane of area A, and if the flux density is constant throughout the area, then this equation reduces to

$$\phi = BA. \tag{2.24}$$

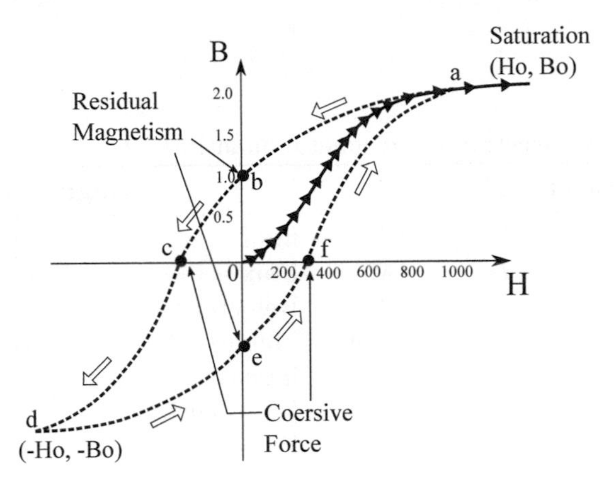

FIGURE 2.5
Hysteresis

Thus, the total flux in the core in Figure 2.8 due to the current i in the winding is

$$\phi = BA = \frac{\mu N i}{l_c}, \tag{2.25}$$

where A is the cross sectional area of the core.

An important effect called hysteresis is presented when the magnetic field increases from positive to negative value. Hysteresis is characterized by the residual magnetism and the coercive force that appear in the B-H curve (see Figure 2.5). This magnetizing curve is determined according to the ferromagnetic material.

To study magnetic circuits, analogies between magnetic and electric circuits can be created. Those analogies allow us to describe the magnetic circuit and solve it using the same laws as for electric circuits. Table 2.1 shows a comparison of electric and magnetic circuits that was presented by [24]. In this table, the reluctance is one of the most important quantities that is determined using the material permeability, the mean length, and the area. The reluctance is an analog to the resistance.

Using the information of Table 2.1, an electric circuit can be created by the magnetic circuit (see Figure 2.6). Electric circuits can represent complex magnetic circuits, so the laws of current and voltages can be used for solving such circuits.

Example 2.1. *Consider Figure 2.7 to represent a quadrangular piece of material with $B = 1$ [T], L_c=0.8 [m], $L_g = 0.8 \times 10^{-3}$ [m], μ_R=3000, and a transversal area $A_g = A_c = 16 \times 10^{-4}$ [m^2]. Calculate the current necessary to hold B.*

First we can calculate the reluctances of the metallic part of the circuit and

TABLE 2.1
Analogies between electric and magnetic quantities [24]

Electrical			Magnetic		
Voltage	V	[V]	mmf	$\mathscr{F}$	[A-t]
Current	I	[A]	Flux	ϕ	[Wb]
Resistance	R	[Ω]	Reluctance	$\mathscr{R}$	[H^{-1}]
Electric field intensity	E	[V/m]	Magnetic field intensity	H	[A-t/m]
Current density	J	[A/m^2]	Flux density	B	[T]
Conductivity	σ	[S/m]	Permeability	μ	[H/m]
Ohm's law	$V = IR$			$\mathscr{F} = \phi\mathscr{R}$	
Flow density	$J = \sigma E$			$B = \mu H$	
Admitance	$R = \frac{l}{\sigma A} = \frac{1}{G}$			$\mathscr{R} = \frac{l}{\mu A} = \frac{1}{\mathscr{P}}$	

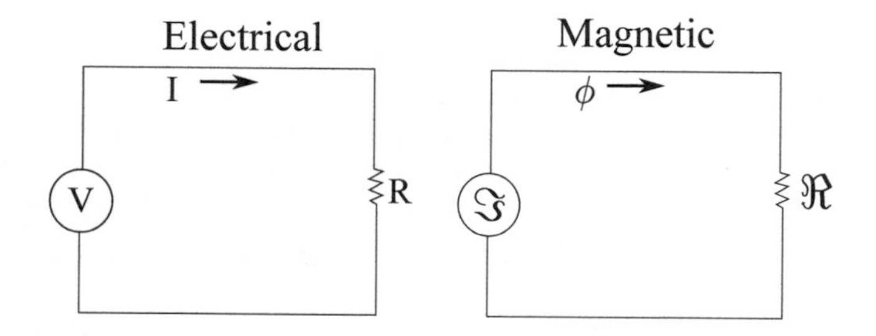

FIGURE 2.6
Analogy between electric and magnetic circuits

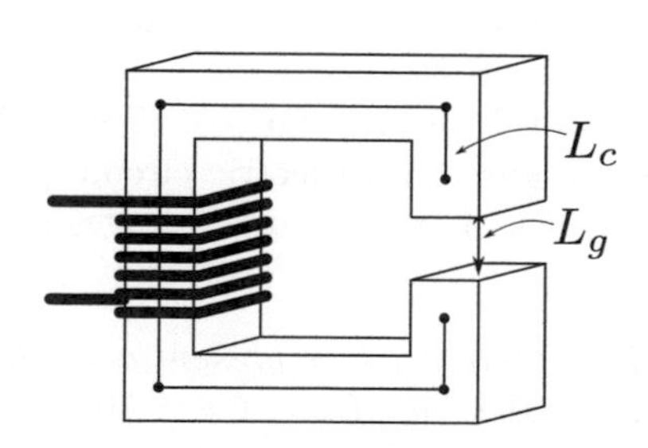

FIGURE 2.7
Schematic for Example 2.1

the air-gap:

$$\mathscr{R}_g = \frac{L_g}{\mu_0 A_g} = \frac{0.8 \times 10^{-3}}{(4\pi \times 10^{-7})\,(16 \times 10^{-4})} = 3.98 \times 10^5,$$

$$\mathscr{R}_c = \frac{L_c}{\mu_0 \mu_R A_c} = \frac{0.8}{(4\pi \times 10^{-7})\,(3000)\,(16 \times 10^{-4})} = 1.33 \times 10^5.$$

The flux can be directly calculated as follows:

$$\Phi = B \cdot A = (1)\left(16 \times 10^{-4}\right) = 1.6 \times 10^{-4}\ [Wb],$$

and the mmf can be obtained as the flux was previously computed:

$$\mathscr{F} = (\mathscr{R}_g + \mathscr{R}_c)\,\Phi = 84.88\ [A - t].$$

With the previous results, the current can be directly computed as:

$$I = \frac{\mathscr{F}}{N} = \frac{84.88\ [A - t]}{600\ [t]} = 0.1415\ [A]. \tag{2.26}$$

2.2.3 Magnetically coupled circuits

Magnetically coupled electric circuits are central to the operation of transformers and electric machines. In the case of transformers, stationary circuits are magnetically coupled for the purpose of changing the voltage and current levels. Circuits in relative motion are magnetically coupled for the purpose of transferring energy between mechanical and electrical systems. Since magnetically coupled circuits play such an important role in power transmission and conversion, it is important to establish the equations that describe their behavior and to express these equations in a form convenient for analysis.

Two stationary electric circuits are magnetically coupled as shown in Figure 2.8. The two coils consist of turns N_1 and N_2 wound on a common core, which is a ferromagnetic material with large permeability relative to that of air. This scheme of the circuit represents the main function of a transformer in which coupled inductors are assumed to have perfect coupling and zero power loss. The transformer is a device that uses magnetic coupling between two inductors. These devices are used to step up or step down AC voltages and currents, commonly used to reduce the voltages and to isolate one circuit from another.

The permeability of free space is $4\pi \times 10^{-7}$ $[H/m]$. This permeability is represented by μ_0, in which the permeability of other materials can be expressed as $\mu = \mu_r \mu_0$ where μ_r is the relative permeability. In the case of transformer steel the relative permeability may be as high as 2000 to 4000.

The flux produced by each coil, for better analysis, can be separated into two components: a leakage component denoted with an l and a magnetizing

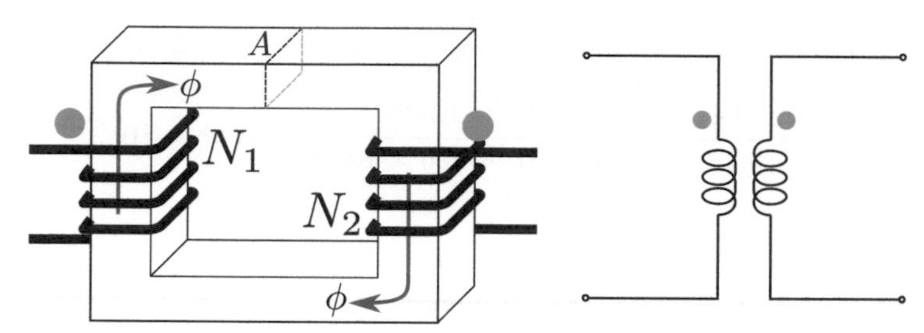

FIGURE 2.8
Magnetically coupled circuit

component denoted by an m. Each of the components are depicted by a single streamline with the positive direction determined by applying the right-hand rule to the directions of current flow in the coil. It is very important to take into account that transformer analysis, i_2, is selected positive out of the top of coil 2 and a dot placed at that terminal. The flux linking each coil may be expressed as

$$\Phi_1 = \Phi_{l1} + \Phi_{m1} + \Phi_{m2}, \tag{2.27}$$
$$\Phi_2 = \Phi_{l2} + \Phi_{m2} + \Phi_{m1}. \tag{2.28}$$

The leakage flux Φ_{l1} is produced by current flowing in coil 1 and it links only the turns of coil 1. Likewise, the leakage flux Φ_{l2} is produced by current flowing in coil 2 and it links only the turns of coil 2. The magnetizing flux Φ_{m1} is produced by current flowing in coil 1 and it links all turns of coils 1 and 2. Similarly, the magnetizing flux Φ_{m2} is produced by current flowing in coil 2 and it also links all turns of coils 1 and 2.

With the selected positive direction of current flow and the way in which the coils are wound, the magnetizing flux produced by positive current in one coil adds to the magnetizing flux produced by positive current in the other coil. In other words, if both currents are actually flowing in the same direction, the magnetizing fluxes produced by each coil are in the same direction, making the total magnetizing flux or the total core flux the sum of the instantaneous magnitudes of the individual magnetizing fluxes. If the actual currents are in opposite directions, the magnetizing fluxes are in opposite directions. In this case, one coil is said to be magnetizing the core, the other demagnetizing.

All of the leakage flux may not link all the turns of the coil producing it. Likewise, all of the magnetizing flux of one coil may not link all of the turns of the other coil. The number of turns is considered an equivalent number rather than the actual number. The inductances of the electric circuit resulting from the magnetic coupling are generally determined from tests. The voltage

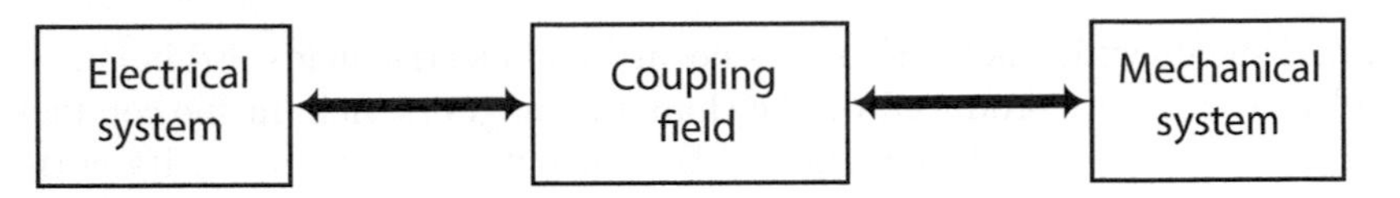

FIGURE 2.9
Block diagram of an electromechanical system

equations are expressed in matrix form as

$$v = ri + \frac{d\lambda}{dt}, \tag{2.29}$$

$$(f)^T = [f_1 f_2], \tag{2.30}$$

where $r = \text{diag}[r_1\ r_2]$, and f represents voltage, current of flux linkage.

The resistance r_1 and r_2, and the flux linkages λ_1 y λ_2 are related to coils 1 and 2. Since it is assumed that Φ_1 links the equivalent turns of coil 1 and Φ_2 links the equivalent turns of coil 2, the flux linkages are written as

$$\lambda_1 = N_1\Phi_1, \tag{2.31}$$

$$\lambda_2 = N_2\Phi_2, \tag{2.32}$$

where Φ_1 and Φ_2 are given by Equations (2.27) and (2.28) respectively.

2.2.4 Electromechanical energy conversion and losses

Electromechanical devices are used in many systems, and machines are the most common. A block diagram of an elementary electromechanical system is shown in Figure 2.9.

Electromechanical systems are comprised of an electrical system. The electrical and mechanical systems can interact. Interaction can take place through any and all electromagnetic and electrostatic fields that are common to both systems, and energy is transferred from one system to the other as a result of this interaction. Both electrostatic and electromagnetic coupling fields may exist simultaneously, and the electromechanical system may have any number of electrical and mechanical systems.

However, before considering an involved system, it is helpful to analyze the electromechanical system in a simplified form. An electromechanical system with one electrical system, one mechanical system, and one coupling field is shown in Figure 2.9. Electromagnetic radiation is neglected, and it is assumed that the electrical system operates at a frequency sufficiently low so that the electrical system may be considered a lumped-parameter system.

Losses occur in all components of the electromechanical system. Heat loss will occur in the mechanical system due to friction and the electrical system will dissipate heat through the resistance of the current-carrying conductors. Eddy currents and hysteresis losses occur in the ferromagnetic material of all magnetic fields while dielectric losses occur in all electric fields. Eddy currents are electric currents produced by the changing magnetic field in the conductor while hysteresis is the dependence of the output of a system on its current input and past inputs.

If W_E is the total energy supplied by the electrical source and W_M the total energy supplied by the mechanical source, then the energy distribution could be expressed as

$$W_E = W_e + W_{eL} + W_{eS}, \tag{2.33}$$
$$W_M = W_m + W_{mL} + W_{mS}. \tag{2.34}$$

In Equation (2.33), W_{eS} is the energy stored in the electric or magnetic fields that are not coupled with the mechanical system. The energy W_{eL} is the heat loss associated with the electrical system. These losses occur because of the resistance of the current-carrying conductors as well as the energy dissipated from these fields in the form of heat due to hysteresis, eddy currents, and dielectric losses. The energy W_E is the energy transferred to the coupling field by the electrical system. The energies common to the mechanical system are defined in a similar manner.

In Equation (2.34), W_{mS} is the energy stored in the moving member and compliances of the mechanical system, W_{mL} is the energy loss of the mechanical system in the form of heat, and W_m is the energy transferred to the coupling field. It is important to note that with the convention adopted, the energy supplied by either source is considered positive. Therefore, $W_E(W_M)$ is negative when energy is supplied to the electrical source (mechanical source).

If W_F is defined as the total energy transferred to the coupling field, then we obtain

$$W_F = W_f + W_{fL}, \tag{2.35}$$

where W_f is the energy stored in the coupling field and W_{fL} is the energy dissipated in the form of heat due to losses within the coupling field. In this case, these losses could be the hysteresis, eddy currents, or dielectric losses, as mentioned before. The electromechanical system must obey the law of conservation of energy, thus the next equation is expressed as

$$W_f + W_{fL} = (W_E - W_{eL} - W_{eS}) + (W_M - W_{mL} - W_{mS}). \tag{2.36}$$

From Equations (2.33), (2.34), and (2.36) we simplify, so Equation 2.37 can be written:

$$W_f + W_{fL} = W_e + W_m. \tag{2.37}$$

This energy relationship is shown schematically in Figure 2.10.

The actual process of converting electrical energy to mechanical energy, or vice versa, is independent of:

- The loss of energy in either the electrical or the mechanical system, expressed as $W_{eL} - W_{mL}$

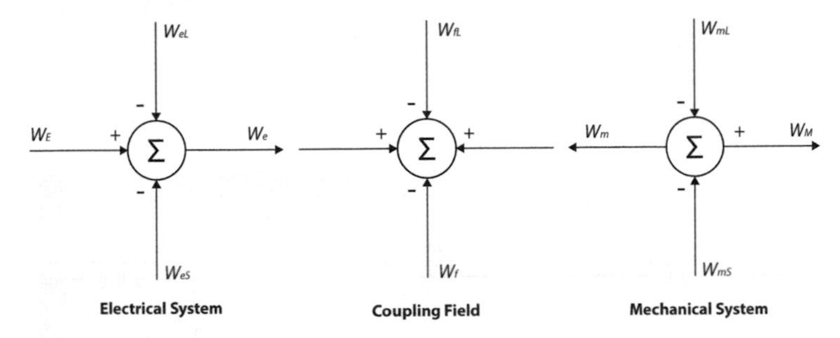

FIGURE 2.10
Energy balance

- The energies stored in the electric or magnetic field that are not common to both systems, denoted by W_{eS}
- The energies stored in the mechanical system, shown as W_{mS}. If the losses of the coupling field are neglected, then the field is conservative and becomes

$$W_f = W_e + W_m. \tag{2.38}$$

The examples of elementary electromechanical systems are shown in Figure 2.11, representing a magnetic coupling field, while the electromechanical system shown in Figure 2.12, employs an electric field as a means of transferring energy between the electrical and mechanical systems.

2.3 Machine parameters

The stator windings of the synchronous machine are embedded in slots around the inside circumference for the stationary member. In the 2-pole machine, each phase winding of the 3-phase stator winding is displaced 120° with respect to each other, as illustrated in Figure 2.13. The field of fd winding is wound on the rotating member. The as, bs, cs, and fd axes denote the positive direction of the flux produced by each of the windings. The as, bs, and cs windings are identical so that each winding has the same resistance and the same number of turns.

When a machine has three identical stator windings arranged as shown in Figure 2.13, it is often referred to as a machine with symmetrical stator windings. The symmetrical induction machine has identical multiphase stator windings and identical multiphase rotor windings. An unsymmetrical induction machine has non-identical multiphase stator windings, which generally are 2-phase, and symmetrical multiphase rotor windings.

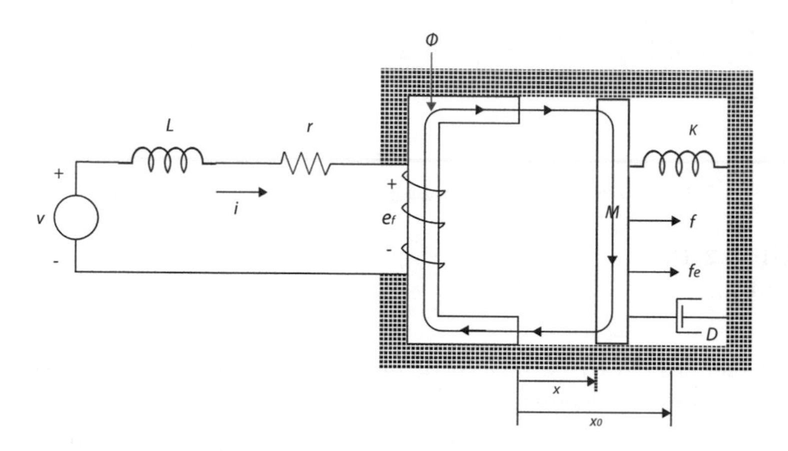

FIGURE 2.11
Electromechanical system with magnetic field

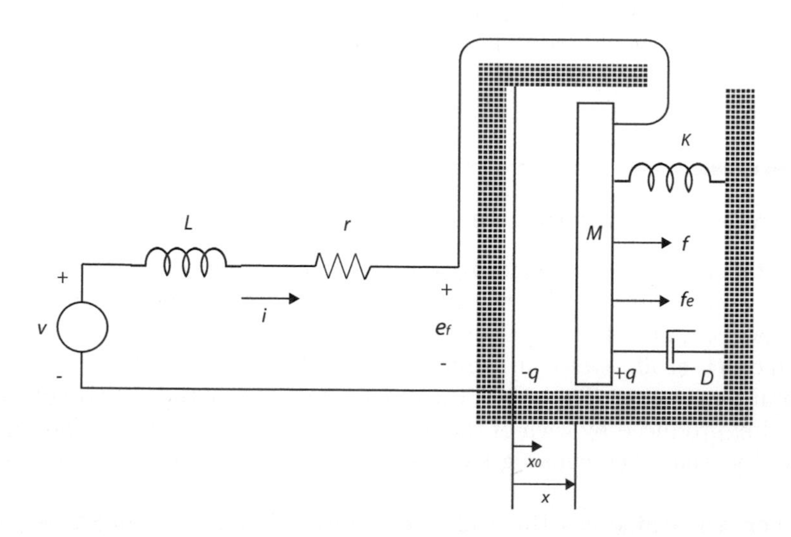

FIGURE 2.12
Electromechanical system with electric field

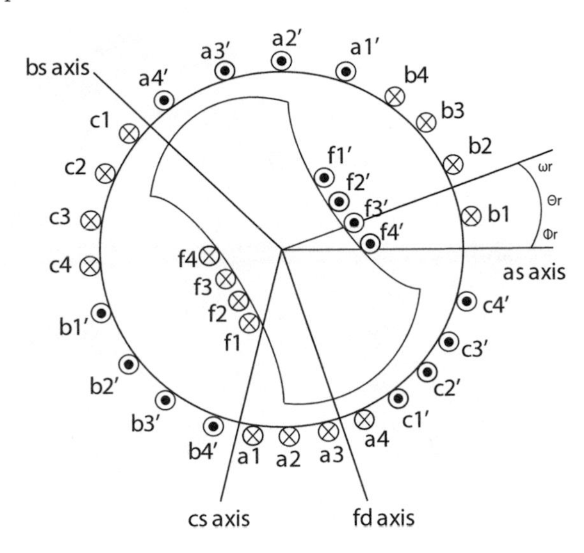

FIGURE 2.13
Elementary 2-pole, 3-phase, Y-connected salient-pole synchronous machine

It is assumed that each coil spans π radians of the stator for a 2-pole machine. One side of the coil (coil side) is represented by $\otimes$, indicating that the assumed positive direction of the current is down the length of the stator (like if it were moving into this page). The point inside the circle, $\odot$, indicates that the assumed positive direction of the current is flowing from the inside of this page. Each coil contains n_c conductors, each of which makes up an individual single conductor coil within the main coil. Thus the number of turns of each winding is determined by the product of n_c and the number of coils' sides carrying current in the same direction.

The winding configuration shown in Figure 2.13 is an oversimplification of a practical machine. The coil sides of each phase winding are considered to be distributed uniformly over 60° of the stator circumference. Generally, the coil sides of each phase are distributed over a larger area, perhaps as much as 120°, in which case it is necessary for some of the coil sides of two of the phase windings to occupy the same slot.

In some cases, the coil sides may not be distributed uniformly over the part of the circumference that the coil occupies. The windings span π radians for the 2-pole machine. This is referred to as full-pitch winding. In order to reduce voltage and current harmonics, the windings are often wound so that they span slightly less than π radians for a 2-pole machine. This is referred to as a fractional-pitch winding.

A salient-pole synchronous machine is selected for consideration since the analysis of this type of machine may be easily modified to account for other machine types. However, a salient-pole synchronous machine would seldom be a 2-pole machine except in the case of small reluctance machines, which are

FIGURE 2.14
A synchronous machine

of the synchronous class but do not have a field winding. Generally, 2- and 4-pole machines are round-rotor machines with the field winding embedded in a solid steel (nonlaminated) rotor. Salient-pole machines generally have a large number of poles composed of laminated steel whereupon the field winding is wound around the poles similar to that shown in Figure 2.13.

After directly measuring stator resistances and finding the coils' distribution, the next step is to determine the self- and mutual inductances of the machine windings. It is advantageous to use the elementary 2-pole, 3-phase synchronous machine to explain these inductance relationships. This procedure can be readily modified to account for additional windings (damper windings) placed on the rotor of the synchronous machine or for a synchronous machine with a uniform air gap (round rotor). These inductance relationships are easily altered to describe the winding inductances of an induction machine.

Figure 2.13 and Figure 2.15 are redrawn with the windings portrayed as sinusoidally distributed windings. In a magnetically linear system, the self-inductance of a winding is the ratio of the flux linked by a winding to the current flowing in the winding with all other winding currents zero. A synchronous machine is shown in Figure 2.14.

Mutual inductance is the ratio of flux linked by one winding due to current flowing in a second winding with all other winding currents zero, including the winding for which the flux linkages are being determined. For this analysis it is assumed that the air-gap length is approximated as

$$g(\Phi_r) = \frac{1}{\alpha_1 - \alpha_1 \cos^2(\Phi_r)}, \tag{2.39}$$

$$g(\Phi_s - \theta_r) = \frac{1}{\alpha_1 - \alpha_1 \cos^2(\Phi_s - \theta_r)}. \tag{2.40}$$

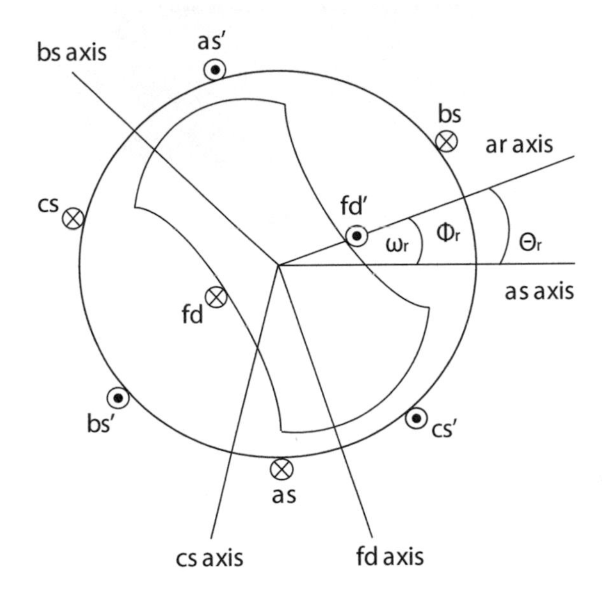

FIGURE 2.15
Elementary 2-pole, 3phase, Y-connected salient-pole synchronous machine

The winding arrangement of a 2-pole, 3-phase, Y-connected symmetrical induction machine is shown in Figure 2.17. The stator windings are identical with equivalent turns N_s and resistance r_s. The rotor windings, which may be wound or forged as a squirrel cage winding, can be approximated as identical windings with equivalent turns N_r and resistance r_r. The air gap of an induction machine is uniform and it is assumed that the stator and rotor windings are approximated as sinusoidally distributed windings. An induction machine is shown in Figure 2.16.

In nearly all industrial applications, the induction machine is operated as a motor with the stator windings connected to a balanced 3-phase source and the rotor windings short-circuited. The principle of operation in this mode is quite easily deduced. With balanced 3-phase current flowing in the stator windings, a rotating air-gap MMF (magnetomotive force) is established, as in the case of the synchronous machine, which rotates about the air gap at a speed determined by the frequency of the stator currents and the number of poles. If the rotor speed is different from the speed of this rotating MMF, balanced 3-phase currents will be induced (thus the name induction)

FIGURE 2.16
Induction machine

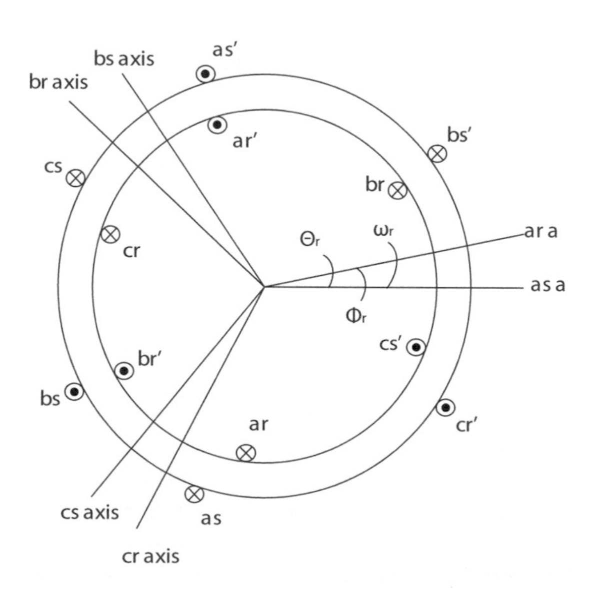

FIGURE 2.17
Two-pole, 3-phase, Y-connected symmetrical induction machine

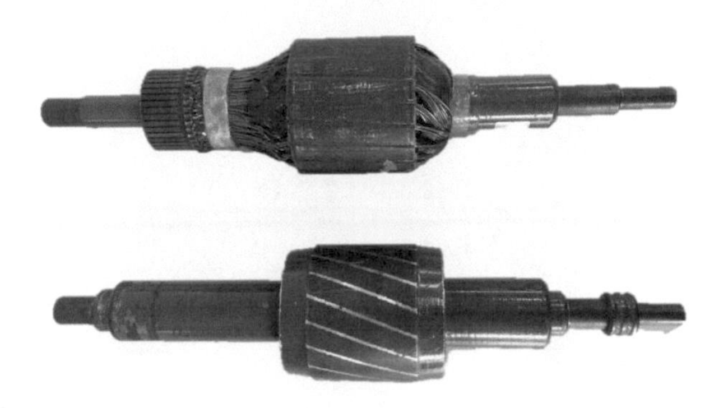

FIGURE 2.18
Wound rotor and squirrel cage rotor

in the short–circuited rotor windings. The frequency of the rotor currents corresponds to the difference in the speed of the rotating MMF due to the stator currents and the speed of the rotor.

The induced rotor currents will in turn produce an air-gap MMF that rotates relative to the rotor at a speed corresponding to the frequency of the rotor currents. The speed of the rotor air-gap MMF superimposed upon the rotor speed is the same speed as that of the air-gap MMF established by the currents flowing in the stator windings. These two air-gap MMFs rotating in unison may be thought of as two synchronously rotating sets of magnetic poles. Torque is produced due to an interaction of these magnetic poles. It is clear, however, that torque is not produced when the rotor is running in synchronism with the air-gap MMF due to the stator currents since in this case currents are not induced into the short-circuited rotor windings.

The winding inductance of the induction machine is expressed from the inductance relationships given for the salient-pole synchronous machine. In the case of the induction machine, the air gap is uniform. Thus, $2\theta_r$ variations in the self- and mutual inductances do not occur. These variations can be eliminated by setting $\alpha_2 = 0$ in the inductance relationship given for a salient-pole synchronous machine.

An induction machine can be a squirrel cage or wound rotor, as shown in Figure 2.18.

The squirrel cage is the most-employed electric machine due to its robustness, high efficiency, and low cost, while the wound rotor allows access to the rotor windings to modify its torque and speed curve, or even feed the electric machine to the rotor in order to increase its speed controllability. The disadvantage is that its maintenance is high due to the weathering of the brushes;

FIGURE 2.19
Frontal view slip-rings induction machine

in addition, friction of the brushes decreases the efficiency of the machine. In order to access a wound rotor, rings of displacement are needed, as shown in Figure 2.19.

The whole induction machine assembly is shown in Figure 2.20.

The test load of an induction machine measures the rotational losses of the motor and provides information about its magnetizing current. The test circuit is shown in Figure 2.21, where wattmeters, voltmeter, and three ammeters are connected. The test is performed on voltage (VTSC) and frequency.

On the other hand, the "locked rotor test" locks the rotor so that it cannot move, a voltage is applied to the motor and the voltage, the current, and the resulting power are measured. Figure 2.22 shows the connections for the test. First, the current flow value to be approximately full-loaded is set. When the current value is at full load, voltage, current, and power flowing in the motor are measured.

This test presents a problem. During normal operation, the stator frequency is the line frequency of the power system (50 or 60 $[Hz]$). In starting conditions, the rotor also has the line frequency. However, in normal operating conditions, the slip in most engines is only 2 to 4% and the resulting frequency of the rotor is in the range of 1 to 3 $[Hz]$. This creates a problem because the frequency of the line does not represent normal operating conditions of the rotor. A typical solution for class B and C motors (resistance versus frequency) is to use a frequency equal to or less than 25% of the nominal frequency. This test is performed to reach nominal current.

FIGURE 2.20
Induction machine assembly

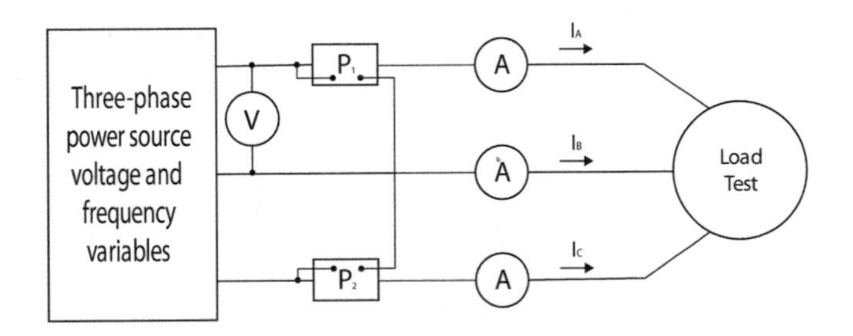

FIGURE 2.21
Load test circuit

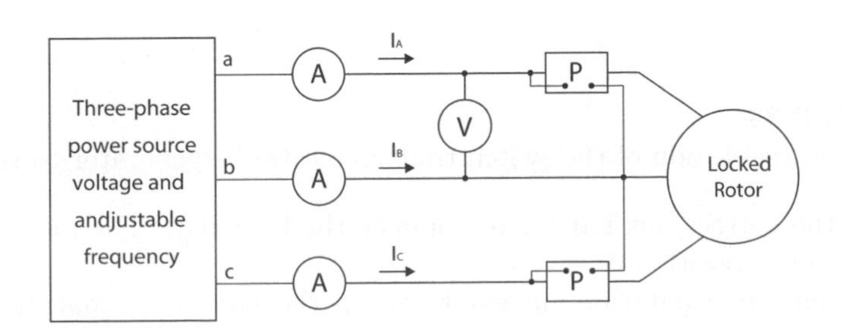

FIGURE 2.22
Locked rotor test circuit

2.3.1 Exercise generator load test

The generator will be connected to a varying load; thus the electrical parameters can be monitored. A manual change on the load will be applied to graphically monitor the electrical parameter changes. Use all the modules shown in the cabling diagram of the exercise and check that the power is off before connecting. The required material is at the end of the exercise and suggested connections appear in Diagram 2.1. Follow the instructions illustrated by Figures 2.23, 2.24, 2.25, and 2.26.

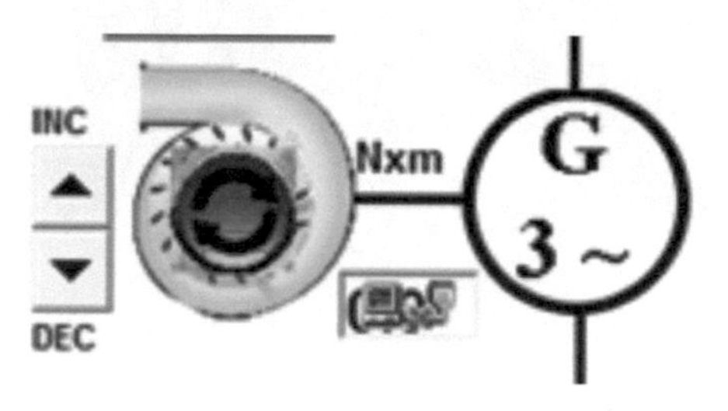

FIGURE 2.23
SCADA representation of the water governor to increase and decrease generator speed

Turn on the brushless driver. Increment the speed of the driver up to 1800 [rpm].

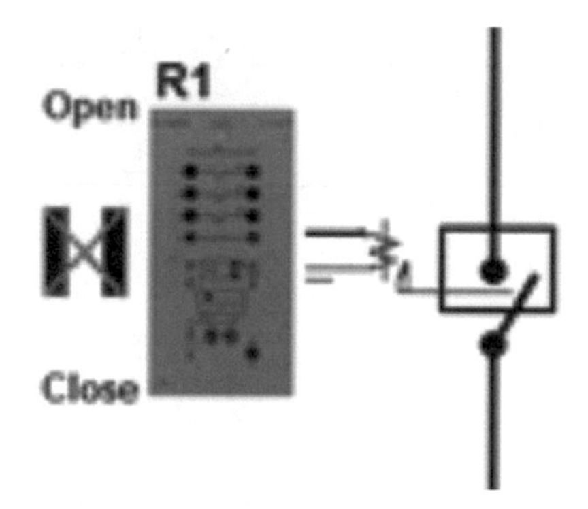

FIGURE 2.24
SCADA representation of the switch that connects the generator to the loads

Close the correct bus bar R1 to connect the load R-L. The load R-L has to be set to the position 0.

Increment the excitation current to reach the nominal armature voltage U_s.

Change the resistive load in a balanced way and observe the variations of VL12, V23, and VL31 due to the variation of load. Increment the excitation current I_e to level the output voltage U_s.

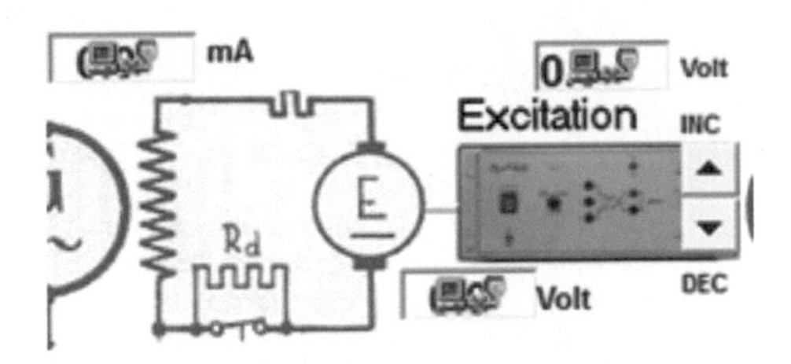

FIGURE 2.25
Excitation voltage

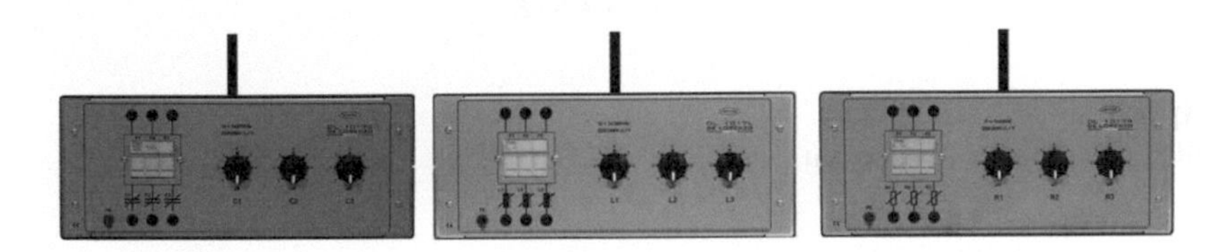

FIGURE 2.26
Loads

Change the inductive load in a balanced way and observe the variations of VL12, V23, and VL31 due to load variation. Increment the excitation current I_e to level the output voltage U_s.

Register the data and draw a graph relating the excitation current I_e to the variation of a purely resistive load and inductive load.

Table 2.2 has been plotted in Graphs 2.1 and 2.2.

Table 2.3 has been plotted in Graphs 2.3 and 2.4.

Suggested material:

- Brushless driver for generating mechanical energy
- Power circuit breaker
- Feeder manager relay
- Module for measuring the electric power
- Three-phase asynchronous generator
- Programmable automatic power supply unit
- RS485 communication
- Resistive load
- Inductive load

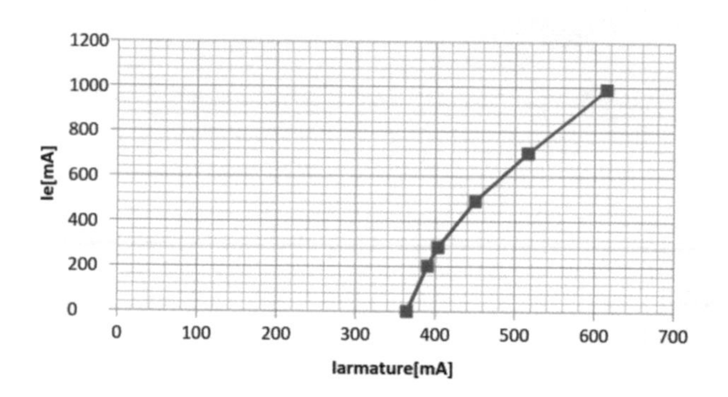

Graph 2.1
Resistive load with constant Us

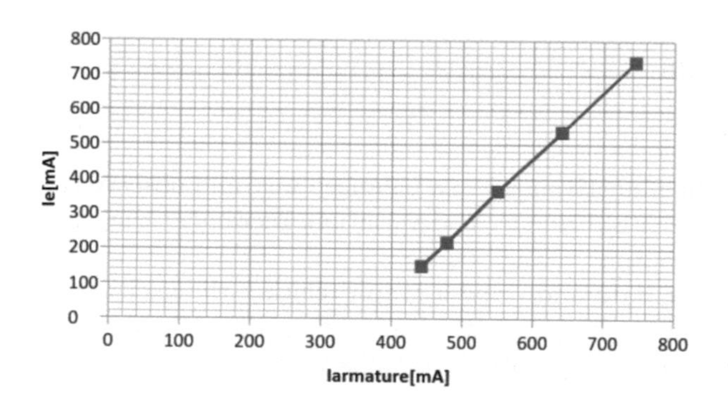

Graph 2.2
Inductive load with constant Us

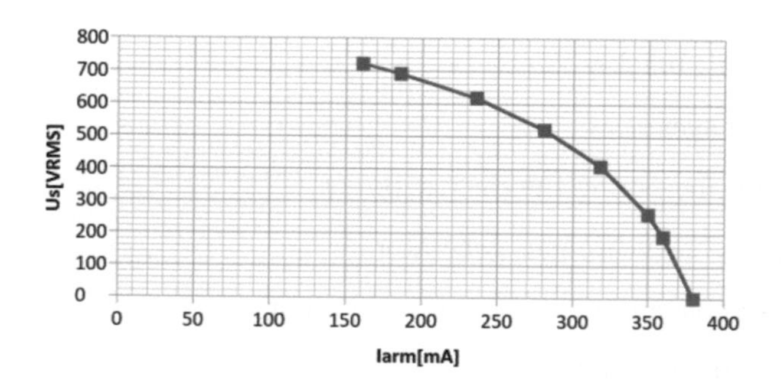

Graph 2.3
Resistive load with constant I_e

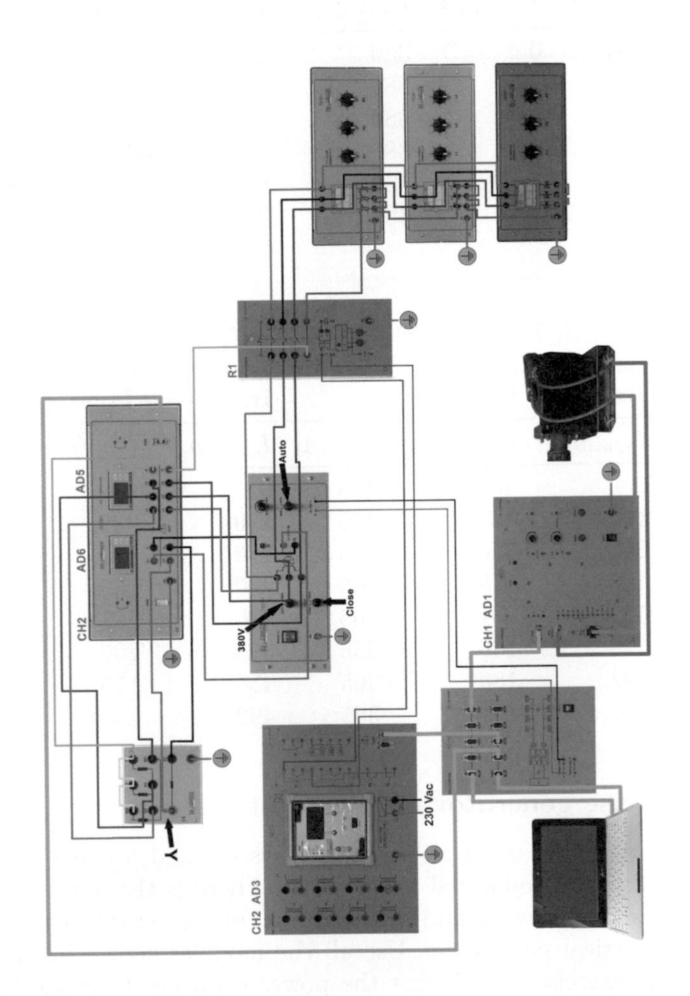

Diagram 2.1
Suggested connection to load test

TABLE 2.2
Data registered with $Us[V_{\mathrm{VRMS}}]$ constant, having I_e as the excitation current, I_{arm} as the armature current, and U_S as the supplied voltage

	Resistive Load			Inductive Load		
Balanced	$I_e[mA]$	$I_{arm}[mA]$	$Us[V_{RMS}]$	$I_e[mA]$	$I_{arm}[mA]$	$Us[V_{RMS}]$
Off	364	0	380	x	x	380
R1	390	202	380	442	150	380
R2	403	285	380	478	218	380
R3	450	490	380	550	366	380
R4	516	704	380	640	536	380
R5	615	990	380	745	740	380

TABLE 2.3
Data registered with $I_e[mA]$ constant, having I_e as the excitation current, I_{arm} as the armature current, and U_S as the supplied voltage

	Resistive Load			Inductive Load		
Balanced	$I_{arm}[mA]$	$Us[V_{RMS}]$	$I_e[mA]$	$I_{arm}[mA]$	$Us[V_{RMS}]$	$I_e[mA]$
Off	0	380	365	0	0	365
R1	192	360	365	130	326	365
R2	262	350	365	176	307	365
R3	410	318	365	260	269	365
R4	522	281	365	326	233	365
R5	618	236	365	390	206	365
R6	692	186	365	458	176	365
R7	722	161	365	492	156	365

2.3.2 Exercise generator no-load test

Start the alternator by using an excitation system and a drive motor. The alternator will not be connected to any load. There is the option to change the speed of the driver motor and the excitation of the generator to verify the behavior of electrical parameters. Use all the modules shown in the cabling diagram of the exercise. Check that the power is off before connecting. The required material is described below and the suggested connection is shown in Diagram 2.2. Follow the instructions illustrated by Figures 2.27 and 2.28. You should obtain similar results as the ones shown in Table 2.4 and Graph 2.5.

Turn on the brushless driver. Increment the speed of the brushless up to 1800 $[rpm]$.

Now, with the alternator set at nominal speed, increment the excitation current to obtain the nominal voltage.

Monitor the different parameters ($V_{armature}$, $I_{armature}$, W, V_{ar}, S, I_{exc},

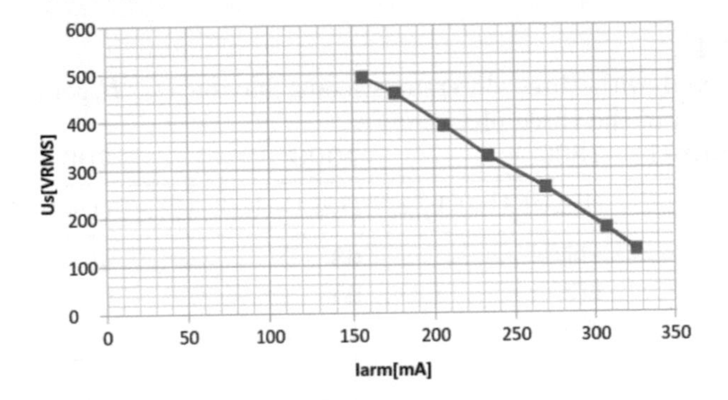

Graph 2.4
Inductive load with constant I_e

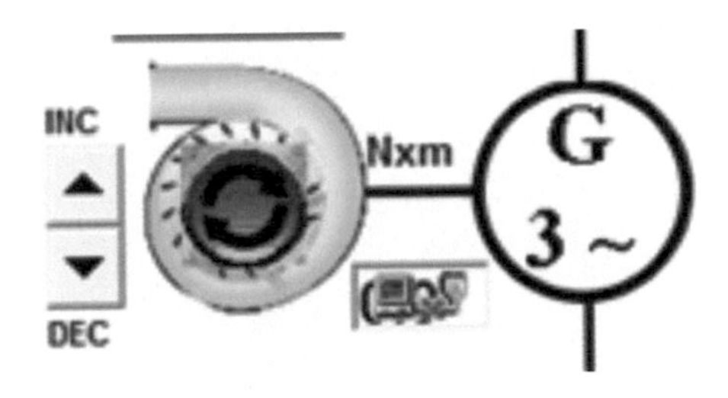

FIGURE 2.27
SCADA representation of the water governor to increase and decrease generator speed

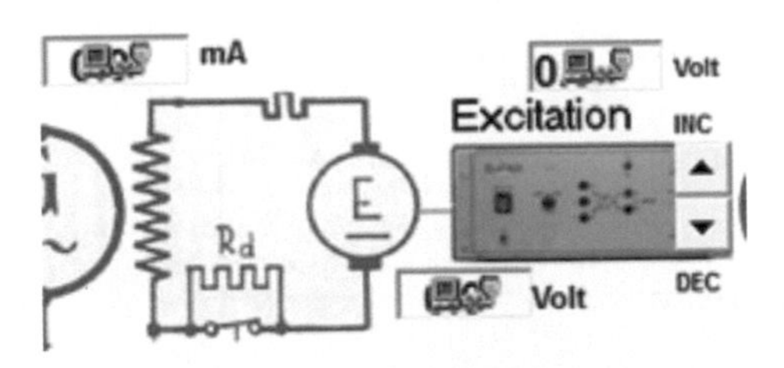

FIGURE 2.28
Excitation of the generator

V_{exc}). Modify the excitation voltage and observe the armature voltage variation. Register the data in a table and draw the graph.

Sugested Material:

- Brushless control
- Power circuit breaker

TABLE 2.4
Data registered with three different speeds, having U_s, U_t, and U_w as the voltage of each of the three phases

Speed (min-1)	1800	1000	700
Ie(mA)	(Us,Ut,Uw) V_{rms}	(Us,Ut,Uw) V_{rms}	(Us,Ut,Uw) V_{rms}
100	117	63	46
150	172	95	69
200	225	124	88
250	275	154	107
300	323	181	126
350	366	202.6	143
400	408	225	158
450	440	243	170
500	467	259	181
550	487	274	192

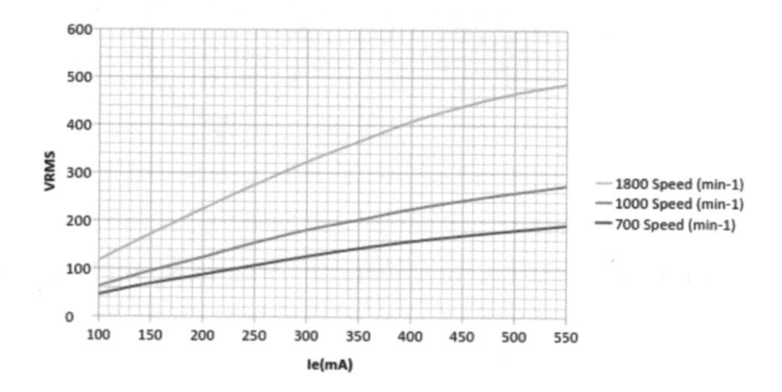

Graph 2.5
Graph from Table 2.4

- Feeder manager relay
- Module for measuring the electric power
- Three-phase asynchronous generator
- Programmable automatic power supply unit
- Resistive load
- Inductive load

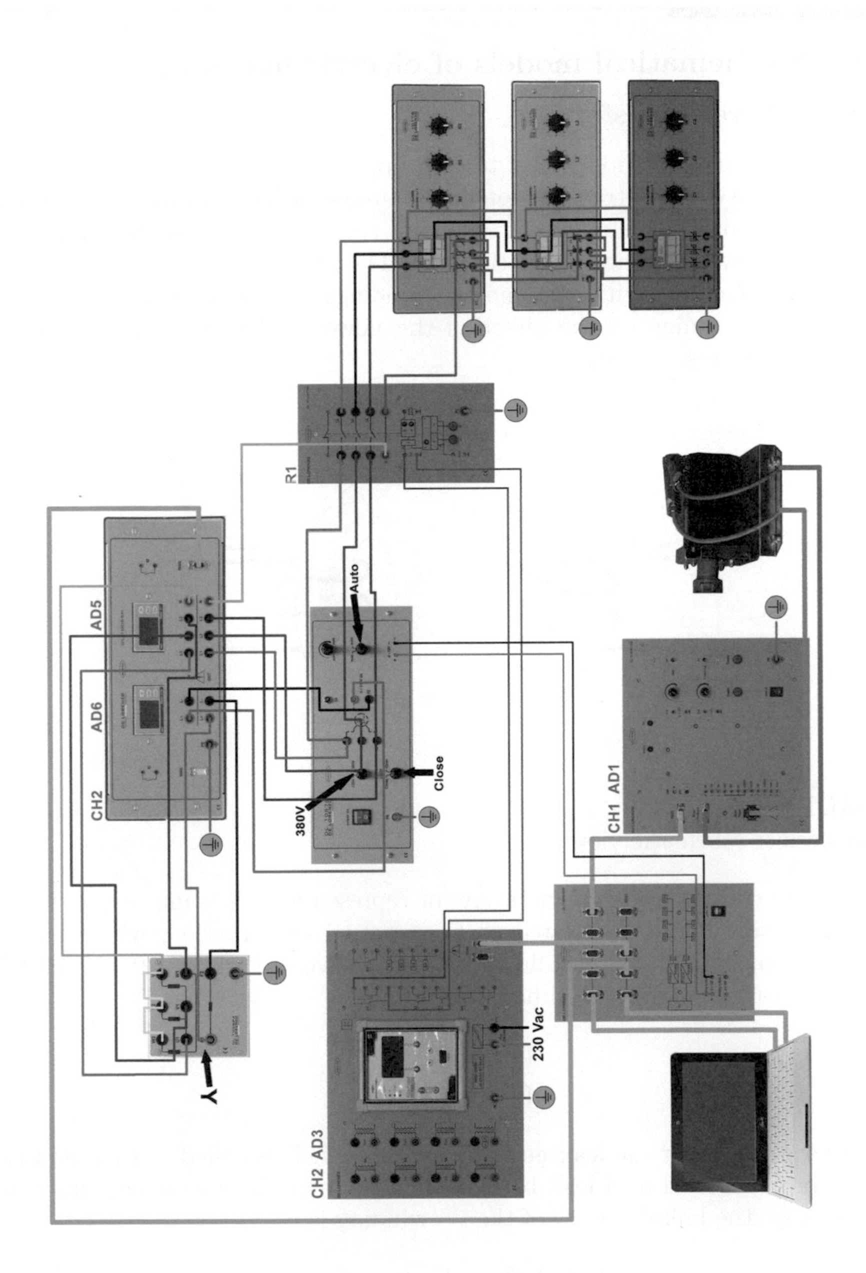

Diagram 2.2
Suggested connection to load test

2.4 Mathematical models of electric machinery

2.4.1 Power transformer

Transformers are commonly used to step up/down a given AC voltage in order to achieve efficient transmission or safe usage. Schematically, they consist of a very simple magnetic circuit involving two windings associated through a common low-reluctance magnetic path. The aforementioned path is commonly manufactured with iron and the windings are classified as primary or secondary depending on the direction the power is desired to flow (always from primary to secondary).

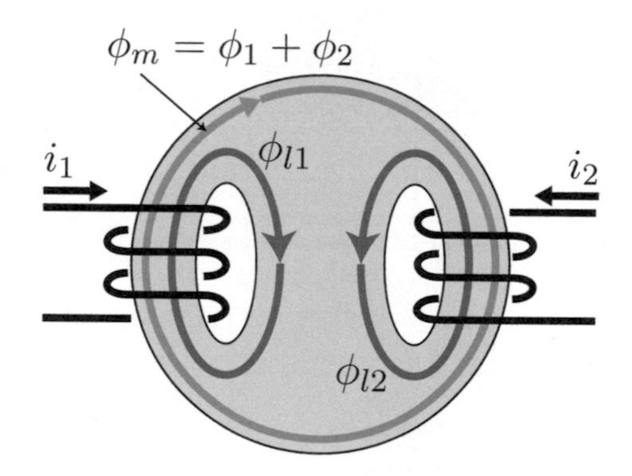

FIGURE 2.29
Transformer schematic view

Besides transformers' magnetic circuit representation's being simple, there are many considerations related to their actual construction, which forces modeling to consider complex flux paths. As shown in Figure 2.29, the actual fluxes of both windings would be given by

$$\begin{aligned}\Phi_1 &= \Phi_{l1} + \Phi_m \\ \Phi_2 &= \Phi_{l2} + \Phi_m,\end{aligned}$$

where Φ_{li} stands for the leakage flux of winding i. It is called *leakage* flux as it can be considered as a loss, because it is not used for core magnetization. In this way, the linked flux λ_i of the ith winding is

$$\lambda_i = N_i \Phi_i = N_i(\Phi_{l1} + \Phi_m),$$

where N stands for the number of turns of the winding. It is possible to substitute the fluxes' magnitudes with their current equivalent, related to magnetic permeability. In this way, currents are solved as

$$\lambda_1 = N_1 \left(\underbrace{N_1 i_1 P_{l1}}_{\Phi_{li}} + \underbrace{(N_1 i_1 + N_2 i_2) P_m}_{\Phi_m} \right) = \left(\underbrace{N_1^2 P_{l1} + N_1^2 P_m}_{L_1 1} \right) i_1 + \underbrace{N_1 N_2 P_m}_{L_{12}} i_2.$$

The linkage flux can be given in terms of system's inductances. The subscripts related to each inductance L_{ij} stand for the relation such inductance keeps towards each coil. For instance, L_{11} focuses on those fluxes going exclusively through winding 1, and is called self-inductance. On the other hand, any $L_{ij}, i \neq j$ takes into account the flux going through two different windings and thus, it is called mutual inductance. Our two-windings transformer can be described by two equations as

$$\begin{aligned} \lambda_1 &= L_{11} i_1 + L_{12} i_2 \\ \lambda_2 &= L_{21} i_1 + L_{22} i_2. \end{aligned}$$

Physically, the self-inductance of both windings is due to the leakage and the actual magnetizing component L_{mi} (only due to current i_i). In order to calculate both inductances, the opposite current is considered to be zero, so

$$L_{11} = \frac{\lambda_i |_{i_j=0}}{i_i} = \frac{N_1(\Phi_{li} + \Phi_{mi})}{i_i} = \underbrace{N_1^2 P_{li}}_{L_{l1}} + \underbrace{N_1^2 P_m}_{L_{lm}}).$$

We can now find the ratio between both magnetization inductances as

$$\frac{L_{m2}}{L_{m1}} = \frac{N_2}{N_1}.$$

So, each linked flux can be given as a function of current through its own magnetizing inductance, as

$$N_1 \Phi_m = N_1 (\Phi_{m1} + \Phi_{m2}) = N_1 \left(\frac{L_{m1} i_1}{N_1} + \frac{L_{m2} i_2}{N_2} \right) = L_{m1} \left(i_1 + \frac{N_1}{N_2} i_2 \right).$$

It is possible to link the transformer equations presented so far with its windings' voltage as the voltage is equal to the linked flux derivative. As in the preceding equations, the mutual inductance can be seen from the primary by using the turns relation:

$$e_1 = \frac{d\lambda_1}{dt} = L_{11} \frac{di_1}{dt} + L_{12} \frac{di_2}{dt} = \underbrace{L_{l1} \frac{di_1}{dt} + L_{m1} \frac{di_1}{dt}}_{\text{self-inductance}} + \underbrace{L_{m1} \frac{d(i_2 N_2 / N_1)}{dt}}_{\text{mutual-inductance}}.$$

This can be associated by solving for the magnetization inductance of the primary

$$e_1 = L_{l1} \frac{di_1}{dt} + L_{m1} \frac{d(i_1 + i_2 N_2 / N_1)}{dt} = L_{l1} \frac{di_1}{dt} + L_{m1} \frac{d\left(i_1 + i_2^{(1)}\right)}{dt},$$

where $i_2^{(1)}$ represents the secondary current referred to the primary. Likewise, the secondary voltage can be found to be

$$e_1 = L_{l2}\frac{di_2}{dt} + L_{m2}\frac{d\left(i_1 N_1/N_2 + i_2\right)}{dt} = L_{l2}\frac{di_2}{dt} + L_{m2}\frac{d\left(i_1^{(2)} + i_2\right)}{dt}.$$

Referred quantities scale their associated magnitudes to preserve equations' balance whether their associated winding was thought to have as many turns as the opposite one. In this way, all variables related to the secondary can, for instance, be referred to the primary and vice versa. Hence, the secondary voltage can be referred to the primary by multiplying all its terms by N_1/N_2.

$$\begin{aligned} e_1 &= L_{l2}^{(1)}\frac{di_2^{(1)}}{dt} + L_{m2}\frac{d}{dt}\left(\left(\frac{N_1}{N_2}\right)^2 i_1 \frac{N_1}{N_2} i_2\right) \\ &= L_{l2}^{(1)}\frac{di_2^{(1)}}{dt} + \underbrace{\left(\frac{N_1}{N_2}\right)^2 L_{m2}}_{L_{m1}}\frac{d}{dt}\left(i_1 \underbrace{\frac{N_2}{N_1} i_2}_{i_2^{(1)}}\right) \\ &= L_{l2}^{(1)}\frac{di_2^{(1)}}{dt} + L_{m1}\frac{d}{dt}\left(i_1 i_2^{(1)}\right). \end{aligned}$$

The induced voltages are not the same as the terminal ones because the windings themselves exhibit some resistance. Thus, the terminal voltage is the algebraic sum of the induced voltage plus the voltage drop at the winding.

$$v_i = i_i r_i + e_i = i_i r_i + L_{li}\frac{di_i}{dt} + L_{mi}\frac{d\left(i_i + i_j^{(i)}\right)}{dt}.$$

So, having the terminal voltage of winding i referred to winding j would be

$$v_i^{(j)} = i_i^{(j)} r_i^{(j)} + L_{li}^{(j)}\frac{di_i^{(j)}}{dt} + L_{mj}\frac{d\left(i_j + i_i^{(j)}\right)}{dt}.$$

If both windings are seen from the primary side, the equations describing the behavior of the transformer are

$$\begin{aligned} v_1 &= i_1 r_1 + L_{l1}\frac{di_1}{dt} + L_{m1}\frac{d\left(i_1 + i_2^{(1)}\right)}{dt}, \\ v_2 &= i_2 r_2 + L_{l2}\frac{di_2}{dt} + L_{m1}\frac{d\left(i_1 + i_2^{(1)}\right)}{dt}. \end{aligned}$$

It is quite clear that the right-hand side of both equations is the same, physically implying that both voltages are influenced by a common inductance. As

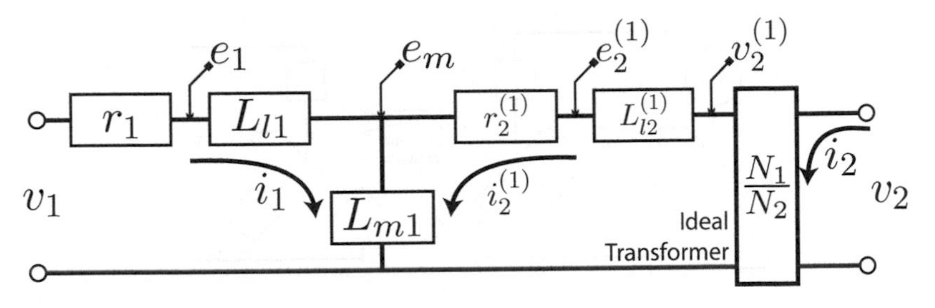

FIGURE 2.30
Transfomer equivalent circuit

a result, it can be seen that the magnetization flux is due to a current $i_1 + i_2^{(1)}$. Both primary and secondary voltages share a common path to ground, provided by L_{m1}. This voltage drop is later followed by two branches with a single series resistive-inductive load, respectively. The preceding description points to the equivalent circuit shown in Figure 2.30.

It is important to notice that the magnetization inductance is not constant as the current rises. As the magnetic core gets saturated, the inductance value will decrease as permeability does. It is also important to notice that the equivalent circuit does not consider a load connected to the secondary terminals. The output voltage can be associated to an electric load as it can be defined as an impedance, so a current can be calculated and used to modify transformer's operating conditions. On the other hand, a full reformulation of the transformer model can be performed for generic impedance considerations as secondary loads; however, this process could lead to unnecessary complexity.

It is noteworthy that a full grid simulation conformed by many subsystems will impose unconsidered variations. However, a full descriptive model of the transformer could consider the electric source connected to the primary to have an associated series impedance (i.e., a Thevenin equivalent), while its output at the secondary would have a generic load. A different approach could lead to a step-by-step simulation where values at each stage are needed to recalculate the system's performance.

A three-phase transformer would require new considerations to be taken as the flux linked by each winding depends on the other two phases. However, if all three phases are balanced, the same monophasic approach used so far can be used to estimate threefold electric-magnetic relations by neglecting new mutual inductances.

It is clear that the application of arbitrary voltages to the primary can be modeled by the simple equivalent circuit shown in Figure 2.30; however, transformers are mostly connected to well-defined three-phase distribution systems, so no arbitrary voltages are applied and the electric relation between phases is determined in a standard manner. For instance, a three-phase transformer

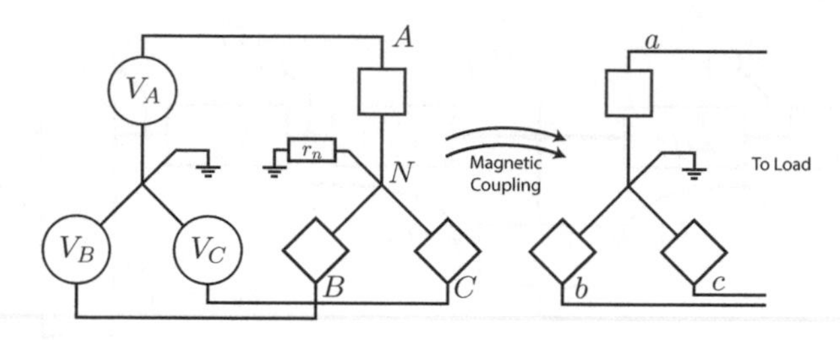

FIGURE 2.31
Y-Y connected transformer

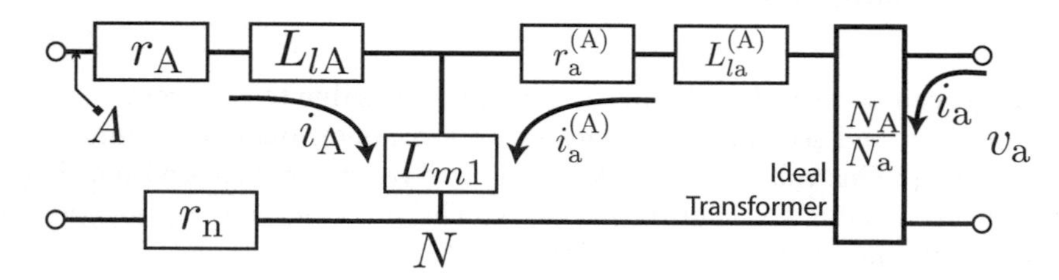

FIGURE 2.32
Equivalent circuit of Y-Y connected transformer for phase A-a

will be connected in delta (Δ) or "wye" (Y) so their behavior can be modeled straightforwardly.

A Y-Y connected transformer schematic is shown in Figure 2.31. The equivalent circuit in Figure 2.32 shows the electric relations found for a single phase if the equivalent circuit is again used. Notice that the neutral connection adds a resistance to the primary loop as the primary is connected to a Y source.

Similarly, a Δ-Y connected transformer will schematically look like Figure 2.33. Notice that no neutral resistance is now added on the primary; however, it could be added in the secondary whenever the load is also Y connected. The equivalent circuit for this case is shown in Figure 2.34, where the currents i_{CA} and i_{BC} are connected to other equivalent circuits. A simulation would be complex in this case as voltage sources and phase coupling is needed in order to attain adequate results.

2.4.2 Asynchronous machine

An induction motor is ideal for use in machine tools because it keeps a fairly constant rate ($0 - 5\%$ slip) within its rated operating range, does not need to be synchronized to connect to the grid, and has a low price relative to other machines. Also, its maintenance can be minimal, since it does not need brushes as DC motors do, and its efficiency is above 80%. Induction motors

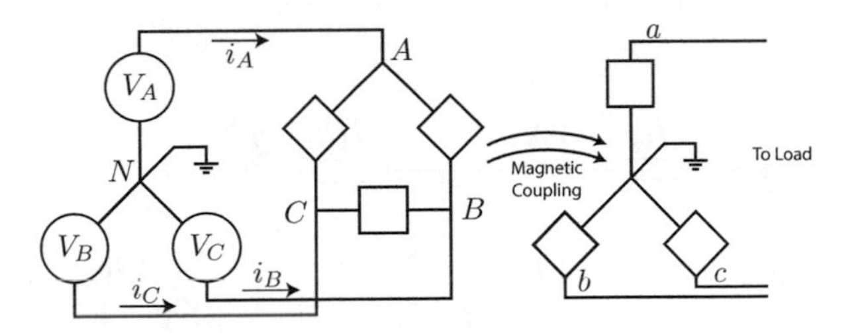

FIGURE 2.33
Δ-Y transformer connection

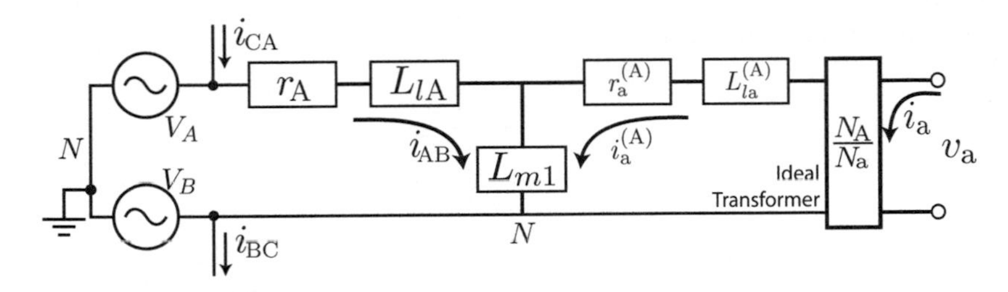

FIGURE 2.34
Equivalent circuit of Δ-Y connected transformer for phase A-a

are classified depending on the number of voltage lines, rated power, NEMA design, and speed (number of poles and operating frequency).

An induction motor can be understood as a transformer with rotating secondary; if the rotor is locked, the motor behaves as a short-circuited power transformer. This implies that the transformer modeling presented in Section 2.4.1 is useful to derive a steady-state model of the motor.

This machine can be represented by an equivalent electrical circuit to describe its behavior (see Figure 2.35). This circuit has a variable load dependent on rotor's slip. The sliding of an engine is the basis of an asynchronous machine operation,

$$s = \frac{n_{sync} - n_m}{n_{sync}}, \tag{2.41}$$

$$n_{sync} = \frac{120 f_e}{p} \ [rpm], \tag{2.42}$$

where n_{sync} is the maximum engine's speed-dependent electrical frequency and number of poles, and n_m is the actual speed of the rotor.

The equivalent circuit of induction motor (shown in Figure 2.36) in Figure 2.35 can be reduced to the Thevenin equivalent circuit in Figure 2.37, neglecting some electric interactions, where

$$V_{TH} \approx V_{\text{in}} \frac{X_m}{X_s + X_m}, \tag{2.43}$$

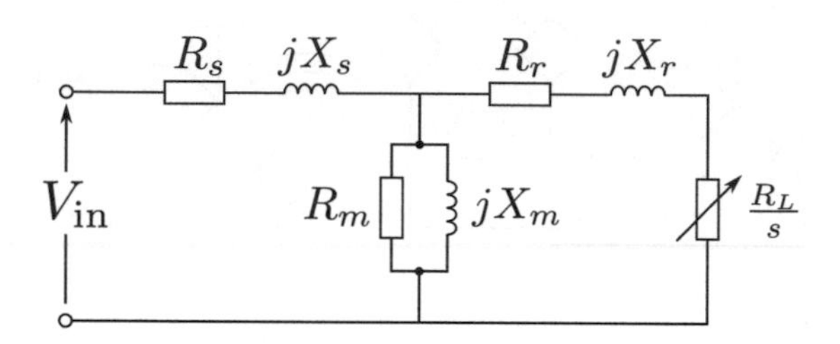

FIGURE 2.35
Equivalent circuit of induction motor

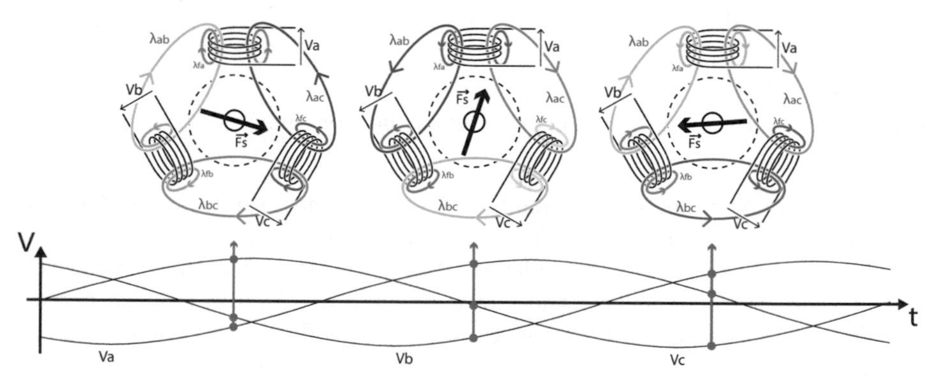

(a) Stator flux linkage at three different voltage stages

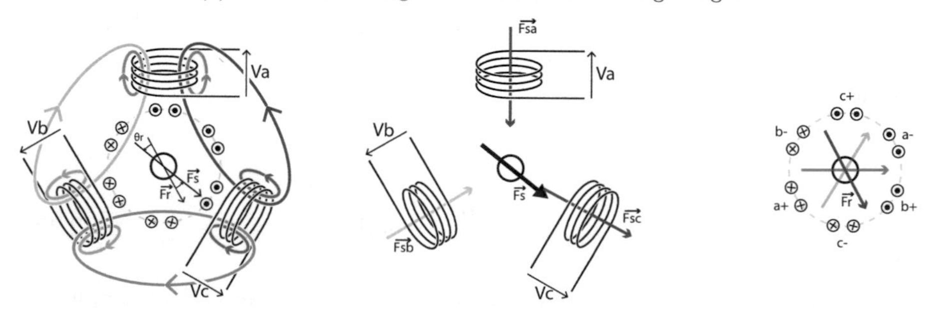

(b) Induced rotor's magnetomotive force

FIGURE 2.36
Graphical representation of asynchronous motor auto-induction

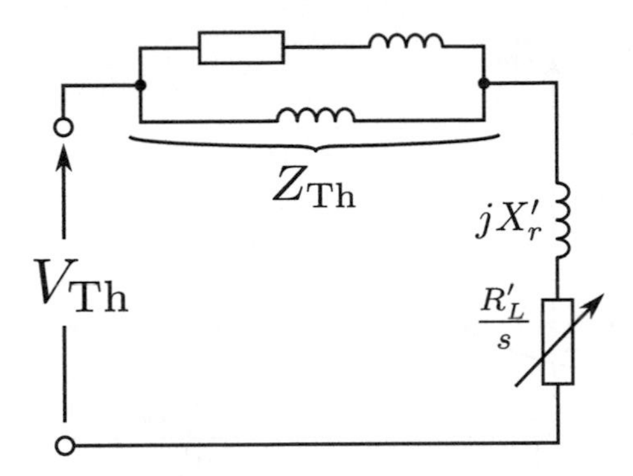

FIGURE 2.37
Thevenin equivalent circuit

$$R_{TH} \approx R_s \left(\frac{X_m}{X_s + X_m} \right)^2, \tag{2.44}$$

$$X_{TH} \approx X_s. \tag{2.45}$$

Then, the torque motor with sliding is calculated as

$$\tau_{ind} = \frac{\frac{3 \times V_{TH}^2 \times R_2}{s}}{s\omega_{sync} \times [(R_{TH} + \frac{R_2}{s})^2 + (X_{TH} + X_2)^2]}. \tag{2.46}$$

By knowing the machine's parameters, it is possible to graph its speed-torque relation, as shown in Figure 2.38. The most important point of the graph is the maximum torque, because as its name indicates, the engine will not be able to exceed this value.

The National Electrical Manufacturers Association (NEMA) has standardized the design of electric motors to establish certain characteristics and requirements that must be met by manufacturers. The main characteristics of the engines as designed are:

Design A The standard design has a current of moderate start, and an equally medium torque, and small slip in the range of nominal operation. The maximum permissible slip is 5% and must be less than that of type B design. The maximum torque is 200 to 300% nominal torque and slippage occurs at less than 20%, while the starting torque for this engine is rated at 60% for large motors and 200% or more for small engines. The main drawback for this type of engine is the large amount of current required to start the line (usually $5 - 7$ times the rated current), and if there is sensitive equipment connected to the same equipment, it can be damaged by voltage drop. Very large motors should be started at reduced voltage to mitigate the problem of voltage drop. Typical applications for this engine are: pumps, machine tools, blowers, compressors, fans.

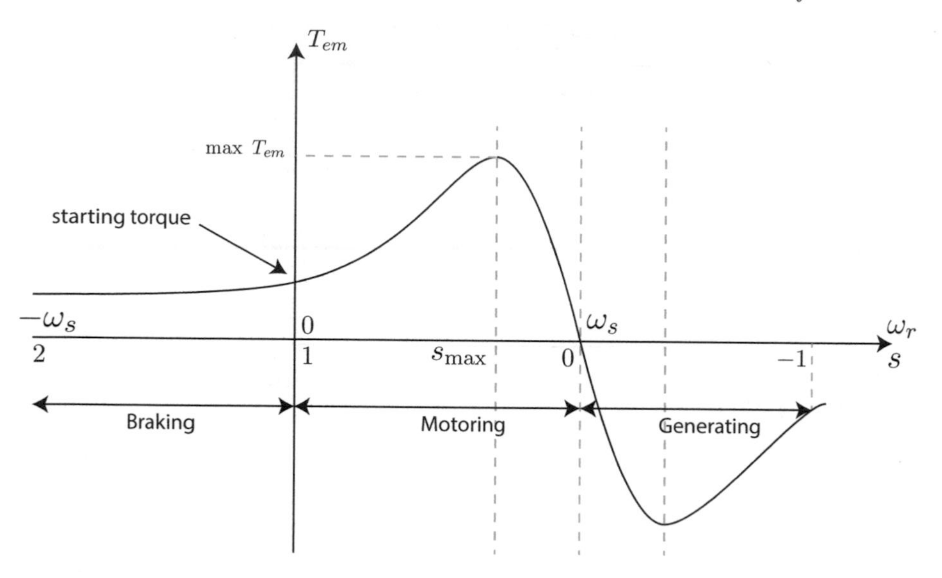

FIGURE 2.38
Induction motor operating modes

Design B This type of engine has a medium starting torque, but with a lower starting current (about 25% less than design A), little slip in the operating range, and minimum torque of 200% nominal torque, although this is less than design A because it has a higher resistance rotor (this also reduces its efficiency). The applications are very similar to design A. Design B has shifted to this design because of its lower starting current.

Design C Design C is for high starting torques (over 250% nominal torque) and low starting current, slippage less than 5% in nominal operation, and maximum torque is slightly lower than design A. As shown in Figure 2.39, the pattern C has a design performance between A and B. The disadvantage of this engine is cost, because achieving this behavior requires manufacturing of dual cage rotors. These engines respond very well for jobs that require motor starting loaded, loaded as pumps, compressors, and conveyors.

Design D These motors have a very large starting torque; some motors even have maximum torque at start-up (275% over the nominal torque) and a small current boot, but they have a huge slide in nominal operation (7 to $11-17\%$) and a very poor efficiency (because it requires a lot of resistance in the rotor to get the boot with little current). These motors are used to start moving large inertial mass; for example, panels that open the floodgates of dams which require a lot of torque to reach full speed. After performing a task, these engines are left off until their next operation is required.

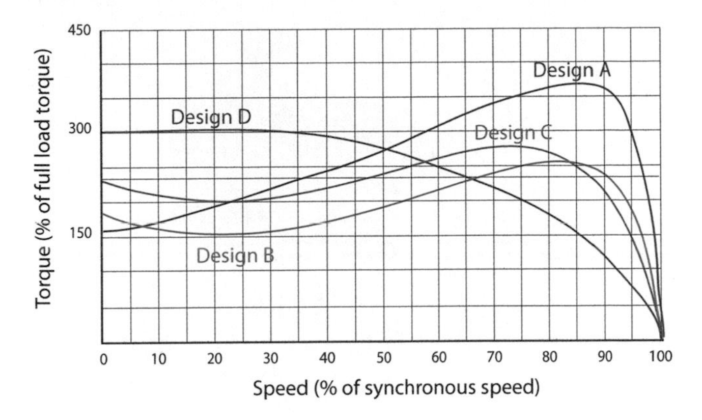

FIGURE 2.39
Torque-speed curves of different NEMA standard motors

TABLE 2.5

NEMA Code	Rotor stopped [kVA/hp]	NEMA Code	Rotor stopped [kVA/hp]
A	0-3.15	L	9.00-10.00
B	3.15-3.55	M	10.00-11.00
C	3.55-4.00	N	11.20-12.50
D	4.00-4.50	P	12.50-14.00
E	4.50-5.00	R	14.00-16.00
F	5.00-5.60	S	16.00-18.00
G	5.60-6.30	T	18.00-20.00
H	6.30-7.10	U	20.00-22.40
J	7.20-8.00	V	Más de 22.40
K	8.00-9.00		

Given the above descriptions, the best design for the implementation of the retrofit design is A, because it has the greatest efficiency in nominal operation and weakness can be mitigated with start control.

To know more precisely what the starting current at the plate will be, NEMA code determines which will be the starting power full motor voltage. Table 2.5 shows the values of the constant that is multiplied by the engine power design for the PTO:

With the design constant, we obtain the initial complex power S,

$$S_{\text{start}} = (\text{rated horsepower})(\text{code letter factor}), \tag{2.47}$$

and the starting current

$$I_{\text{start}} = \frac{S_{\text{start}}}{\sqrt{3} \times V_{\tau}}. \tag{2.48}$$

FIGURE 2.40
Brushless motor

2.4.3 Brushless motor

The construction of brushless motors is very similar to that of AC motors. The rotor is a permanent magnet element, and the stator is made up as an AC motor with several phases. The big difference between these two types of motors is how to detect the position of the rotor in order to know how to find the magnetic poles and generate the control signal using power transistors. These sensing and power electronics elements are commonly integrated (or sold together) with the machine.

The position sensing is normally done with Hall effect sensors. The Hall effect is the appearance of an electric field (the Hall field) in a conductor when it is crossed by a magnetic field. The Hall effect-based sensors comprise a conductive or semiconductive element and a magnet element. Thus, the proximity of an object can be determined if it is a ferromagnetic material. Brushless motors can also use optical sensors or encoders to determine the position at which the rotor is located. A brushless motor is shown in Figure 2.40.

Like all electric machines, a brushless motor is based on two basic principles of electromagnetic force induced in a conducting wire in the presence of a magnetic field (deduction of Lorentz force law),

$$F = Bli, \tag{2.49}$$

where B is the magnetic field density, l is the conductor length, and i is the current intensity flowing through the conductor. The voltage induced in a conductor in the presence of an alternating magnetic field (deduction of Faraday's Law) is

$$e_b = \frac{d\lambda}{dt} = Blv\ [V], \tag{2.50}$$

where v is the relative velocity between the field and the conductor. To determine the torque generated by the engine, you will have to build a circuit to

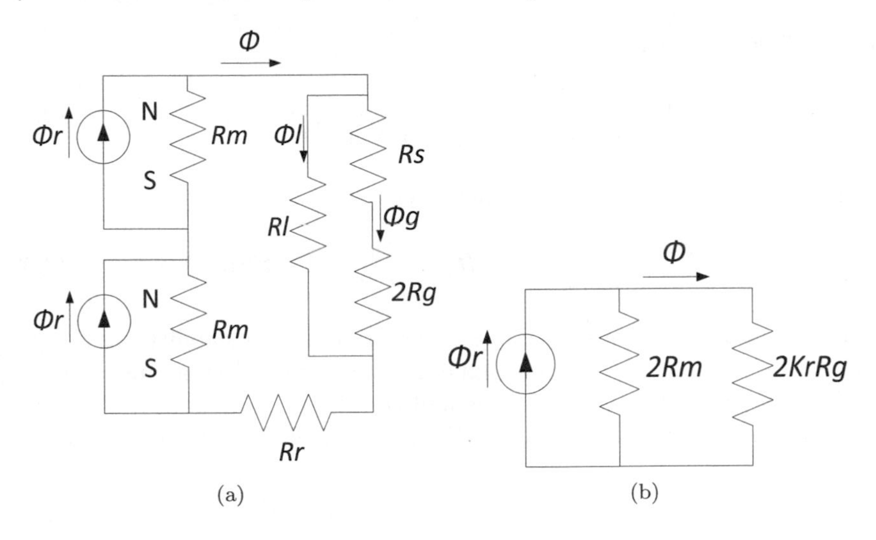

FIGURE 2.41
a) Magnetic circuit model, b) magnetic circuit simplified

determine the magnetic flux linkage, then the generated electromotive force (EMF), and finally the induced torque. The flow exits the north pole of the rotor magnet, passes through the air gap by the stator, and travels through the gap again to reach the south pole of a magnet. Finally it closes the circuit in the inner part of the rotor. A small part of the stream does not flow through the stator, representing the leakage flow in parallel with reluctance (Rl) coupled with the flow through the return gap. This is the reason why the circuit of Figure 2.41(a) has $2Rg$, the magnetic circuit, which is in series with the stator. Figure 2.41(b) represents the first circuit model that describes the direction of the magnetic flux (only one pole pair).

Permanent magnets are represented by sources with their respective reluctance flow in parallel. The metallic inner part of the rotor, the air gap, and the stator are represented as reluctances, and the elements that make the circuit are: Rm, is the reluctance of the magnet, Rs, is the reluctance of the stator, g, is the air gap reluctance, and Rr, is the reluctance of the metal part of the rotor.

The airgap flux can be written in terms of the total flow $\Phi_g = k_l \Phi_r$ where K_l is known as factor binding, usually close to 1. Thus, we can eliminate the air-gap flux and simplify the magnetic circuit by removing R_l, then you can add R_s, R_r, and $2R_G$, since they are in series and the magnets can be represented by one in parallel with a resistor of $2WD$ value, shown in Equation 2.50. Considering also that the reluctance of the air gap is greater than the stator and rotor, then we can simplify these two reluctances representing circuit (R_r and R_s) as factor K_r reluctances, multiplying the air gap reluctance R_g. K_r is typically slightly larger than 1. With the simplified circuit Figure 2.41(b), the flow link is as follows:

$$\Phi = \frac{2R_m}{2R_m + 2K_r R_g} = \frac{1}{1 + K_r \frac{R_g}{R_m}} \; [Wb]. \tag{2.51}$$

If we substitute the value of reluctances

$$R_m = \frac{l_m}{\mu_R \mu_0 A_m}, \quad R_g = \frac{g}{\mu_0 A_g} \; [A - \text{turns}], \tag{2.52}$$

where l_m is the magnetic length, μ_R is the relative permeability, μ_0 the permeability of vacuum, Am is the magnetic cross-sectional area, and g and A_g are the distance of the gap and the area of the air gap, respectively. Also, consider that the airgap flux is a part of the total flux Φ, obtaining the following expression:

$$\Phi_g = K_l \Phi_r = \frac{K_l}{1 + K_r \frac{R_g}{R_m}} \Phi_r \; [Wb]. \tag{2.53}$$

Equation 2.53 describes the link flow engine. Taking into account that there is more than one winding, the expression of the total flow is

$$\lambda = N\Phi_g \; [Wb - turns]. \tag{2.54}$$

Equation 2.53 depends only on the angular position of the motor. In Figure 2.42(a), the magnetic flux of the magnet pole is opposite to that generated by the stator winding sense, so that the flow is equal to $\lambda = -N\Phi_g$. Moreover, in Figure 2.42(b), the stator is 90 electrical degrees, so that the flow goes in and out, resulting in a flow of $\lambda = 0$. Figure 2.42(c) shows that the flux of the rotor magnet coincides with that generated by the stator winding current, so that flow is $\lambda = N\Phi_g$.

Considering that the flow varies linearly with respect to the angular position, we get the graph shown in Figure 2.43; for the induced EMF in Equation (2.53), we apply Faraday's law.

In Figure 2.43(a), the waveform of inductive coupling is shown when it is pursuing a counter electromotive force at π. The electromotive force opposes the passage of electrical current in an inductor, in this case the rotor, which will tend to slow down. In Figure 2.43(b), the square waveform of the electromotive force due to the triangular waveform applied to the inductive coupling is shown at π. If we consider the law of conservation of energy,

$$e_b i = T\omega_n, \tag{2.55}$$

clearing the pair T of the expression,

$$T = \frac{e_b i}{\omega_n} = \frac{K_e \omega_n i}{\omega_n} = K_e i. \tag{2.56}$$

This approach does not consider magnetic saturation in the material or

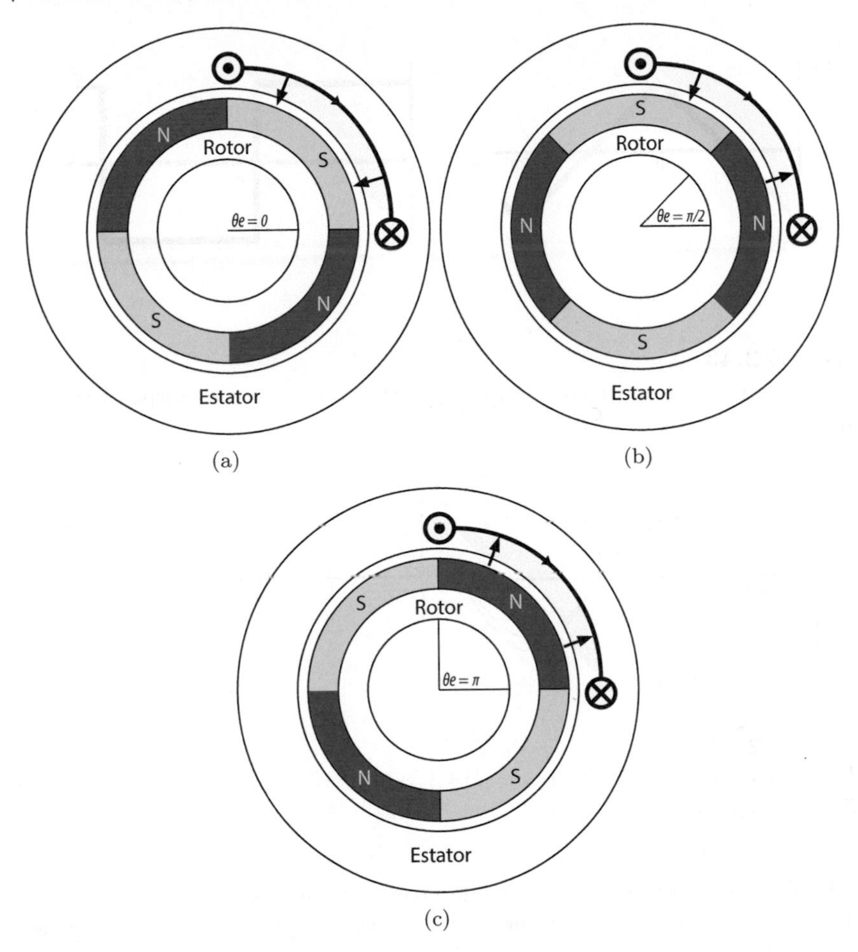

FIGURE 2.42
a) Motor with a fractional step, b) motor with 90° E, c) motor with 180° E

the energy necessary to re-align the magnetic domains, which are different groups of magnetic currents within a magnetic field. A more realistic graph of an engine is shown in Figure 2.44, where it can be seen that actually EMF is trapezoid.

Traditional conductors have an electrical resistance that is proportional to the length of the conductor and inversely proportional to the cross-sectional area through which electric current flows. In the case of electric motors, the windings contain a varnish layer that isolates the cables and prevents shorting of windings in the stator. Thus, the resultant electric resistance is equal to

$$R = \frac{\rho l}{A}, \tag{2.57}$$

where ρ is a constant known as electrical resistivity of the material, l is the

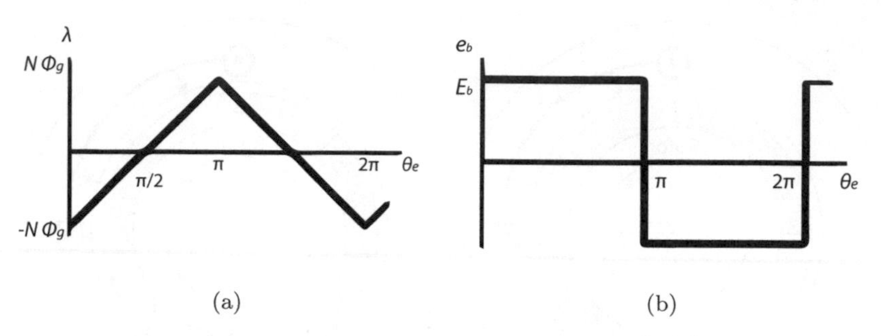

FIGURE 2.43
a) Waveform of the inductive coupling in π, b) square waveform of the electromotive force

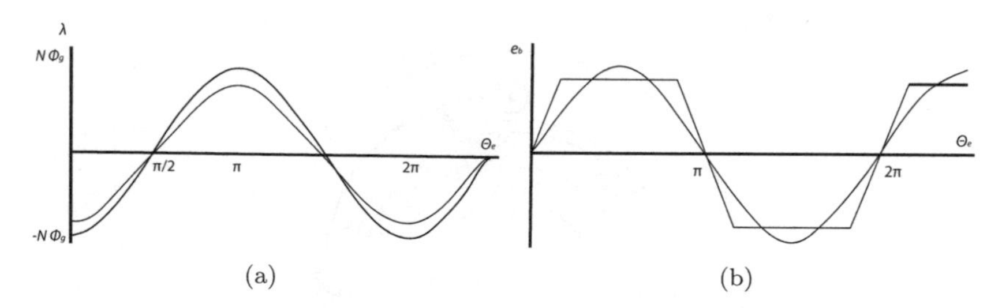

FIGURE 2.44
a) Inductive coupling in its trapezoidal shape, b) electromotive force in its trapezoidal shape

length of the conductor, and A is the cross sectional area of the conductor. The electrical conductivity of the metals varies approximately linearly with respect to temperature, thus

$$\rho(T) = \rho(T_0)\left[a + \alpha(T - T_0)\right], \tag{2.58}$$

where α is the coefficient of thermal resistivity and T0 is the base temperature. The magnetic flux is related to the current through a constant, modeled by an inductance, so it should be included in the model brushless motor,

$$L = \frac{\lambda}{i}. \tag{2.59}$$

If we consider the magnetic circuit of Figure 2.45, we can notice that F_r contains a value point "potential" of 0, then the circuit is balanced. Moreover, each flow Φ is equal to

$$\Phi_1 = \Phi_2 = \Phi_3 = \Phi_4 = \overline{R_g + R_m} \tag{2.60}$$

so that the total flow is

$$\Phi_1 + \Phi N(2 + \Phi_3 + \Phi_4) = \frac{N_m N_i^2}{R_g + R_m}\lambda. \tag{2.61}$$

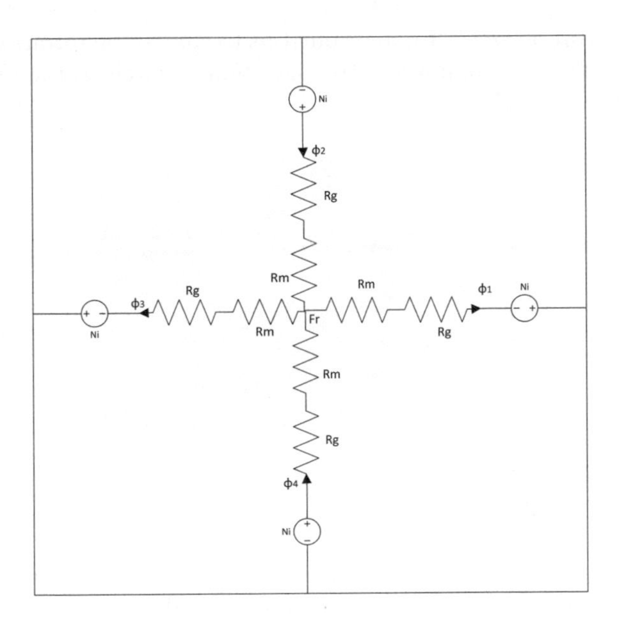

FIGURE 2.45
Magnetic circuit

Thus, the inductance is equal to

$$L = \frac{N_m N_i^2}{R_g + R_m}\lambda. \tag{2.62}$$

2.4.4 Brushless exercise

Using three different methods, this system will be used in the next exercise to simulate a turbine; thus, it will be able to track different speed changes. Check that the power is off before connecting. It is necessary to have a brushless control and a brushless motor.

Switch the Power ON and check the LEDs Mains and OK. If the LED OK is off, check the green cable connection to the brushless motor. Press the combo box "control speed" and select ANALOG, always press "apply" to confirm every change, as shown in Figures 2.46 and 2.47.

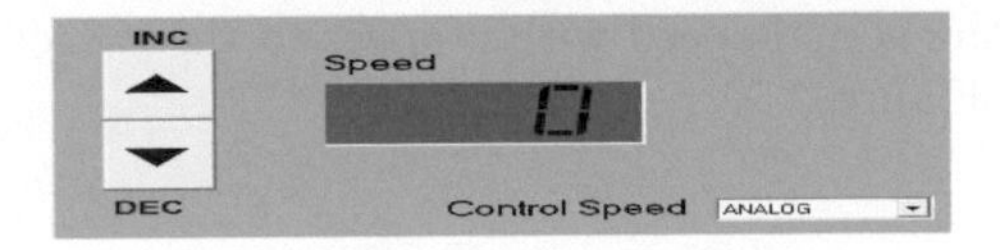

FIGURE 2.46
Digital control speed display

Press the "Torque" and "Enable" buttons to enable the torque to the shaft. Set the knob to the minimum and set the "Volt" switch to the ON position, as shown in Figure 2.47

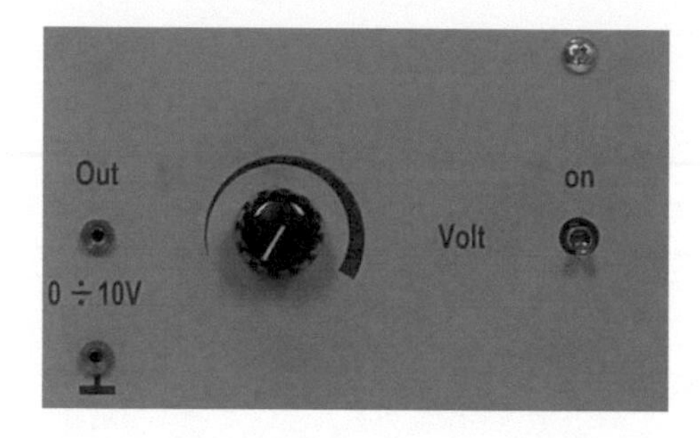

FIGURE 2.47
Analog control speed

Adjust the speed by varying the knob Volt. Check the actual speed on the graphic. Reduce the speed to 0 [*rpm*] and switch OFF the level of the analog control, as shown in Figure 2.47.

FIGURE 2.48
Software driven control speed

Press the combo box "control speed" and select SOFTWARE and then press Apply to confirm (see Figure 2.48).

Press the "Control Type" combo box to select "Speed control". Press the "Set Ramp" combo box to select "Trapezoidal". Set the acceleration/deceleration profiles to 0, which implies a slow speed increase toward the speed set point.

Increment the acceleration by using the buttons INC & DEC. Connect the three-phase induction generator and check the mechanical parameters.

Measure the Mechanical power with the load to determine the power loss. Type in the red block the speed at < 3000 and check the speed plot. Observe the time delay of the actual speed with respect to the target speed. A similar graph is shown in Figure 2.50.

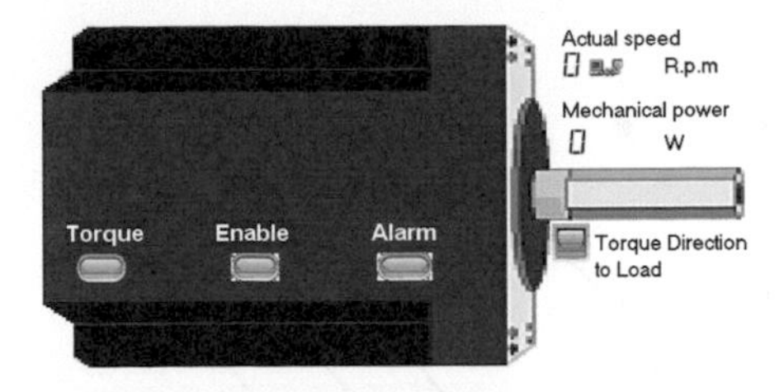

FIGURE 2.49
Actual speed

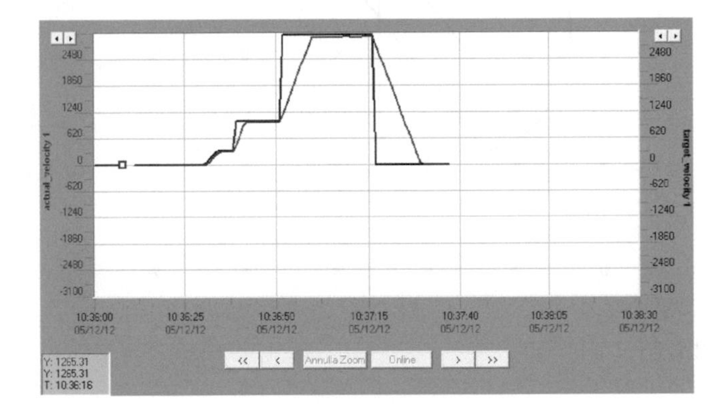

FIGURE 2.50
Speed graph

2.5 Theoretical problems

Problem 2.1. *Figure 2.51 shows a conductor moving with a velocity of* 10 $[m/s]$ *to the right in a magnetic field. The flux density is* 0.5 $[T]$, *out of the page, and the wire is* 1 $[m]$ *in length, oriented as shown. What are the magnitude and polarity of the resulting induced voltage?*

Solution 2.1. *The direction of the quantity $v \times B$ is down. The wire is not oriented on an up-down line, so choose the direction of l as shown to make the smallest possible angle with the direction of $v \times B$. The voltage is positive at the bottom of the wire with respect to the top of the wire. The magnitude of the voltage is*

$$\begin{aligned} e_{ind} &= (v \times B)\dot{I} \\ &= (vBsin90^{\circ})/cos30^{\circ} \\ &= (10m/s)(0.5T)(1m)cos30^{\circ} \\ &= 4.33\ [V]. \end{aligned}$$

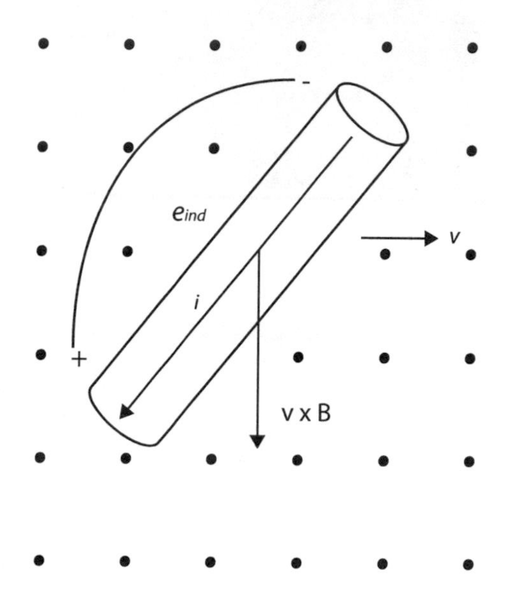

FIGURE 2.51
Movement of a conductor

Problem 2.2. *Figure 2.52 shows a coil of wire wrapped around an iron core. If the flux in the core is given by the equation* $\Phi = 0.05sin377t$ $[Wb]$*, and if there are* 100 $[turns]$ *on the core, what voltage is produced at the terminals of the coil? Of what polarity is the voltage during the time when flux is increasing in the reference direction shown in the figure? Assume that all the magnetic flux stays within the core (assume that the flux leakage is zero).*

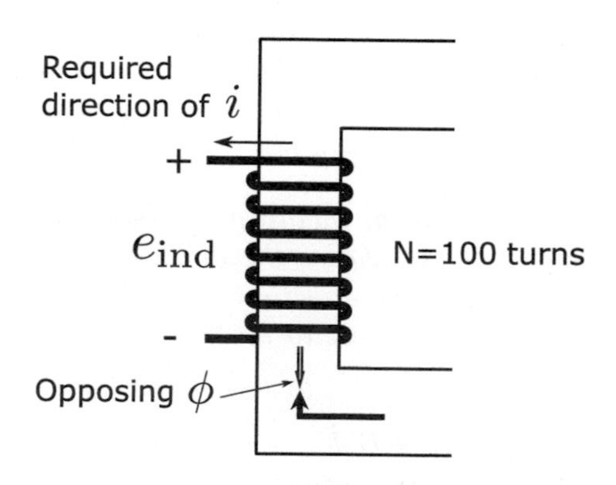

FIGURE 2.52
Wire wrapped around an iron core

Solution 2.2. *The direction of the voltage while the flux is increasing in the reference direction must be positive to negative. The magnitude of the voltage is given by:*

$$\begin{aligned} e_{ind} &= N\frac{d\Phi}{dt} \\ &= 377t0.05sin(100turns)\frac{d}{dt} \\ &= 1885cos377t, \end{aligned}$$

or alternatively,

$$e_{ind} = 1885sin(377t + 90°)\ [V].$$

Problem 2.3. *Find the relative permeability of the typical ferromagnetic material at* $H = 50\ [A - turns/m]$.

Solution 2.3. *The permeability of a material is given by*

$$\mu = BH,$$

and the relative permeability is given by

$$\mu_r = \frac{\mu}{\mu_0}.$$

Thus, it is easy to determine the permeability at any given magnetizing intensity. At $H = 50\ [A - turn/m]$ *and* $B = 0.25\ [T]$, *so*

$$\mu = BH = \frac{0.25\ [T]}{50\ [A - turn/m]} = 0.005\ \left[\frac{H}{m}\right],$$

and

$$\mu_r = \frac{\mu}{\mu_0} = \frac{0.005\ \left[\frac{H}{m}\right]}{4\pi \times 10^{-7}\ \left[\frac{H}{m}\right]} = 3980.$$

Problem 2.4. *A square magnetic core has a mean path length of* $55\ [cm]$ *and a cross-sectional area of* $150\ [cm^2]$. *A* $200\ [turn]$ *coil of wire is wrapped around one leg of the core. The core is made of a material that has a magnetizing intensity of* $H = 115\ [A - turn/m]$. *How much current is required to produce* $0.012\ [Wb]$ *of flux in the core?*

Solution 2.4. *The required flux density in the core is*

$$B = \Phi A = \frac{1.012\ [Wb]}{0.015\ [m^2]} = 0.8\ [T].$$

The magnetomotive force needed to produce this magnetizing intensity is

$$F = HL_c = (115\ [A - turn/m])(0.55\ [m]) = 63.25\ [A - turn].$$

So the required current is

$$i = F/N = 63.25\frac{[A - turn]}{200\ [turn]} = 0.316\ [A].$$

Problem 2.5. *From Problem 2.4, calculate the core's relative permeability at that current level and its reluctance.*

Solution 2.5. *The core's permeability at this current is*

$$\mu = BH = \frac{0.8T}{115A - turns/m} = 0.00696\frac{H}{m}.$$

Therefore, the relative permeability is

$$\mu_r = \frac{\mu}{\mu_0} = \frac{0.00696H/m}{4\pi \times 10^{-7}H/m} = 5540.$$

The reluctance of the core is

$$R = \frac{F}{\Phi} = \frac{63.25A - turns}{0.012Wb} = 5270 \left[\frac{A - turns}{Wb}\right].$$

2.6 Homework problems

Problem 2.6. *Suppose there is a one-phase transformer that exhibits the following results after performing open-circuit and short-circuit tests:*

- *OC:* $V = 120\angle 0^\circ$ $[V_{rms}]$, $I = 0.169\angle -89.98^\circ$ $[A_{rms}]$
- *SC:* $V = 5\angle 0^\circ$ $[V_{rms}]$, $I = 12.501\angle -16.259^\circ$ $[A_{rms}]$, *and* $I_{out}^{(1)} = 12.5\angle - 16.248^\circ$ $[A_{rms}]$

Estimate all transformer parameters with respect to its primary side. The short-circuit measurements are given regarding the primary side as well. Preserve calculation precision up to the thousandth.

Solution 2.6. *Note: Rounding heavily affects these results.*

The open circuit test reveals a direct relation for the first mesh as follows: $\frac{120}{0.1693\angle -89.98^\circ} = R_1 + j(X_1 + X_m) = 0.248 + j710.059$ $[\Omega]$.

So, $R_1 = 0.248$ $[\Omega]$ *and* $X_1 + X_m = 10.0597$ $[\Omega]$.

The short-circuit measured magnitudes can provide the missing information through the following mesh analysis:

$$\begin{aligned} 5 &= i_1(R_1 + jX_1) + jX_m(i_1 - i_2^{(1)}) \\ 0 &= i_2(R_2^{(1)} + jX_2^{(1)}) + jX_m(i_2^{(1)} - i_1) \\ X_1 + X_m &= 710.059. \end{aligned}$$

So, using the first and third equation together with short-circuit data:

$$L_1 = 146\ [\mu H], \quad and \quad L_m = 1.883\ [\mu H].$$

The voltage through Xm can be easily calculated as $V_{X_m} = jX_m(i_1 - i_2^{(1)}) = 1.846\angle 6.367°$, *so the second mesh impedance can be easily computed as:* $R_2^{(1)} + jX_2^{(1)} = V_{X_m} i_2^{(1)} = 0.136 - j0.056$ $[\Omega]$.

So: $R_2^{(1)} = 0.136$ $[\Omega]$, *and* $L_2^{(1)} = 151$ $[\mu H]$.

Problem 2.7. *Take the transformer from Problem 2.6 and assume that, together with the open-circuit and the short-circuit measured values, the DC resistances were also acquired:*

- $R_1 = 0.25$ $[\Omega]$, $R_2 = 0.268$ $[\Omega]$

Estimate the missing impedances and compare the results with those from Problem 2.6, considering the transformer's ratio is 1:2.

Solution 2.7. *The open-circuit test can yield the value of* X_1 *and* X_m *directly. As the resistance is known, only the imaginary part of the equation provides new information:* $Im(\frac{120}{i_1}) = 710.059$ $[\Omega] = X_1 + X_m$. *The real part does not match with the measured resistance. Such remainder is due to numeric approximation issues.*

For the short-circuit situation, the equivalent circuit can be solved through:

$$\begin{aligned} 5 &= i_1(R_1 + jX_1) + jX_m(i_1 - i_2^{(1)}) \\ 0 &= i_2(R_2^{(1)} + jX_2^{(1)}) + jX_m(i2^{(1)} - i_1) \\ X_1 + X_m &= 710.059 \end{aligned}$$

So, all the three inductances can be solved directly (remember to use $R_2^{(1)}$ *instead of* R_2*)* $X_1 = 659$ $[\mu\Omega] + j\omega 204.8$ $[\mu H]$, $X_m = 659$ $[\mu\Omega] + j\omega 1.88$ $[H]$, $X_2(1) = 362$ $[\mu\Omega] + j\omega 204.9$ $[\mu H]$.

Regarding the first problem, there are some differences regarding the values of the inductances. Notice that those differences are due to numerical precision, so conclusively, the common tests performed on transformers require high measuring precision and calculation consistency.

Problem 2.8. *Assuming a transformer with the following parameters, draw the RMS value of the output voltage, from 1/3 to 3 times its nominal power value.*

- $V = 120$ $[V_{rms}]$, $V_{out} = 240$ $[V_{rms}]$, $P_n = 1500$ $[VA]$
- $R_1 = 0.25$ $[\Omega]$, $L_1 = 148.5$ $[\mu H]$, $L_m = 1.88$ $[H]$
- $R_2(1) = 0.134$ $[\Omega]$, $L_2(1) = 148.5$ $[\mu H]$

Estimate the missing impedances and compare the results with those from the first problem, considering the transformer's ratio is 1:2.

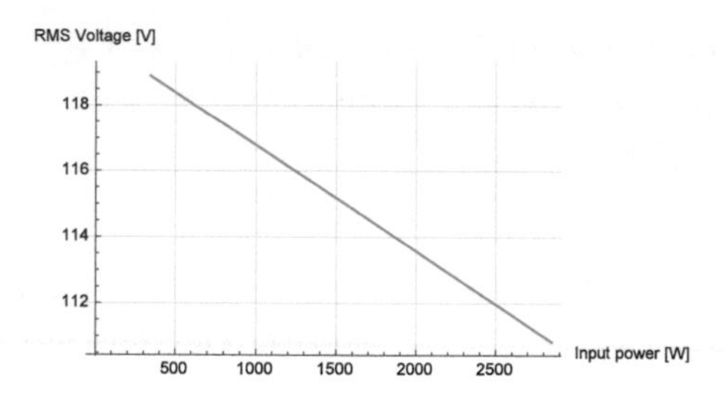

FIGURE 2.53
Voltage - Power curve

Solution 2.8. *The circuit equivalent of the transformer can be used to find its efficiency for some set of output loads. The equations that depict such interaction are as follows:*

$$V = i_1(R_1 + j\omega L_1) + j\omega L_m(i_1 - i_2^{(1)}),$$
$$0 = i_2^{(1)}(R_2^{(1)} + R_L^{(1)} + j\omega L_2^{(1)}) + j\omega L_m(i_2^{(1)} - i_1),$$

where $R_L^{(1)}$ is the load resistance seen from the primary. By sweeping through different resistance values, the following curve can be drawn:

Notice that there is a loss of about 5 $[V]$ *when the transformer is consuming* 1500 $[W]$. *Such loss is seen as a* 10 $[V]$ *loss if seen from the secondary winding.*

Problem 2.9. *Some self-made inductance has a circular sectional area of* 49 $[mm^2]$ *and is made of iron* (6.3×10^{-3} $\left[\frac{H}{m}\right]$), *and a mean path length of* $l = 5$ $[cm]$. *How many turns are required to generate a* 3 $[mH]$ *inductance?*

Solution 2.9. $B = \frac{\Phi}{A}$, *and* $L = \frac{N\Phi}{i}$, *so* $B = \frac{Li}{NA}$.

In addition, $F = Ni = Hl$, *so*

$\frac{B}{H} = \frac{Ll}{N^2 A} = \mu$, *and finally*

$$L(N) = \frac{\mu N^2 A}{l} = \frac{(6.3\times10^{-3})(49\times10^{-6})N^2}{0.05} = 6.17 \times 10^{-6} N^2.$$

So, for a 3 $[mH]$ *inductance, only* 22 $[turns]$ *are needed.*

Problem 2.10. *How strong will the flux be if a current of* 3 $[A]$ *is passed through a* 3 $[mH]$ *cylindrical inductance* ($r = 7$ $[mm]$, $l = 5$ $[cm]$), *made with electrical steel* (5×10^{-3} $\left[\frac{H}{m}\right]$)? *Do not consider saturation.*

Solution 2.10. $B = \frac{\Phi}{A}$, $L = \frac{N\Phi}{i} \therefore B = \frac{Li}{NA}$.

$F = Ni = Hl \therefore H = \frac{Ni}{l}$.

$\mu = \frac{B}{H} = \frac{Li}{NAH} = \frac{Ll}{N^2 A}$.

As $N = \frac{Li}{\Phi}$, $\mu = \frac{Ll}{(\frac{Li}{\Phi})^2 A} = \frac{l\Phi^2}{Li^2 A} \therefore \Phi = \sqrt{\frac{\mu L i^2 A}{l}}$.

Substituting: $\Phi = \sqrt{\frac{(5\times10^{-3})(3\times10^{-3})(9)(154\times10^{-6})}{0.05}} = 644.8 \times 10^{-6}$ $[Wb]$.

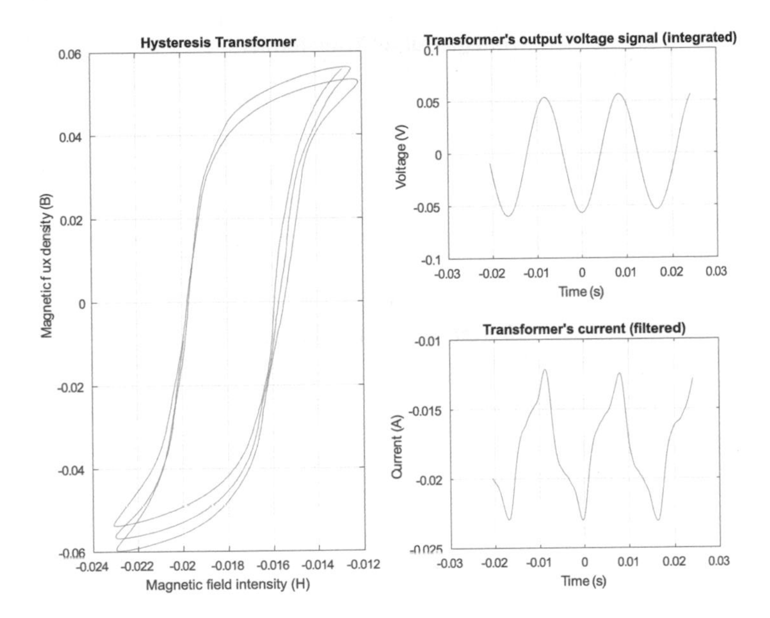

FIGURE 2.54
Resulting hysteresis curve

2.7 Simulation

2.7.1 Transfomers programs

Problem 2.11. *Hysteresis test*

Plot the hysteresis phenomena of a monophasic transformer, with the following features: $V = 120$ $[V]$, $V = 24$ $[V]$, *and* $I = 1.2$ $[A]$. *The measurement of voltage and current of transformer are in the files "CurrentTransformerFilPas.csv" and "VoltageTransformaderInteger.csv," respectively. The MATLAB program has been developed to get hysteresis phenomena in a plot and the code is shown below.*

Solution 2.11. *After running the code, the resulting plot should be as shown in Figures 2.54 and 2.55.*

```
clc, clear;
% Read Data
data_corriente = csvread('CurrentTransformerFilPas.
    csv', 9, 0)
data_voltage = csvread('VoltageTransformaderInteger.
    csv', 9, 0)
```

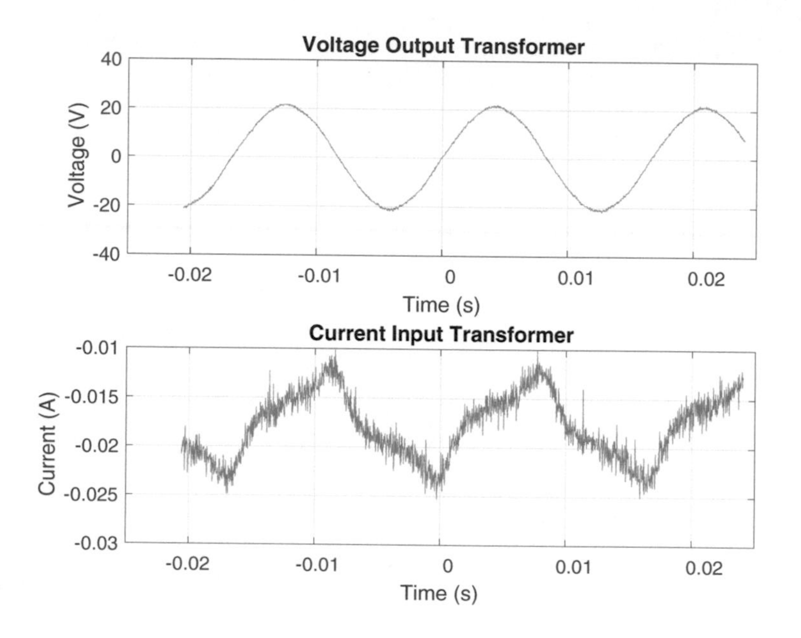

FIGURE 2.55
Resulting voltage and current

```
var1 = data_corriente(170:1956,1);
var2 = data_corriente(170:1956,2);
var3 = data_corriente(170:1956,3);
var4 = data_corriente(170:1956,4);
var5 = data_voltage(170:1956,1);
var6 = data_voltage(170:1956,2);
var7 = data_voltage(170:1956,3);
var8 = data_voltage(170:1956,4);

% Plot Hystesis
figure
ax1 = subplot(2,1,1);
ax2 = subplot(2,1,2);
% Voltage Output Transformer
plot(ax1,var1,var2)
title(ax1,'Voltage Output Transformer')
ylabel(ax1,'Voltage (V)')
xlabel(ax1,'Time (s)')

% Current Input Transformer
```

```
plot(ax2,var1,var3)
title(ax2,'Current Input Transformer')
ylabel(ax2,'Current (A)')
xlabel(ax2,'Time (s)')

% Plot Results
figure
axy1 = subplot(2,2,[1,3]);
axy2 = subplot(2,2,2);
axy3 = subplot(2,2,4);

% Hysteresis Transformer
plot(axy1,var4,var8)
title(axy1,'Hysteresis in Transformer')
ylabel(axy1,'Voltage (V)')
xlabel(axy1,'Current (A)')

% Integration of signal Voltage Output of Transformer
plot(axy2,var5,var8)
title(axy2,'Integration of signal Voltage Output of
   Transformer')
ylabel(axy2,'Voltage (V)')
xlabel(axy2,'Time (s)')
```

Problem 2.12. *Parameters' determination test*

Through a short-circuit test and open-circuit test, get the parameters of a transformer that operates with windings and core at normal operating temperatures, these tests are to $480-240$ $[VA]$ *and* 25 $[kVA]$*. The next MATLAB program has been developed to get the parameters of the transformer.*

```
% open-circuit & short-circuit test
clear;
V1=480; V2=240; n=V1/V2;        % Rated voltage values
Vsc=37.2; Isc=51.9; Psc=750;     % Test data
Voc=240; Ioc=9.7; Poc=720;
% dc resistance of windings - if not known, set R1dc=
   R2dc=1
R1dc=0.110; R2dc=0.029;
Voc=n*Voc; Ioc=Ioc/n;
Req=Psc/Isc^2; R1=R1dc; R2=Req-R1dc;
Zsc=Vsc/Isc; X1=sqrt(Zsc^2-Req^2)/2; X2=X1/n^2;
Rc=Voc^2/Poc; Xm=Voc/sqrt(Ioc^2-(Voc/Rc)^2);
disp(' '); disp(['   TRANSFORMER EQUIVALENT CIRCUIT
   PARAMETERS - ',...
   date]);
```

```
disp('                Classical Data Reduction');
disp(' '); disp(' ')
disp([blanks(3) 'R1(ohm)' blanks(8) 'R2(ohm)' blanks
    (8) 'X1(ohm)' ...
      blanks(8) 'X2(ohm)']);
disp([blanks(3) num2str(R1) blanks(9) num2str(R2) ...
      blanks(8) num2str(X1) blanks(9) num2str(X2)]);
disp(' ');
disp([blanks(3) 'Rc(ohm)' blanks(8) 'Xm']);
disp([blanks(4) num2str(Rc) blanks(8) num2str(Xm)]);
% Refinement of data reduction
thetoc=acos(Poc/Voc/Ioc); Eoc=abs(Voc-(R1+j*X1)*Ioc*
    exp(-j*thetoc));
Poc=Poc-Ioc^2*R1; Qoc=Voc*Ioc*sin(thetoc)-Ioc^2*X1;
Rc=Eoc^2/Poc; Xm=Eoc^2/Qoc;
thetsc=acos(Psc/Vsc/Isc); Esc=abs(Vsc-(R1+j*X1)*Isc*
    exp(-j*thetsc));
Psc=Psc-Esc^2/Rc; Qsc=Vsc*Isc*sin(thetsc)-Esc^2/Xm;
Req=Psc/Isc^2; R1=Req*R1dc/(R1dc+n^2*R2dc); R2=(Req-
    R1)/n^2;
X1=Qsc/Isc^2/2; X2=X1/n^2;
disp(' '); disp(['   TRANSFORMER EQUIVALENT CIRCUIT
    PARAMETERS - ',...
    date]);
disp('                Refined Data Reduction');
disp(' '); disp(' ')
disp([blanks(3) 'R1(ohm)' blanks(8) 'R2(ohm)' blanks
    (8) 'X1(ohm)' ...
      blanks(8) 'X2(ohm)']);
disp([blanks(3) num2str(R1) blanks(9) num2str(R2) ...
      blanks(8) num2str(X1) blanks(9) num2str(X2)]);
disp(' ');
disp([blanks(3) 'Rc(ohm)' blanks(8) 'Xm']);
```

Problem 2.13. *Cost analysis*

Analyze the difference between a low-cost transformer and a loss-optimized transformer, making the following assumptions:

- *The transformers are operated continuously.*
- *The transformers operate at partial load, but this partial load is constant.*
- *Additional cost and inflation factors are not considered.*
- *Demand charges are based on* 100% *load*

The total cost of owning and operating the transformer for one year is thus defined as follows:

- *Capital cost (C_c), taking into account the purchase price (C_p), the interest rate (p), and the depreciation period (n).*
- *Cost of no-load loss (C_{P0}), based on the no-load loss (P_0)and energy cost (C_e).*
- *Cost of load loss (C_{Pk}), based on the load loss(P_k), the equivalent annual load factor (a) and energy cost (C_e).*
- *Cost resulting from demand charges (C_d), based on the amount set by the utility and the total kW of connected load.*

The next MATLAB program have been developed for analyzing a low-cost transformer and a loss-optimized transformer.

```
clc;
clear;
% Transformer Loss Evaluation
% Distribution tranformer whith a low-cost
   transformer
% Data:
%        Depreciation period                        n = 20
   years
%        Interest rate                              p = 12 %
   p.a.
%        Energy charge                              Ce = 0.25
    Euro/kWh
%        Demand charge                              Cd = 350
   Euro/(kW.year)
%        Equivalent annual load factor              alpha =
   0.8

% Cc     Capital Cost
% where:
%          Cp     Purcharse price
%          r      Depreciation factor
%          q      Interest factor
%          p      Interest rate
%          n      Depretation period

n = 20;                % Years
p = 12;                % 12% p.a.
Ce = 0.25;             % Euro/kWh
Cd = 350;              % Euro/(kW.year)
```

```
alpha = 0.8;        % Equivalent annual load factor
Po = 19;          % kW No-load loss
Pk = 167;         % kW Load loss
Cp = 521000;      % Purcharse price  ($)

q = p/(100)+1
r = (p*(q^n))/((q^n)-1)

Cc = (Cp*r)/100  % Capital cost per year ($/year)

% Cpo   Cost of no-load loss
% Ce    Energy Cost
% where 8760 are hours in a year (hr/year)
% Po    No-load loss

Cpo = Ce*8760*Po    % Cost of no-load loss ($/year)

% Cpk   Cost of load loss
% Pk    Load loss
% a     annual load factor

Cpk = Ce*8760*(alpha^2)*Pk  % Cost of load loss ($/
   year)

% CD    Cost resulting from demand charge
% Cd    Demand charges amount/(kW*year)

CD = Cd*(Po+Pk)    % Cost resulting from demand
   charge ($/year)

% Total cost of owning and operating this transformer

Total_Low = Cc+Cpo+Cpk+CD.
```

```
clc;
clear;
% Transformer Loss Evaluation
% Distribution tranformer whith a loss-optimized
   transformer
% Data:
%       Depreciation period                  n = 20
   years
%       Interest rate                        p = 12 %
   p.a.
```

```
%       Energy charge                           Ce = 0.25
    Euro/kWh
%       Demand charge                           Cd = 350
   Euro/(kW.year)
%       Equivalent annual load factor           alpha =
   0.8

% Cc    Capital Cost
% where:
%        Cp    Purcharse price
%        r     Depreciation factor
%        q     Interest factor
%        p     Interest rate
%        n     Depretation period

n = 20;             % Years
p = 12;             % 12% p.a.
Ce = 0.25;          % Euro/kWh
Cd = 350;           % Euro/(kW.year)
alpha = 0.8;          % Equivalent annual load factor
Po = 16;            % kW No-load loss
Pk = 124;           % kW Load loss
Cp = 585000;        % Purcharse price  ($)

q = p/(100)+1
r = (p*(q^n))/((q^n)-1)

Cc = (Cp*r)/100  % Capital cost per year ($/year)

% Cpo   Cost of no-load loss
% Ce    Energy Cost
% where 8760 are hours in a year (hr/year)
% Po    No-load loss

Cpo = Ce*8760*Po    % Cost of no-load loss ($/year)

% Cpk   Cost of load loss
% Pk    Load loss
% a     annual load factor


Cpk = Ce*8760*(alpha^2)*Pk  % Cost of load loss ($/
   year)

% CD    Cost resulting from demand charge
```

```
% Cd     Demand charges amount/(kW*year)

CD = Cd*(Po+Pk)     % Cost resulting from demand
   charge ($/year)

% Total cost of owning and operating this transformer

Total_Loss = Cc+Cpo+Cpk+CD
```

2.7.2 Synchronous generator

It is important to be familiar with those electrical effects dependent on electric machinery which, in a broader sense, are required to understand their performance while connected to an electric network. A synchronous machine (see Figure 2.56), for example, can be evaluated as a stand-alone device to understand the effects of varying the input mechanical power or the voltage field in terms of output voltage. If the machine is later connected as a generator whose voltage is leaded by the infinite bus, no variations of voltage will occur, but of power.

If the generator's behavior is known, its integration to a complex grid will be transparent to the electrical engineer and undesired effects will be avoided.

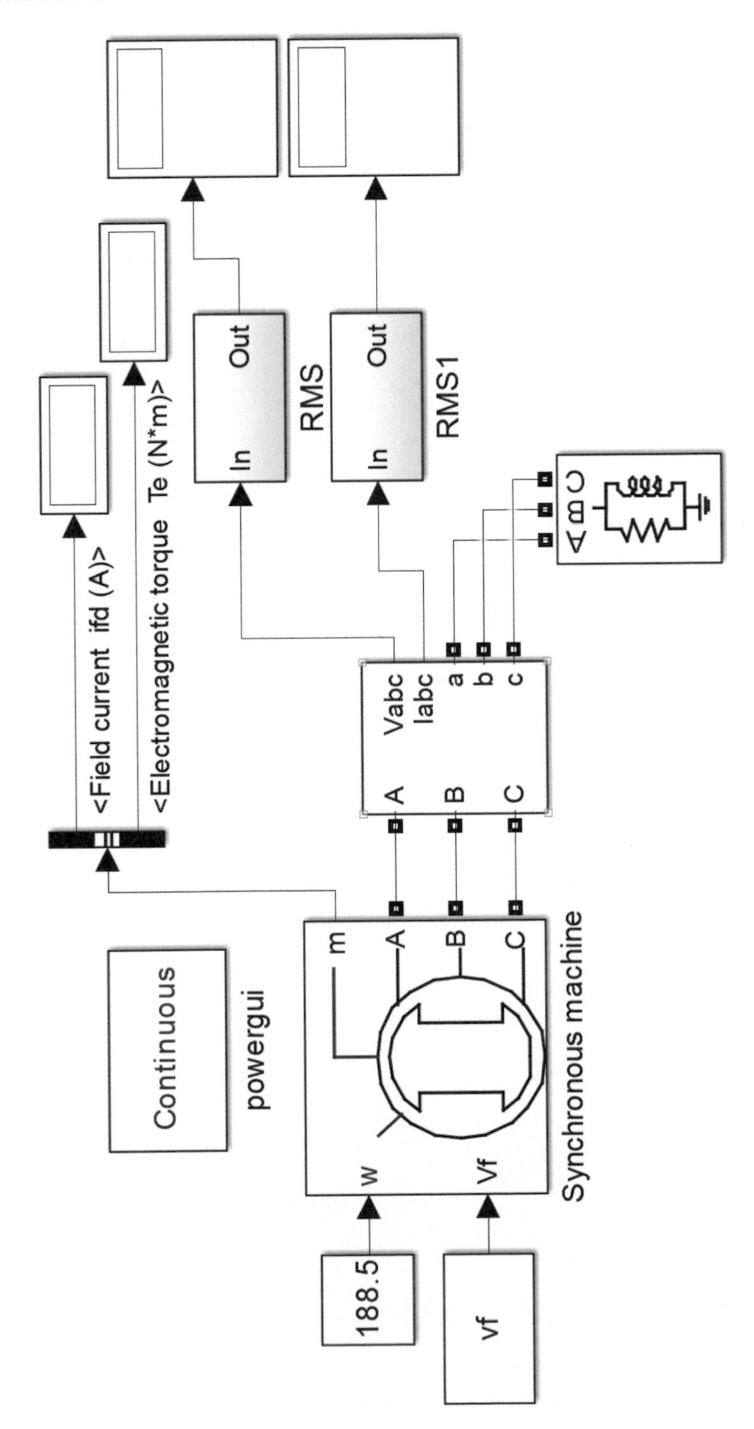

FIGURE 2.56
Power manipulation through a synchronous generator

3

Hydroelectricity

Generation of electricity based on water power is called hydroelectricity. Water potential or kinetic energy can be used to this purpose due to water height or flow, respectively [33]. About 99% of world energy is produced from primary sources, which are water, fossil fuel, and nuclear fuel [32, p. 53], making hydroelectricity paramount in overall energy usage. Hydroelectric generation can be broadly separated into conventional hydroelectric technologies and emerging hydrokinetic technologies; new techniques involve ocean tidal currents, wave energy, and thermal/salinity gradients in water, and represent a small but promising full-renewable energy source. On the other hand, conventional methods guarantee high energy production but have negative consequences on the ecosystem [57, p. 98].

As energy generation depends on the water cycle, numerous factors affect a hydroelectric power plant: rainfall intensity, the location's topography, etc. [33, p. 177]. Water is also required for crops, cattle, human consumption, and ecosystem balance, so a power plant's establishment and management imply a high-resources exhaustive-planning project. Besides those restrictions, hydroelectric plants have an average lifetime of 50 years, require no fuel, have low operation/maintenance costs, represent a negligible contribution to air pollution, and allow parallel operation, such as navigation, flood control, irrigation, etc. [33, p. 180]. Conventional hydroelectricity is one of the least expensive power plants; however, the change in environmental conditions have disabled many prime water sources [57, p. 98].

Hydroelectricity is considered to be renewable, as it depends on the everlasting water cycle; nonetheless, many variables must be taken into account since changes in water supply for a given location can decrease performance. Environmental drawbacks are varied and some of them are severe; the effects due to a hydroelectric power plant are not consistent with sustainability premises, and activists look forward to forcing removal of dams and returning rivers to their free-running conditions [57, p. 103].

Only conventional hydroelectric power plants are considered in this chapter, as they provide large amounts of energy to actual energy distribution installations, while emerging technologies are mostly low-power proposals that are rarely integrated in real distribution systems [57, pp. 99–103].

3.1 Overall characteristics and operation

Hydroelectric plants' basic operating principle implies the utilization of the accumulated potential energy due to a raised water body as a high-pressure flow when released [33, p. 177]. Whenever water is allowed to run from its initial position to a lower one, potential energy (proportional to the height difference between initial to final points) will be converted into kinetic energy; a high-energy flow can then propel an electrical turbine permitting electric generation from mechanical energy [32, p. 53]. Some variations can also achieve energy generation from low height differences by managing a river flow [32, p. 54].

Hydroelectric power plants can be classified under three different categories [32, pp. 53–54]:

- *Impoundment hydroelectric*: Most common type, large generation capacity, needs a big raised reservoir behind the dam
- *Diversion hydroelectric*: Manages the river flow (tide) to propel the turbine, does not need a reservoir, low generation capacity
- *Pump storage hydroelectric*: Similar to impoundment hydroelectric, pumps water back to the reservoir when demand is low for later use

They can also be classified by their power generation capacity as [35, p. 103]:

- *Micro*: Generates less than 100 $[kW]$
- *Small*: From 100kW to 30 $[MW]$
- *Large*: Greater than 30 $[MW]$

Other alternative classifications are made about flow control, load conditions, and head altitude [33, pp. 214–218].

Impoundment hydroelectric generators (Figure 3.1) use the most basic operation principle, as they are based on converting potential energy of raised accumulated water into a quick flow to move a generator's propeller. There are six fundamental components in this type of power plant: dam, reservoir, penstock, turbine, generator and governor [32, p. 55]. Other secondary elements are machine room, internal pipelines, and oscillation tanks [33, p. 183].

The dam is the concrete wall that holds the water in the reservoir and allows potential energy to be accumulated. In order to move the turbine attached to the generator, the governor (control valve) must allow the water to flow through the penstock; once the water has completed its trajectory, it is released to the outlet, which is an open water body. The head represents the height difference between the reservoir surface and the water outlet. This

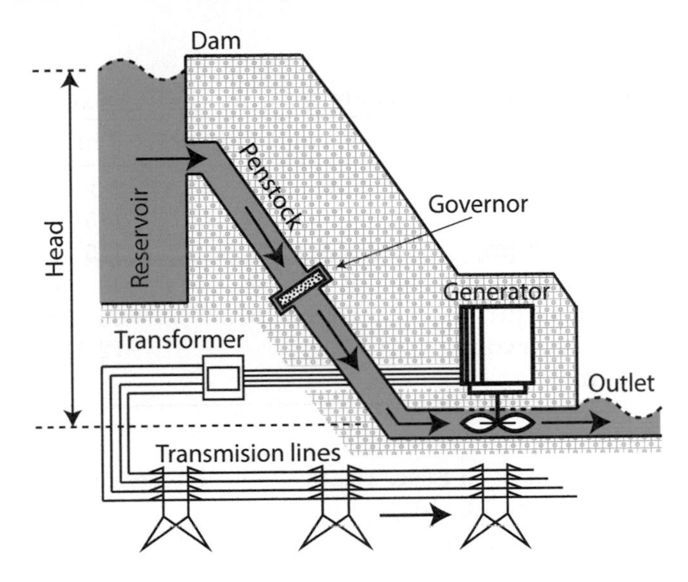

FIGURE 3.1
Impoundment hydroelectic power plant representation

measurement is useful to calculate total potential energy; however, due to energy losses in the penstock, an effective head is often considered to include effects of friction and turbulence [35, p. 101].

The way in which rivers and lakes are connected along the whole water path from the mountain top to the sea offers many opportunities to install hydroelectric plants as it goes downhill. Total water height (and thus, total potential energy) can be seized by successive dams in order to take full advantage of the same water body. Terrain's topography and path's conditions are paramount if dams are planned to be located this way. Generated electric energy of an impoundment power plant depends on some key parameters, such as the water behind the dam, reservoir capacity, flow rate of the water inside the penstock, and the efficiencies of penstock, turbine, and generator [32, p. 57].

Example dimensions of two large impoundment hydroelectric plants are shown in Table 3.1. The Three Gorges dam in China is 22.5 $[GW]$ power capacity rated, while the Grand Coulee dam in Washington is 7.0 $[GW]$. Both plants are large; nonetheless, their physical sizes are quite different.

Short term projects (2020) include the expansion of conventional hydropower and the technology deployment of design-enhanced turbines to improve fish passage and water quality; furthermore, medium term projects (2035) foresee a higher use in emerging technologies focusing on avoiding ecosystem damages, while in the long term (>2035), depending on technology improvements, large-scale installations of emerging technologies are expected, as well as their integration to the main grid; however, large scale projects are

TABLE 3.1
Example features of two large impoundment hydroelectric plants [32, p. 56]

Feature	Grand Coulee	Three Gorges
Length of dam [km]	1.6	2.34
Height of dam [m]	170	185
Width at river base [m]	150	115
Width at crest [m]	9	40
Volume of concrete [m^3]	9.16×10^6	28×10^6
Surface area of reservoir behind dam [km^2]	320	72,128

those implying deployment in issues related to workforce, capital, and other industrial matters [57, pp. 101–103].

A hydroelectric power plant needs two main analyses to plan and manage electric load capabilities. Hydrology allows the total water flow into the reservoir to be known, considering rainfall, evaporation, and water runoff. Hydrography studies water flow variations due to rainfall patterns, and the location's geology and topography, vegetation, and atmospheric conditions [33, pp. 189–191].

$$\text{Water runoff} = \text{total rainfall} - \text{evaporation} \tag{3.1}$$

As the reservoir's head depends on water runoff, its flow $[m^3/time]$ must be analyzed for at least 20 years to precisely know its energy capacity. The minimum flow and reservoir's size determine the primary available power, which is the power the plant can provide constantly. Secondary power represents the possible energy flow under runoff surplus and determines the outlet drain dimensions. Average water flow helps in deciding plant's nominal power [33, p. 190].

Figure 3.2 represents hydrographic analyses' results compared to expected demand line. The months April, May, October, and November present runoff deficiencies that would make the plant provide suboptimal operation, and the energy demand would be unreached. This is the main reason why a dam is built: to provide an artificial water reservoir that can accumulate water surpluses from other months. The total reservoir capacity, then, must be calculated based on expected deficiencies and a risk factor that provides a safety gap for continuous operation.

The reservoir must provide enough water to cover deficiencies; however, it does not need a capacity for holding the total surplus as the water can be allowed to run free in case of overpassing. According to the examples given in Figure 3.2 and Figure 3.3, the power plant will need water reserves 33.3% of the year. Both deficiency blocks imply a lack of 400 $[m^3/s]$ and 500 $[m^3/s]$, respectively; nonetheless, they can be covered with past months'

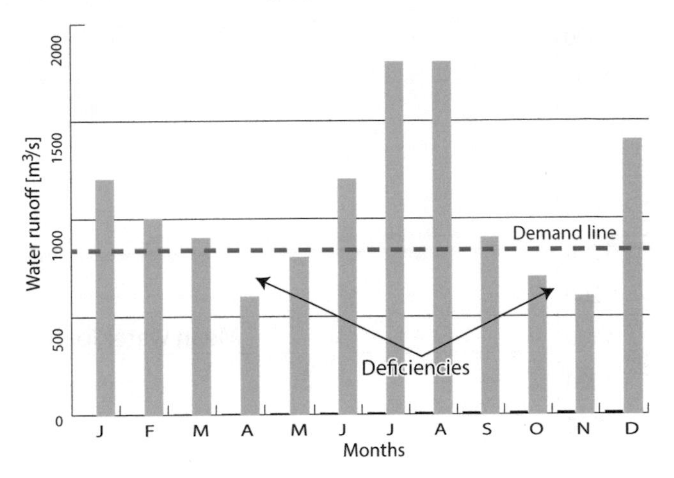

FIGURE 3.2
Water runoff example graph

surpluses. The reservoir must at least hold 200 [m^3/s-month] for October and 300 [m^3/s-month] for November, which results in a total minimum capacity of 1.31×10^9 [m^3]. However, power consumption is not constant and in this case has been only estimated as a mean demand value; moreover, hydrographical data is not the same every year and the reservoir must be prepared to hold a bigger amount of water. Previous calculations were only demonstrative.

Considering the mean water load value of the previous example, a preliminary nominal power value can be assigned to the impoundment hydroelectric plant. Assume the plant's head is 90 [m], with 20% in losses due to friction, turbulence, and generator efficiency,

$$W_n = \overline{m}gh\eta = \delta\overline{Q}gh\eta, \tag{3.2}$$

where $\overline{m}$ is the mean mass flow, calculated as water density by its mean flow ($\delta\overline{Q}$), g stands for gravity acceleration, h is the dam's head, and η the efficiency. Using Equation (3.2), a nominal average power of 759 [MW] can be constantly provided by the plant during a whole year. Calculations shown before are very simple and obviously cannot represent the actual characteristics of a plant; nevertheless, they can approximately offer some values to show how large a plant is and how powerful. The previous procedure is similar to the one shown by Enríquez [33, pp. 193–199].

The power plant energy losses are due to penstock friction and turbulence, and turbine and generator efficiencies. Particularly speaking, turbines are normally rated over 90% efficient under full charge condition and can run from 100 [rpm] and 1000 [rpm] [33, p. 219]. Depending on their construction, they can be classified as impulse or reaction turbines. Impulse turbines (Figure 3.4(a)) rotate thanks to water receiving cups located all the way around

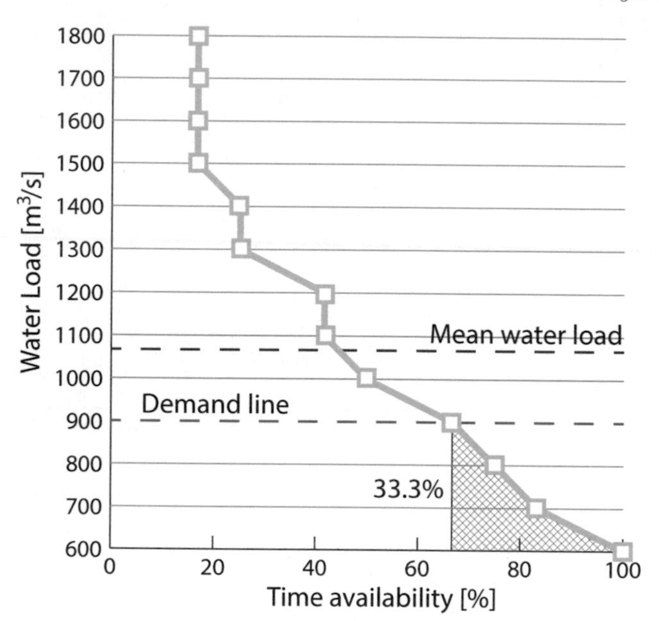

FIGURE 3.3
Water load through time example graph

the turbine axis. The momentum gained by each cup makes the runner turn at constant speed if mechanical load equals the gained momentum through time; regulation must be made by the governor valve in order to modify the amount of water and its speed through turbine nozzles.

In order to prevent the water jet from arbitrarily splashing in the cup and canceling the axial momentum of the runner, a different design, which considers two attached cups (Figure 3.5), can be used so the reflected water takes a known path. The harnessed momentum equals $m_i(v_i - v_c)$, where subscript i represents the water jet and c, the cup. On the other hand, the remaining momentum (lost) is $m_r(v_r + v_c)$, where subscript r stands for reflected water. Notice that both speeds are added as their direction is opposite; at last, the total momentum acquired by the cup is the difference between the gained momentum and the lost one. The mass of incoming water must be equal to that of the reflected one as well as its speed [32, pp. 58–59].

$$\Delta M = 2m_i(v_i - v_c) \tag{3.3}$$

From Newton's second law, the equation above can yield to turbine power as $F = \Delta M/t$ and $P = Fv_c$. So, the power obtained from each cup can be estimated under steady state conditions due to water pressure and power plant design. If the power equation is derived in terms of the incoming water speed, an optimal operation point can be obtained if $v_i = 2v_c$ [32, p. 60]. Optimal

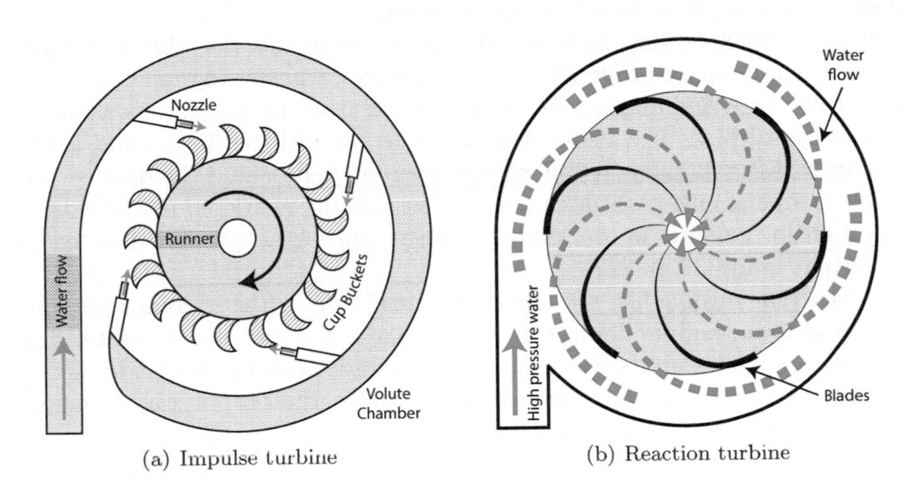

(a) Impulse turbine (b) Reaction turbine

FIGURE 3.4
Turbine types schematics

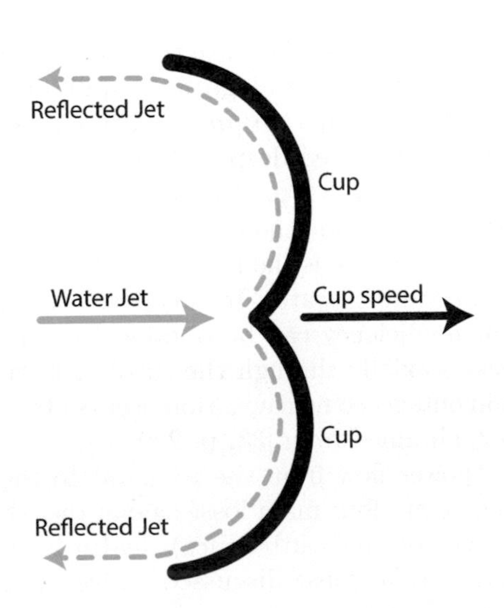

FIGURE 3.5
Improved cup design and water jets indication

operation can be achieved at constant speed, also related to output frequency, consistent with synchronous machine operation.

Impulse turbines take the water into the volute chamber, so that water jets of pressure are injected from all of several nozzles, transferring momentum to the turbine cups and moving the runner attached to the generator's shaft.

Impulse turbines like Pelton turbines operate with big heads and a low amount of water as the potential energy accumulated behind the dam is converted to kinetic energy through the nozzles, i.e., the restricted nozzle output permits high pressures in the volute chamber. Once used, water is directed to the output drain [33, p. 222].

A reaction turbine like the one shown in Figure 3.4(b) is completely immersed in water and its drain is located at the center, forcing the water flux to move the blades on its way down. The energy taken by the blades is the difference between the input water energy minus the output water energy, assuming this energy gap is absorbed by the blades [32, p. 62].

Many simplifications are done to calculate energy harnessing of a reaction turbine. First, changes in volume or mass are considered to be zero, as well as the water speed inside the turbine (as it is immersed in the water), so kinetic energy is small compared to potential energy; secondly, output water is considered to be at a very low pressure, letting the blade absorb $E = P_r V$, where P_r stands for the incoming pressure from the penstock and V for the volume of water in which the turbine is immersed. In this way, the developed power can be written as $P = P_r V/t = P_r f$, where f is the water flow [32, pp. 63–65].

Some known reaction turbine designs are the Francis and Kaplan turbines, the first one being suitable for mid to high heads (80 $[m]$ – 500 $[m]$) and the second one for low heads (1.5 $[m]$ – 80 $[m]$) [32, p. 65]. The output drain of a reaction turbine is always placed deep in the water outage so stability of water disposal can be achieved [33, p. 227]. Francis turbines are sensitive to cavitation as the water spins inside the turbine body.

Another reaction turbine type is the Kaplan turbine, which is designed for low heads and large amounts of water. It is capable of adjusting the blades' orientation so that high efficiency can be obtained under different load conditions; the water passes axially through the turbine, hitting the blades as it goes down to the drain outage, so no cavitation occurs [33, p. 229]. A summary is offered in Table 3.2, obtained from [33, p. 229].

The hydroelectric power flow from the reservoir to the generator can be seen in Figure 3.6. There are four main losses along the whole process, which are penstock losses (friction and turbulence), hydro losses (turbulence and viscosity), turbine losses (like those discussed above), and generator losses. Every step in the power flow process has an inherent efficiency that captures the effect of the total losses; consequently, the power delivered by the plant considering an initial potential energy coming from the reservoir behind the dam is [32, p. 68].

TABLE 3.2
Turbine types summary

Type	Max. Head [m]	Max. Power [MW]	Max. Diameter [m]
Francis	2000	>250	5.5
Pelton	500	>750	10
Kaplan	70	>225	10

$$P_{\text{out}} = P_{\text{in}}\eta_p\eta_h\eta_t\eta_g. \tag{3.4}$$

3.1.1 Exercises

Based on the nominal supply of 380 $[V_{\text{L-L}}]$ 60 $[Hz]$, achieve a steady no-load operating point for the hydroelectric generator. Attain desired frequency by modifying the governor position using the SCADA system and set the output voltage to 380 $[V]$ manually (without an automatic controller). Use Diagram 3.1 for connections.

Speed [rpm]	Excitation Voltage [V]	Attained Frequency [Hz]	Attained Voltage [V]
1820	80	60.1	380

Recommended equipment and connections:

- Synchronous generator
- Motor-driven power supply
- Module for measuring the electric power

If the generator is to be connected alone to some changing load (separated from the infinite bus), variations in performance are expected as load is modified. Connect the hydraulic generator to resistive loads Y-Y and describe the operating point changes due to resistive variations; assume a normal operation of 380 $[V_{LL}]$ 60 $[Hz]$. (First, take the generator to steady operation and modify its excitation through the manual knob in the controller if needed.)

Recommended equipment and connections

- Synchronous generator
- Motor-driven power supply
- Module for measuring the electric power

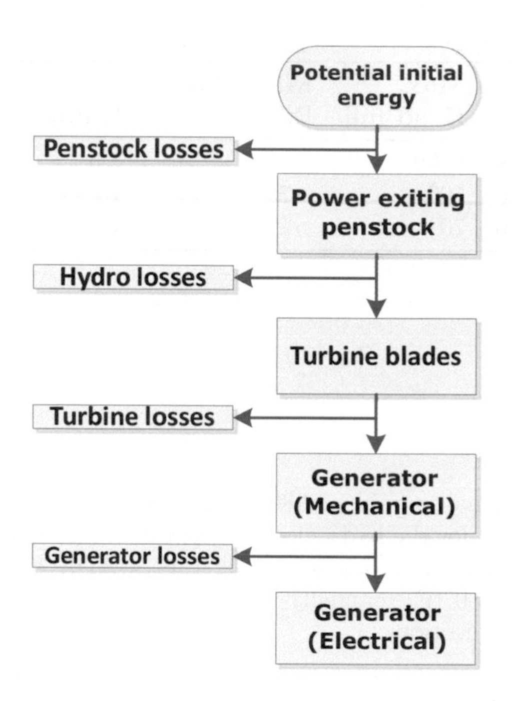

FIGURE 3.6
Power flow of a hydroelectric plant

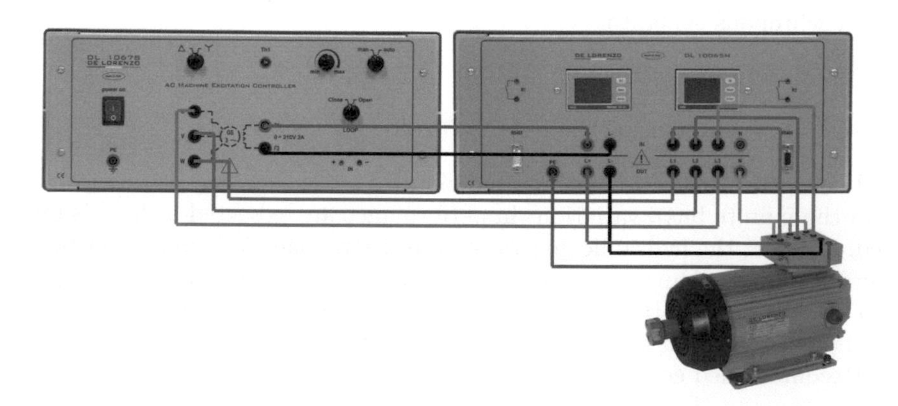

Diagram 3.1
Testing synchronous generator's void operation

Load level	Current [mA]	Output Volt. [V_{RMS}]	Excitation Volt. [V]	Torque [Nm]
No load	0	380	80	0.6
1	200	380	90.6	1.5
2	280	380	93	1.8
3	500	380	103	2.6

What happens if the current is taken over 500 [mA]? Why?
The generator decelerates and stops. Simulated water flow cannot provide enough torque to operate, so the relation between input mechanical power, efficiency, and output electrical power cannot be attained.

- Resistive load

Make the hydraulic generator operate steadily at 380 [V_{L-L}] 60 [Hz] at full load conditions (lowest resistive load value possible) and measure delivered power. Use Diagram 3.2 for connections. If the didactic generator is scaled 1:1000 and represents a hydroelectric power plant, answer and justify the type of plant it would be, the needed water flow at full-load conditions, and the potential energy needed at the dam to attain an operation of 1 month (30 days) at maximum rating (estimate total losses to be of 20% and the head of 50 [m]).

Max. Power [W]	× 1000	Type	Water flow [m^3/s]	Energy [GJ]
332	332 [kW]	Small	0.846	860

Recommended equipment and connections

- Synchronous generator
- Motor-driven power supply
- Module for measuring the electric power
- Resistive load
- Maximum demand meter

Make the hydraulic generator operate steadily at 380 [V_{L-L}] 60 [Hz] at average load conditions and measure delivered power. Use Diagram 3.3 for connections. If the didactic generator is scaled 1:1000 and represents a hydroelectric power plant, answer and justify the amount of water needed at the reservoir to operate 3 months (90 days) without water runoff, the type of turbine needed for such a plant, and the average water flux. Considering

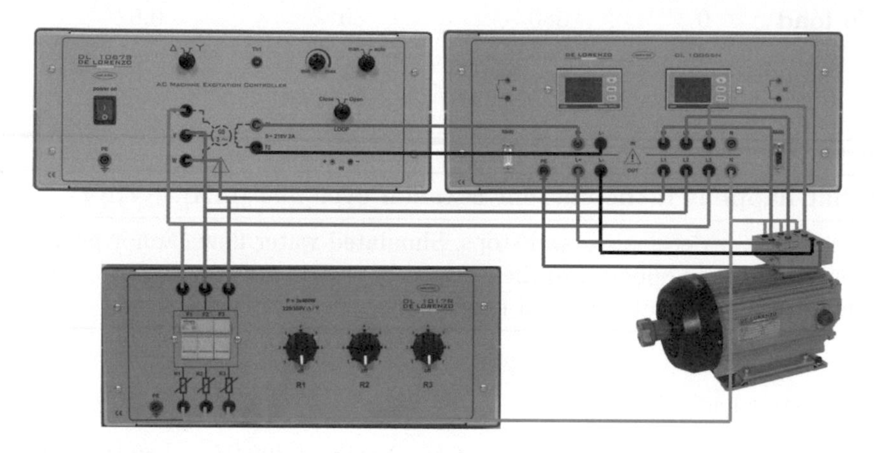

Diagram 3.2
Synchronous generator with resistive loads

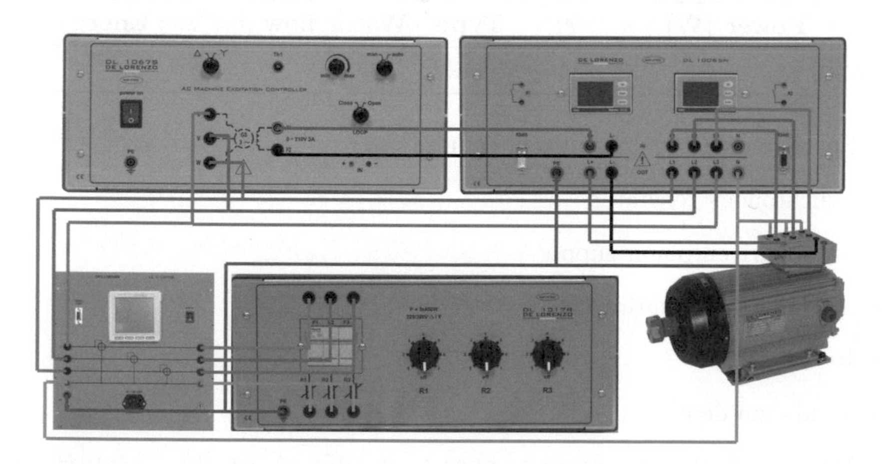

Diagram 3.3
Instrumented operation

the average demand line, estimate the needed mean water load if a risk factor of 20% is desired. Use a mean resistive value as the average consumption. (Estimate total losses to be of 20% and the head of 50 $[m]$.)

Without risk factor:

Avg Power [W]	× 1000	**Reservoir** $[m^3]$	**Turbine**	**Avg. water flux** $[m^3/s]$
189	189 $[kW]$	3.74×10^6	Kaplan	0.481

With risk factor:

Avg Power [W]	× 1000	**Reservoir** $[m^3]$	**Turbine**	**Avg. water flux** $[m^3/s]$
226.8	226.8 $[kW]$	4.49×10^6	Kaplan	0.577

3.2 Integration with the infinite bus

Electric generators are rarely connected to a single load; in fact, they are parallel added to a supply system called an infinite bus or grid [68, p. 299]. The infinite bus is formed by all generators connected together through transmission lines to load centers and distribution transformers; obviously, all generators integrated into the power grid provide the same voltage magnitude and frequency, provoking the infinite bus to have very low variations as it is sourced by many generators [68, p. 299]. It is present at main transmission lines at high voltages of hundreds of kilovolts to maintain high efficiency power distribution at 60 $[Hz]$ (America) or 50 $[Hz]$ (Europe).

Synchronous generators, as those in hydroelectric power plants, work at tens of kilovolts, so transformers are needed to match the infinite bus voltage; likewise, load centers like domestic or industrial facilities typically need 600/480/230/115 $[V]$, so various conversion transformer steps are needed to lower the voltage provided. A representation of the infinite bus is shown in Figure 3.7.

If a generator is to be added to the infinite bus, it must have the same voltage amplitude, frequency, phase sequence, and phase [68, p. 300]. If any of the preceding conditions is not met and the connection is done anyway, the generating nature of the newly added alternator will not be respected, so it can be seen as a load by the infinite bus, and high risk is expected at generator facilities, with disastrous consequences.

Phase sequence "is the time order in which the voltages pass through their

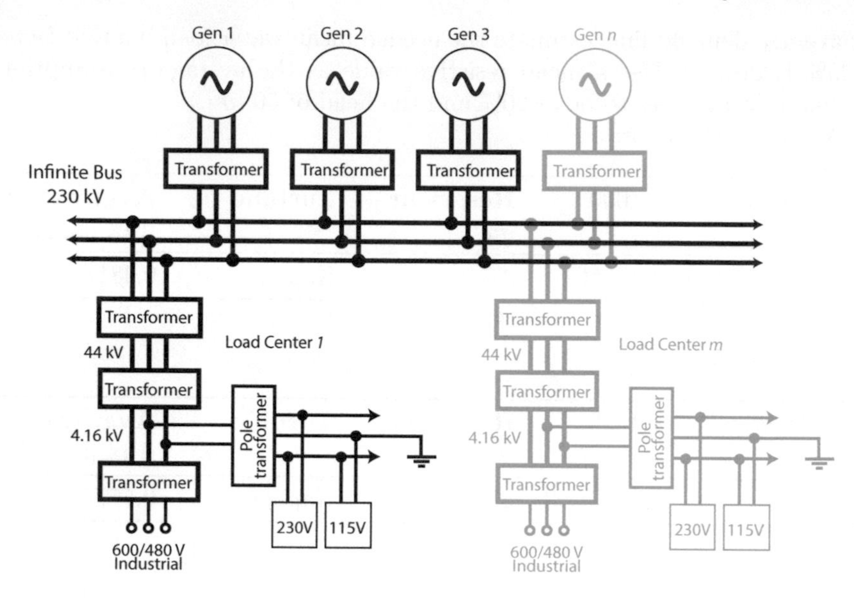

FIGURE 3.7
Infinite bus representation

respective maximum values" in a polyphase AC system [14, p. 507]. If the phasor diagram of a three-phase system is assumed to rotate CCW (as its phase voltage changes periodically) as shown in Figure 3.8, whether or not an observer is placed in the position o, the maximum observed values will be sequentially abc for the positive phase sequence and acb for the negative phase sequence. The preceding reasoning implies that for a rotating generator, the phase sequence can be modified only by interchanging two of its phases.

However, as polyphase systems' phases are commonly labeled arbitrarily, the connection of a generator to the infinite bus cannot be made just by matching phases' names, and previous checking process is needed before connection.

There is a method that easily shows if the previously mentioned conditions are attained in order to achieve successful connection to the infinite bus. The synchronization lamps are literally lamps connected in parallel to the circuit breaker, as shown in Figure 3.9.

Although specific equipment is used to confirm the generator's valid configuration in real plants, the method proposed in Figure 3.9 is an effective laboratory arrangement for connection validation. Different equipment like a synchronoscope (obsolete) or an automatic breaker can provide enough information to decide the moment when the connection is made; some equipment can actually connect the generator automatically when the sensed conditions are suitable.

In order to attain proper connection conditions, different synchronous gen-

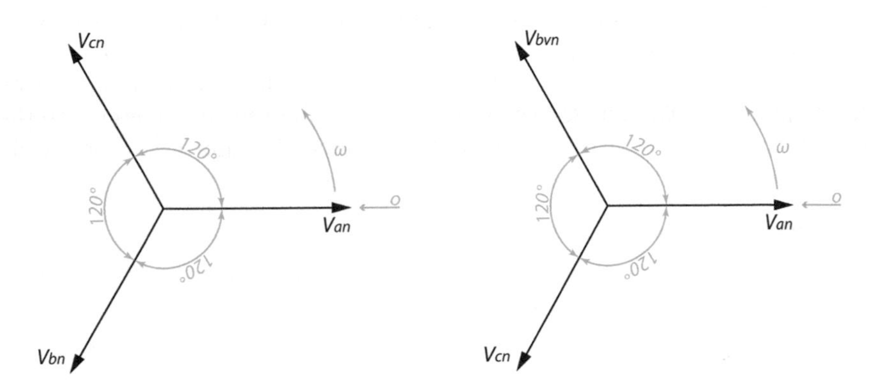

FIGURE 3.8
Positive and negative phase sequences

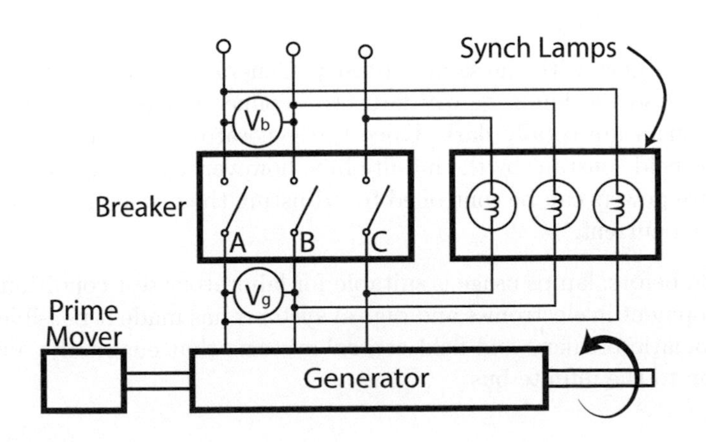

FIGURE 3.9
Schematic of integration of a generator to the infinite bus

erator parameters need to be adjusted: shaft speed, field current, and phase sequence. The shaft speed shall be adjusted so the output voltage frequency is 60/50 $[Hz]$; similarly, the field current must be set so both V_b and V_g read the same; if the phase sequence or phase are not correct, the lamps will never hold a brightness level [68, p. 301].

There are four possible scenarios considering Figure 3.9 as described by Sen [68, pp. 301–303]. The following list assumes all other parameters are the same; obviously, if all voltages are exactly the same all lamps would be totally off.

1. Voltages are different

   All lamps' brightness will be steady and equal. Adjustments must be made in the generator's field current.

2. Frequencies are different

   All lamps brighten and darken simultaneously (change speed depends on how different the frequencies are). Shaft speed must be modified by using the prime mover (Coaxial electric machine connected to the shaft of the generator — Figure 3.9) of a different breaking/accelerating device. If shaft speed is changed, the output voltage also will change, so adjustments in the field current will also be necessary.

3. Phase sequences are different

   Lamps will glow and fade independently. Phase connections must be interchanged between two of them, e.g., A with B.

4. Phases are different

   All lamps glow with the same intensity. Generator speed must be slightly modified so the lamps slowly fade; the connection must be made when all of them are totally dark. Once the generator is connected, its speed will be held constant by the infinite bus; however, the delivered active and reactive powers can be controlled by adjusting the prime mover power and the field current.

As told before, lamps usage is suitable for laboratory test conditions. Modern development in electronics and digital systems has made it possible to produce automatic breakers and field control systems that enable the automatic connection to the infinite bus.

3.2.1 Exercises

A synchronous generator is to be incorporated to the infinite bus. Achieve a successful connection through the generator synchronizing relay and measure the generator's contribution to the grid under various R-L load conditions and water flow. Use Diagram 3.4 for connections.

Start the generator and make its voltage through the power transformer equal to the grid's voltage, and attain a speed such that output frequency is 60 $[Hz]$. Varying the brushless servo speed in very little amounts will slowly make the phase of both grid and generator match, and the automatic relay will close the connection automatically. **Be very careful with phase sequence; make sure the rotation direction is consistent with the grid's sequence before starting.**

L stands for line — G stands for generator

LOAD	Speed [RMS]	PL vs PG [W]	QL vs QG [VAr]	SL vs SG [VA]
R2	1818	110.08 / 144.8	-556.1 / 632	566.88 / 648.3
R2	1845	92.2 / 157.2	-498.38 / 573.3	506.82 / 594.2
R4	1845	280.86 / 160.8	-350.56 / 552.4	449.64 / 575
⋮				

Recommended equipment and connections

- Synchronous generator
- Motor-driven power supply
- Module for measuring the electric power
- Power circuit breaker
- Feeder manager relay
- Synchronizing relay
- Programmable automatic power supply unit
- Synchronization indicator
- Phase sequence indicator
- Three-phase transformer

3.3 Theoretical problems

Problem 3.1. *Broadly propose the parameters of a new hydroelectric plant based on Figure 3.2. Suppose the risk factor has been set at 50%, the head is* 90 $[m]$, *and there is a 20% loss.*

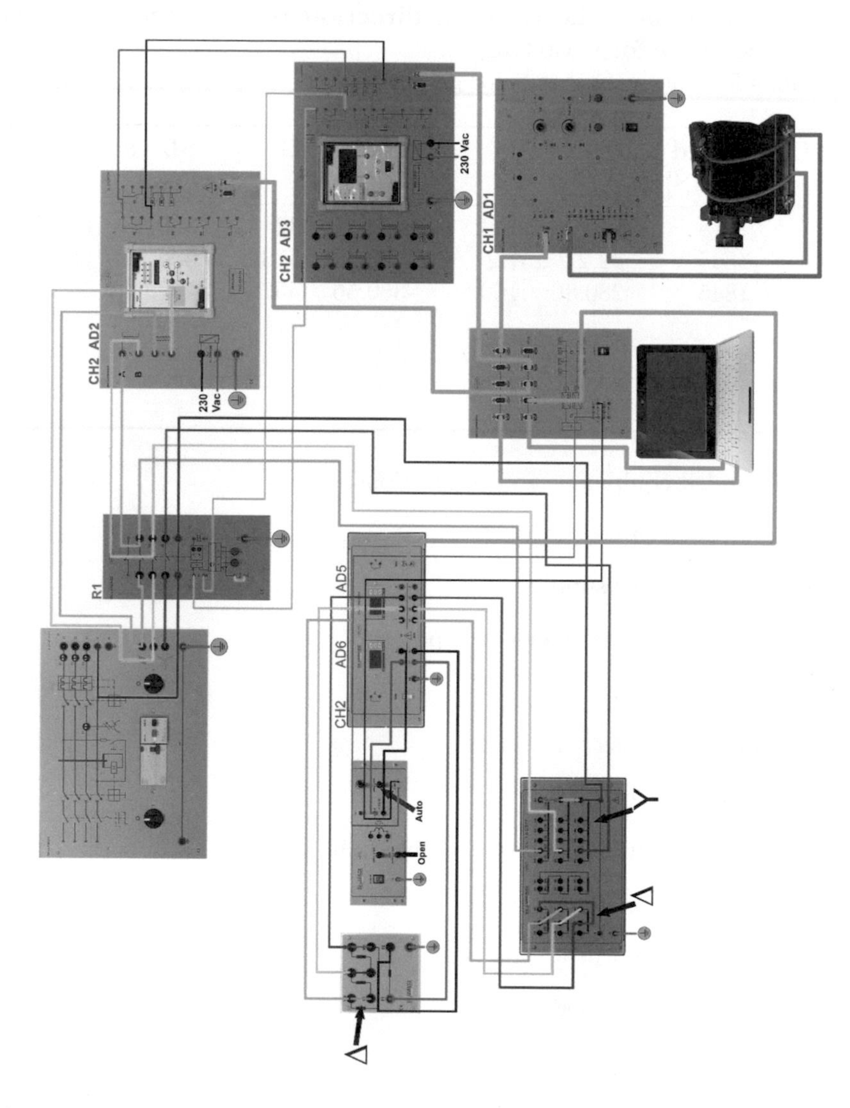

Diagram 3.4
Hydroelectrical power plant synchronization connections

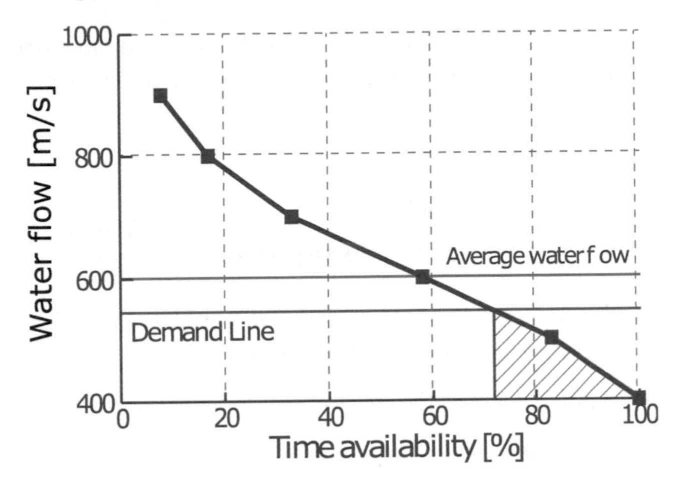

Graph 3.1
Water load vs time availability graph

Solution 3.1. *Average power can be calculated as follows:*

J	*F*	*M*	*A*	*M*	*J*	*J*	*A*	*S*	*O*	*N*	*D*
1200	*1000*	*900*	*650*	*800*	*1200*	*1750*	*1750*	*900*	*750*	*650*	*1400*

$$1000 \left[\frac{m^3}{s}\right] \times 2[month] \left[30\frac{day}{month}\right] \left[24\frac{h}{day}\right] \left[60\frac{min}{h}\right] \left[60\frac{s}{min}\right] \times 1.15 = 1.79 \times 10^9 \; [m^3].$$

Due to power capacity, this plant is large and needs a Pelton turbine due to head restrictions. So results are:

Avg Power [W]	*Reservoir* [m^3]	*Type*	*Turbine*	*Water flow* [m^3/s]
600×10^6	1.79×10^9	*Large*	*Pelton*	*850*

Problem 3.2. *Plot the water load vs time availability for the conditions described below. Plot the mean water load and demand lines (550 [m^3/s]). All values in [m^3/s]*

J	*F*	*M*	*A*	*M*	*J*	*J*	*A*	*S*	*O*	*N*	*D*
500	*600*	*600*	*400*	*600*	*800*	*900*	*700*	*700*	*500*	*400*	*500*

Solution 3.2. *Should look like Graph 3.1*

Problem 3.3. *How tall does the head need to be so an average power of* 300 $[MW]$ *could be taken from a hydroelectrical plant, considering* 350 $[m^3/s]$ *mean water flow and an overall efficiency of 75%?*

Solution 3.3.

$$\frac{300[MW]}{1000\left[\frac{kg}{m^3}\right] \times 350\left[\frac{m^3}{s}\right] \times 9.81\left[\frac{m}{s^2}\right] \times 0.75} = 116[m].$$

Problem 3.4. *Considering the data provided in Problem 3.2, what is the total amount of deficiencies expected in a whole year?*

Solution 3.4. *Whole year* $= -450\ [m^3/s]$ *and the longest gap* $= -300\ [m^3/s]$.

Problem 3.5. *Considering the data provided in Problem 3.2, calculate the reservoir capacity considering a demand line of* 600 $[m^3/s]$.

Solution 3.5. *The longest gap deficiency is* $-500\ [m^3/s]$*, in a period of 4 months, so:*

$$500\left[\frac{m^3}{s}\right] \times 4\,[month] \times \left[30\frac{day}{month}\right]\left[24\frac{h}{day}\right]\left[60\frac{min}{h}\right]\left[60\frac{s}{min}\right] = 5.18 \times 10^9\left[m^3\right].$$

3.4 Homework problems

Problem 3.6. *Consider a location where the water runoff has been measured to be (in* $[m^3/s]$*):*

J	F	M	A	M	J	J	A	S	O	N	D
700	750	900	800	900	1200	1300	1200	1000	800	700	650

Calculate the overall required efficiency of the plant if the average demand is about 935 $[MW]$ *and the head is* 140 $[m]$.

Solution 3.6. *The mean can be easily computed to be* 908.33 $[m^3/s]$*. Considering a head of* 140 $[m]$*, the power generation can be expressed as* $W_n = \delta\overline{Q}gh\eta$*, implying:*

$$935\ MW = \left(100\ \left[\frac{kg}{m^3}\right]\right)\left(908.33\ \left[\frac{m^3}{s}\right]\right)\left(9.81\ \left[\frac{m}{s^2}\right]\right)(140\ [m])\,\eta.$$

So, $\eta = 0.749$.

Problem 3.7. *Suppose there is a small town fed by a single hydroelectric plant. This town increases its peak power consumption over time, while the water runoff has been measured to be decreasing. Both relations are shown below. It is known that the plant is being used at 22.8% of its capacity, and that when it was installed 10 years back, it was used at 2.5% of its total capacity.*

Find the time at which the town will be at risk of a power shortage.

- $W_t(t) = 20e^{(0.6t)}$ $[MW]$, *with t in years*
- $\frac{\Delta \overline{Q}}{\Delta t} = -2t \left[\frac{m^3}{s}\right]$, *with t in years*

Solution 3.7. *At the present time* $(t = 0)$ *the town takes* 20 $[MW]$ *from the plant, implying the plant's rated power is* $W_n(0) = 87.72$ $[MW]$ *at the present water runoff. On the other hand, the town used to consume* 2.7 $[MW]$ *ten years back, which was known to be 2.5% of the plant's capacity given the water runoff at that time.*

The plant's specific parameters are unknown; however, the water's density, gravity acceleration, head, and efficiency can be considered constant over time, so $W_n = k\overline{Q}$, *and* $W_n(t) = k(\overline{Q_0} - 2t)$.

So, $W_n(-10) = \frac{2.7\ [MW]}{0.025} = 108\ [MW] = k(\overline{Q_0} + 20)$, *leading to*
$108\ [MW] = k(\overline{Q_0} + 20)$, *and* $87.72\ [MW] = k\overline{Q_0}$.
So, it can be solved that $\overline{Q_0} = 86.5\ [m^3/s]$, *and* $k = 1.014 \times 10^6 \left[\frac{kg}{ms^2}\right]$.
Now both power equations can be solved together:
$W_n(t_f) = k(\overline{Q_0} - 2t_f) = W_t(t_f) = 20e^{(0.6t_f)}$, *leading to*
$t_f = 24.13$ *years* $W_t = 38.78\ [MW]$.

Problem 3.8. *Suppose a hydroelectric plant exhibits the following performance (first row shows water runoff* $[m^3/s]$, *while the second one shows required average power during that month [MW]):*

J	F	M	A	M	J	J	A	S	O	N	D
700	750	900	800	900	1200	1300	1200	1000	800	700	650
620	500	440	490	580	720	800	700	650	600	600	710

If its head is 90 $[m]$ *and its overall efficiency is 0.8, build a table to show how much water must be released during each month, considering the initial reserved water to be* 500×10^6 $[m^3]$, *and the total dam capacity to be* 1×10^9 $[m^3]$. *Report whether a power shortage is possible.*

Solution 3.8. *It would be useful to calculate the available power per month, considering each month separately. In addition, calculate the difference regarding the available power and finally, the compensating runoff.*

The generated power and equivalent runoff can be easily computed by following the equation $W = \delta\overline{Q}gh\eta$.

Dam status can be easily calculated considering $C = \overline{Q}(3600\ [h/s] \times 24\ [day/h] \times 30\ [month/day])$. *All aforementioned computations are reported in Table 3.3.*

The dam capacity is not enough to compensate for December consumption, so the plant will fail to deliver the required power.

Problem 3.9. *Consider the problem described in Problem 3.8 and calculate the required reservoir size to enable continuous operation throughout the year.*

Solution 3.9. *The required average power throughout the year is of* $618\ [MW]$. *The equation* $W_n = \delta\overline{Q}gh\eta$ *can be used over the monthly remainder. Such remainders imply the following requirements:*

- *A water reserve of at least* $774 \times 10^6\ [m^3]$ *is needed before January starts*
- *A water reserve of at least* $192 \times 10^6\ [m^3]$ *is needed to compensate April*
- *A water reserve of at least* $1230 \times 10^6\ [m^3]$ *is needed to compensate the last three months of the year*

January, February, and the last three months of the year need to be seen as a single unit, so having $2 \times 10^9\ [m^3]$ *is required in the reservoir before October. This can be effectively taken from the preceding surplus of* $3.18 \times 10^9\ [m^3]$, *collected from May to September. In addition, April's required surplus of* $192 \times 10^6\ [m^3]$ *can be taken partly from March (*$66.7 \times 10^6\ [m^3]$*) but* $125 \times 10^6\ [m^3]$ *will still be missing.*

This implies that $2.13 \times 10^9\ [m^3]$ *is the maximum amount of required water for a single time period during the year. If a 15% risk factor is considered, the reservoir must be of* $2.45 \times 10^9\ [m^3]$, *and must be entirely filled during the summer.*

Problem 3.10. *Plot a curve representing the size or the reservoir versus the overall efficiency for a dam with a maximum deficiency of* $30\ [MW]$ *for 2 months, and a head of* $h = 140\ [m]$.

Solution 3.10. *Both dam equations can be combined as follows:*

$$W = \delta\overline{Q}gh\eta,$$
$$\frac{W}{\delta gh\eta} = \overline{Q},$$
$$C_{month} = \frac{W}{\delta gh\eta}(3600 \times 24 \times 30 \times 2),$$
$$C(\eta) = W\frac{5.18 \times 10^6}{\delta gh\eta} = \frac{113.15 \times 10^6}{\eta}.$$

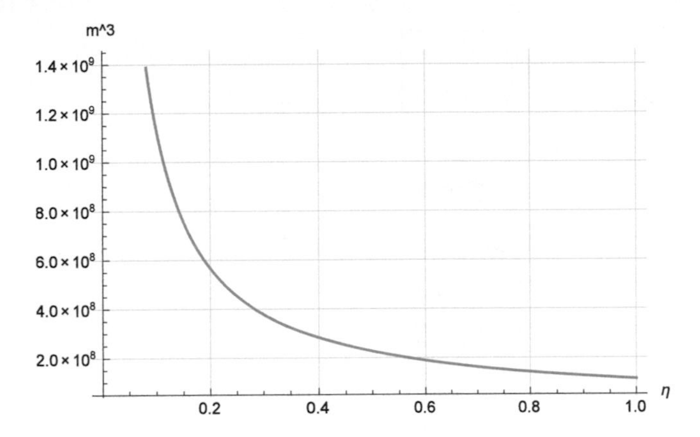

3.5 Simulation

Once a hydraulic power plant is connected to the infinite bus, the behavior and characteristics of the generator will depend on clear parameters. As the voltage is led by the bus because it dominates the generator (in terms of power capacity), variations of input mechanical power (water power) or field voltage will not exhibit a direct effect on output voltage. However, they will enable the power flow to/from the infinite bus.

A synchronous machine can provide its own magnetization between stator and rotor so it can provide reactive power to the network. As the relation between active/reactive power is clearly dominated by field current and mechanical power, respectively, this generator can quickly manage both types of grid variations. The diagram of such simulated system is shown in Figure 3.10.

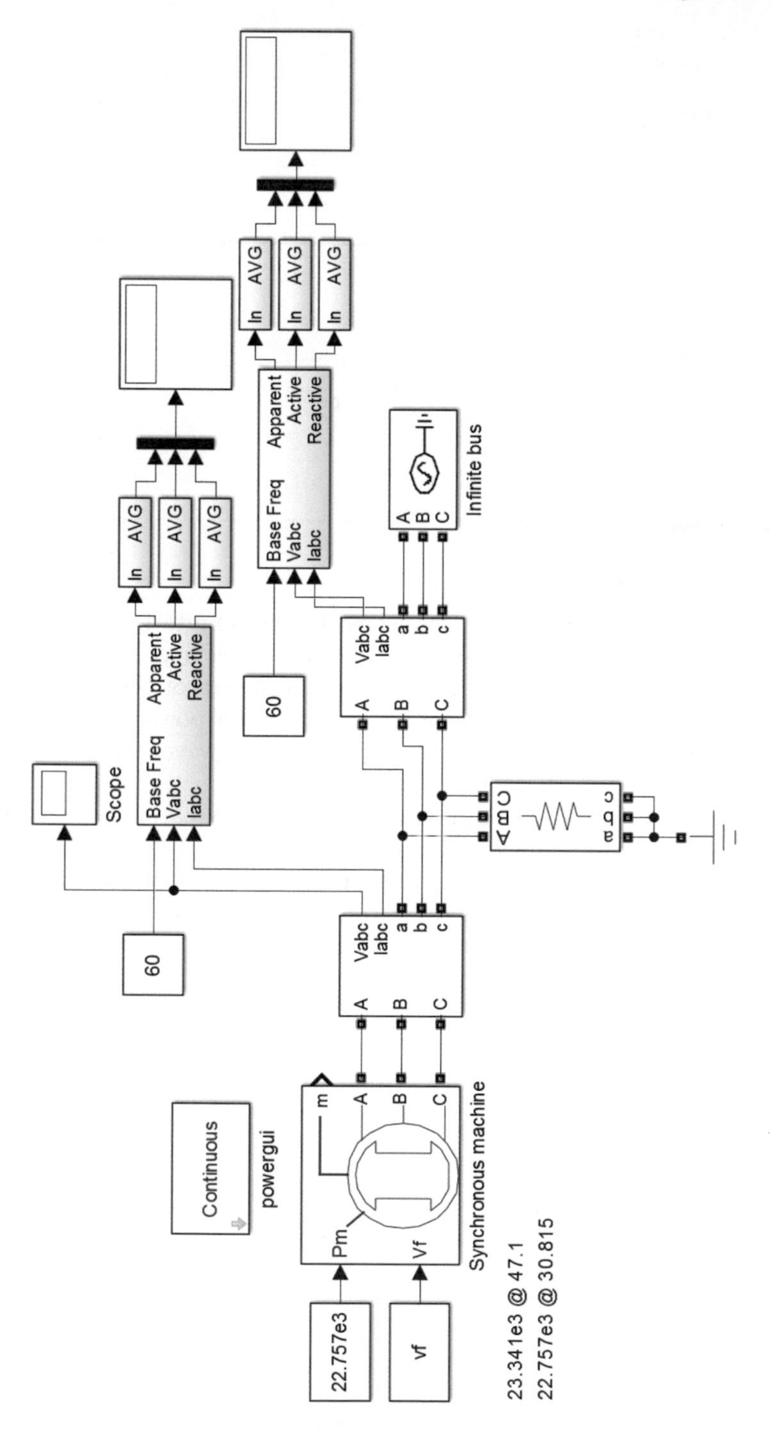

FIGURE 3.10
Generator interconnection to the infinite bus simulation

TABLE 3.3
Computations to solve Problem 3.8

	J	**F**	**M**	**A**	**M**	**J**	**J**	**A**	**S**	**O**	**N**	**D**
Runoff $[m^3/s]$	700.00	750.00	900.00	800.00	900.00	1200.00	1300.00	1200.00	1000.00	800.00	700.00	650.00
Gen Power $[MW]$	494.42	529.74	635.69	565.06	635.69	847.58	918.22	847.58	706.32	565.06	494.42	459.11
Req Power $[MW]$	620.00	500.00	440.00	490.00	580.00	720.00	800.00	700.00	650.00	600.00	600.00	710.00
OverPower $[MW]$	−125.58	29.74	195.69	75.06	55.69	127.58	118.22	147.58	56.32	−34.94	−105.58	−250.89
OverRunoff $[m^3/s]$	−177.79	42.11	277.05	106.26	78.84	180.63	167.37	208.95	79.74	−49.47	−149.47	−355.21
Compensate $\times 10^8 [m^3]$	−4.61	1.09	7.18	2.75	2.04	4.68	4.34	5.42	2.07	−1.28	−3.87	−9.21
Dam status $\times 10^8 [m^3]$	3.92	1.48	8.66	1.00	1.00	1.00	1.00	1.00	1.00	8.72	4.84	−4.36
Release $\times 10^8 [m^3]$	0.00	0.00	0.00	−1.42	−2.04	−4.68	−4.34	−5.42	−2.07	0.00	0.00	0.00

4

Wind Energy

Wind energy has been exploited for thousands of years. Its principal ancient applications are extracting water from wells, cutting lumber, grinding grain for making flour, and other mechanical power applications. Nowadays wind is employed mostly for electric power generation. Wind generators are also called wind turbines, because their function is similar to the gas and steam turbines that are used for electric energy generation. Since the end of the nineteenth century, attempts at electric generation from wind power have been made; nevertheless, combustion-based turbines had much greater development and applications due to their higher reliability and the possibility of being used whenever necessary. Due to oil crises in the 1970s, wind energy has been growing and has been the subject of major investment in Europe and North America, and more recently, in China. In the 1980s, the power electronics technology and aerodynamics of wind turbines were sufficiently mature to begin large-scale manufacturing of wind turbines for commercial use. Today, wind energy is one of the most cost-effective methods of electricity generation and it is the most important in terms of power capacity installed as a renewable energy source.

4.1 Wind turbine basic structure

A wind turbine is a device that transforms the kinetic energy coming from the air stream into electrical energy; this conversion requires a mechanical step and then an electrical one. The element that extracts the energy from the wind is the rotor, and the rotor drives the electrical machine, usually through a gearbox. After that, the generator converts the mechanical power into electrical power.

Many designs of wind turbines have been developed over the years. Here we are not studying the historical survey of the wind conversion systems, but an interesting resume can be found in [11]. Basically there are two configurations: vertical axis and horizontal axis wind turbines, the latter ones being more popular due to their better efficiency. Today for onshore and offshore applications almost all of the manufactured turbines are horizontal axis with two or three blades. Figure 4.1 presents a horizontal axis wind turbine and

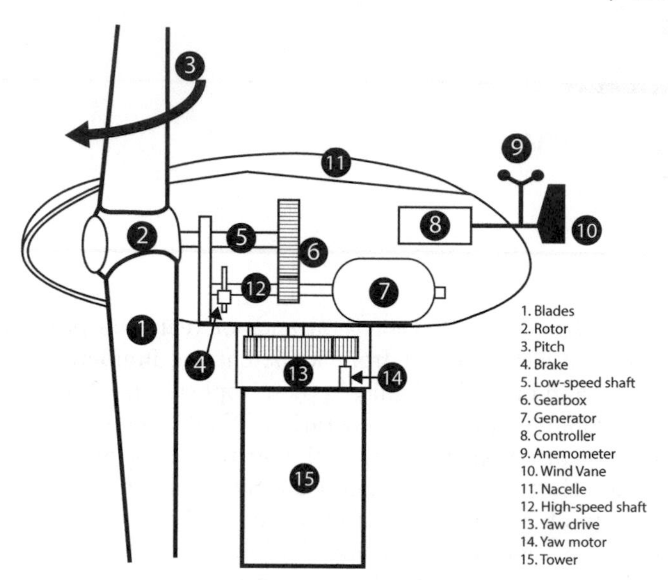

FIGURE 4.1
Wind turbine structure

its elements. The wind speed increases with the height, and therefore a tower is used to elevate the turbine. The energy conversion chain can be organized into:

- Aerodynamic system, consisting of the turbine rotor, which is composed of blades, supported by a hub and mounted on the rotor.
- A transmission system composed of a low-speed shaft, connected directly to the rotor, a gearbox that increases the speed of the low-speed shaft, and the high-speed shaft, which drives the power to the electrical generator. Some designs avoid the transmission system by incrementing the number of poles of the electrical generator.
- The electromagnetic system formed by an electrical machine (synchronous or induction generator).
- The electric system, which includes the grid connection.
- The yaw mechanism, which moves the nacelle so the blades are perpendicular to the wind direction.
- A control system that manages the rest of systems via an anemometer for measuring the speed of the wind, and a wind vane that indicates the direction of the wind.

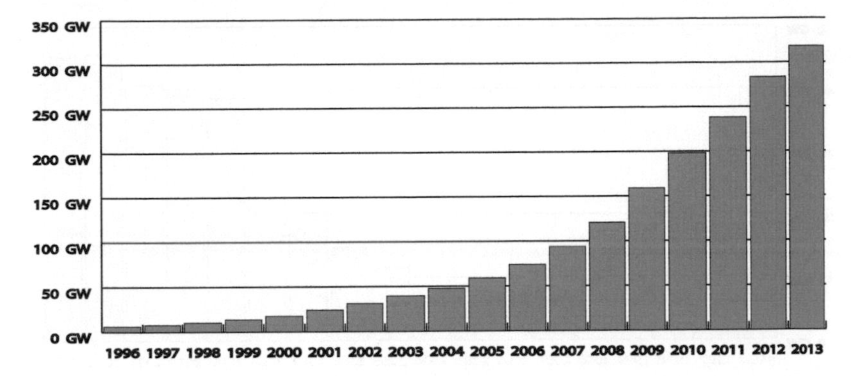

FIGURE 4.2
Global cumulative installed wind capacity from 1996 to 2013, modified from [29]

4.2 Worldwide eolic energy production

Wind energy has become an important issue for European nations, and political measures are implemented in many nations in order to encourage its use. In 2007 the European Union established the target to achieve 20% of energy production from renewable energy sources, while in the United States the renewable portfolio standard (RPS) has established an objective of obtaining at least 15% of total energy from renewable sources [16].

Figure 4.2 shows the annual growth of installed wind capacity, with data taken from the Global Wind Energy Council's report of 2013. The global capacity has grown from 6,100 $[MW]$ installed in 1996 to 318,110 $[MW]$ in 2013. Figure 4.3 shows the new installed capacity; it can be seen that up to today the trend is to continue installing new eolic generators. The enthusiasm of the countries that produce a greater percentage of their consumption by wind energy comes from government reforms that encourage the population to invest in renewable production for injecting energy to the grid.

4.3 Basics of wind energy

The kinetic energy contained in any material element is

$$E = \frac{1}{2}mv^2, \tag{4.1}$$

where m is the mass of the element, and v is the speed. If we consider that mass is flowing thought a cylinder equal to the wind turbine area, it is convenient

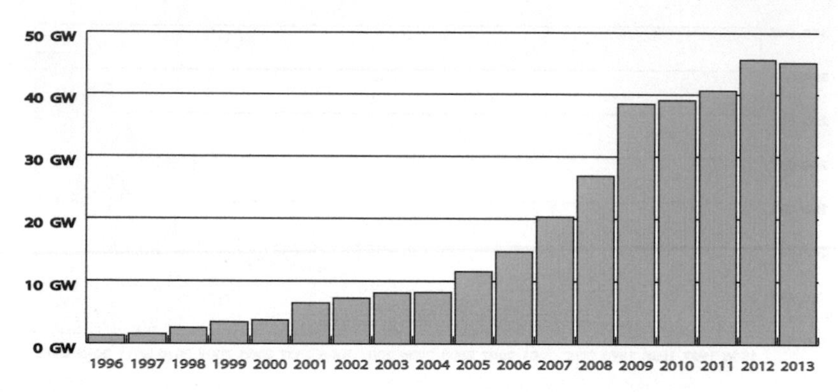

FIGURE 4.3
Global new installed wind capacity from 1996 to 2013, modified from [29]

to express the kinetic energy equation in terms of power flow,

$$\dot{E} = \frac{1}{2}\dot{m}v^2. \tag{4.2}$$

If we substitute mass flux as $m = \rho Av$, where ρ is the density of the fluid, A is the cross-sectional area of the turbine, and v is the velocity of the fluid, total power available on the wind is

$$\dot{E} = P = \frac{1}{2}\rho Av^3. \tag{4.3}$$

It is not possible to extract all this energy from wind. In 1919 the German physicist Albert Betz demonstrated that the maximum percentage of energy that can be captured by a turbine is $16/27 \approx 0.593$, basing his analysis on mass conservation and momentum of the air stream:

From Newton's second law, the rate of change of momentum in a control volume is equal to the sum of all forces acting on it. Consider the wind turbine of Figure 4.4, with the following assumptions:

- The pressure forces at point 1 and point 4 are equal
- There are no shear forces on the x direction
- Only on points 1 and 4 is there momentum loss or gain

So, the equation of momentum for the large volume control (from point 1 to point 4) becomes

$$F_R = \dot{m}(V_4 - V_1). \tag{4.4}$$

Consider the volume of control from point 2 to point 3. $A_2 = A_3$, and $V_2 = V_3$, because the turbine is thin and the velocity cannot change its magnitude dramatically. Nevertheless, the energy extracted is evidenced by the pressure

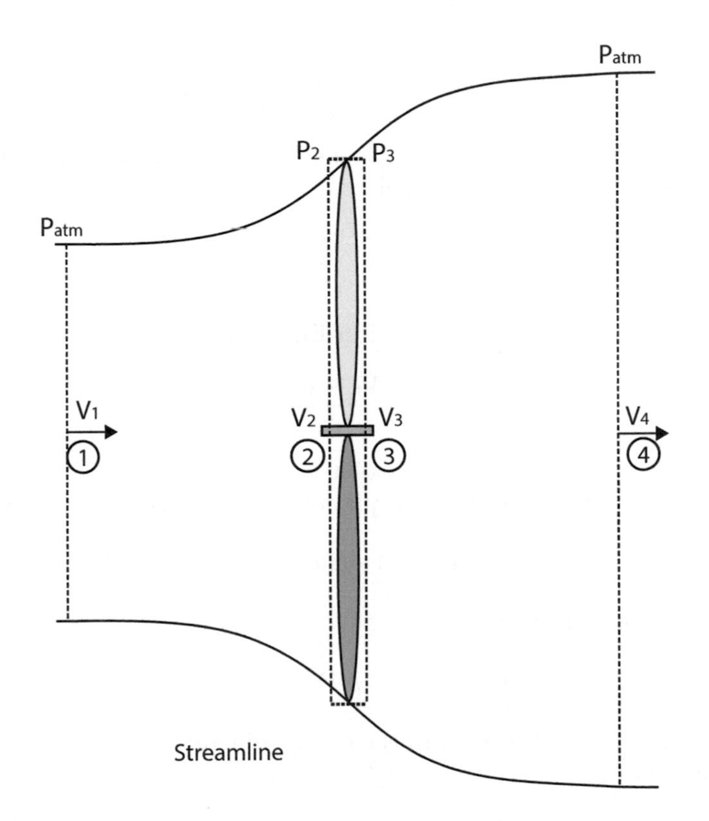

FIGURE 4.4
Control volumes for the analysis of a wind turbine

gradient. Conservation of momentum applied to volume of control from 2 to 3 yields

$$F_R = P_3A - P_2A = (P_3 - P_2)A. \tag{4.5}$$

The Bernoulli equation is valid when there is no energy change, so we can apply it from point 1 to point 2, and from point 3 to point 4 as well:

$$\frac{P_1}{\rho g} + \frac{V_1^2}{2g} + z_1 = \frac{P_2}{\rho g} + \frac{V_2^2}{2g} + z_2, \tag{4.6}$$

and

$$\frac{P_3}{\rho g} + \frac{V_3^2}{2g} + z_3 = \frac{P_4}{\rho g} + \frac{V_4^2}{2g} + z_4. \tag{4.7}$$

If we add Equations (4.6) and (4.7) considering $z_1 = z_2 = z_3 = z_4$, $V_2 = V_3$, and $P_1 = P_4 = P_{atm}$, this is obtained:

$$\frac{V_4^2 - V_1^2}{2} = \frac{P_3 - P_2}{\rho}. \tag{4.8}$$

If $\dot{m} = \rho A V_2 = \rho A V$ is substituted in Equation 4.8,

$$V_2 = \frac{V_1 + V_4}{2}. \tag{4.9}$$

Now let's consider the speed of point 2 as a fraction of the speed: $V_2 = (1 - 2a)V_1$; if we substitute this in Equation 4.9 it yields

$$V_4 = (1 - 2a)V_1, \tag{4.10}$$

while the mass flux is computed as

$$\dot{m} = \rho A V_2 = \rho A V_1(1 - a). \tag{4.11}$$

The power extracted is equal to the energy difference between point 1 and point 4, that is

$$P = \frac{\dot{m}(V_1^2 - V_4^2)}{2} = \frac{AV_1(1-a)[V_1^2 - V_1^2(1-2a)^2]}{2} = 2\rho A V_1^3 a(1-a)^2, \tag{4.12}$$

$$\eta_{turbine} = \frac{P}{P_{max}} = \frac{2\rho A V_1^3 a(1-a)^2}{\frac{1}{2}\rho A v^3}. \tag{4.13}$$

Efficiency is maximum when $a = \frac{1}{3}$ if this value is substituted in the efficiency equation (4.13), $\eta_{max} = \frac{16}{27} \approx 0.593$. Maximum efficiency of modern wind turbines is normally about 0.3.

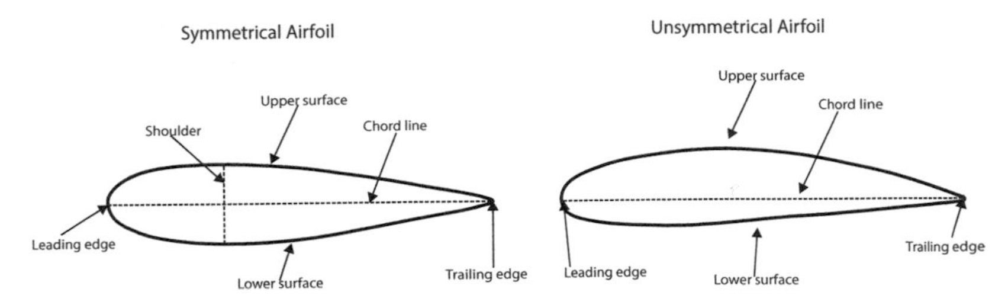

FIGURE 4.5
Symmetrical and unsymmetrical airfoil terminology and shape

4.4 Aerodynamics of wind turbines

Wind is unpredictable, and it is very hard to establish a complete mathematical model for a turbine; normally, experimental results arranged and organized with dimensionless parameters are used for calculations. Yet it is very important to understand the basic physical principles for lifting and dragging forces on airfoils, for understanding many design characteristics of blades.

4.4.1 Airfoils

An airfoil is the cross-sectional area of a turbine's blade. There are many designs of airfoils, depending on the characteristics of the operating environment. Figure 4.5 shows a scheme of an airfoil. The chord line is an imaginary line that connects the leading edge with the trailing edge. The shoulder is the point of higher thickness.

On a symmetrical airfoil, when the airflow comes parallel to the chord line, the lift force equals zero, because as the upper and lower surfaces are identical, the air will flow at the same speed and the pressure gradients are equal. On the other hand, unsymmetrical airfoils have a larger area and are designed to have a greater velocity of air on the upper surface; this provokes a decrease of pressure on the upper side that tends to lift the airfoil (see Figure 4.6).

The angle of the chord line with respect to the airflow is known as the angle of attack (α). Lift and drag force are functions of the angle of attack; as a convention, lift force is perpendicular to the direction of air and drag is parallel to it. The physical explanation of lifting is the curvature of streamlines around the curve, as indicated in Figure 4.7. From fluid mechanics a pressure gradient is necessary to curve the streamlines: $\frac{\delta p}{\delta r} = \frac{PV^2}{r}$, where r is the curvature of streamlines, V is the velocity of the fluid, and ρ is the density. The pressure difference between the upper surface and the lower surface gives a force acting like a centripetal force from the circular motion of the particles. The dragging force at low attack angles is mainly caused by the friction of the surface with the blade.

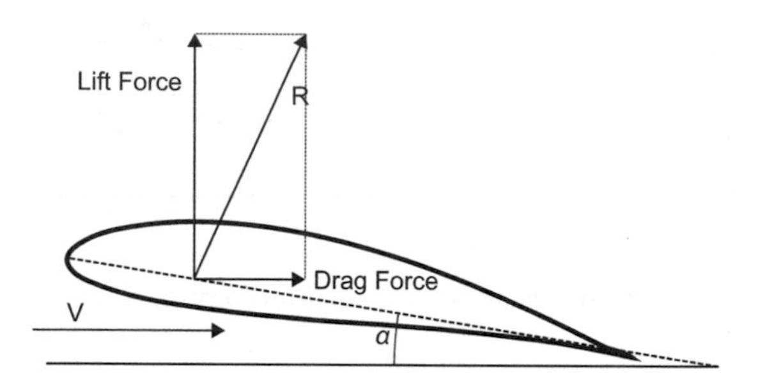

FIGURE 4.6
Lifting and dragging force in an airfoil

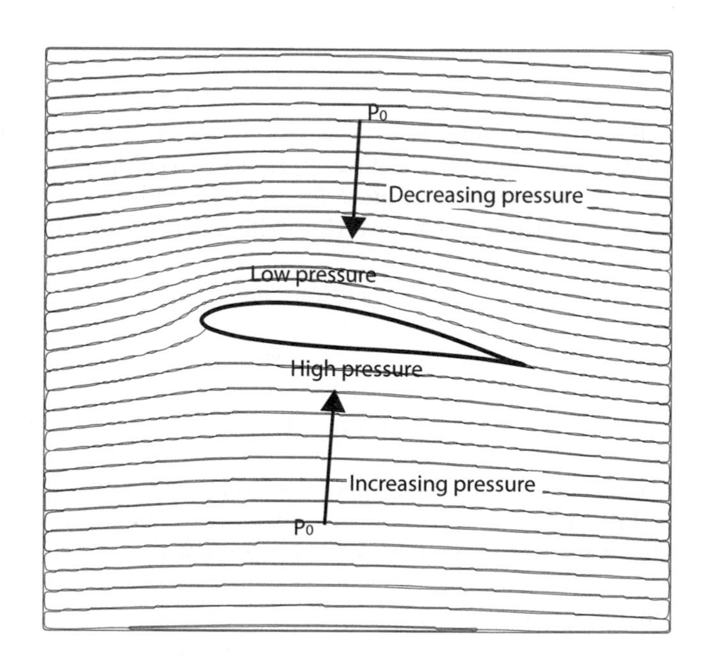

FIGURE 4.7
Generation of lift

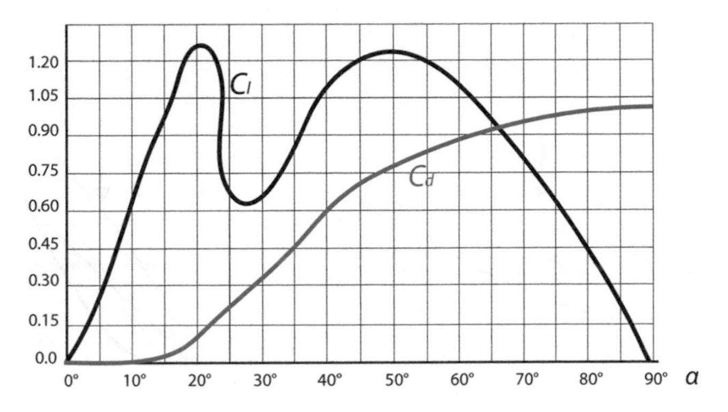

FIGURE 4.8
Lift and drag coefficient as functions of attack angle

A dimensionless coefficient relating the lift generated to the density of the fluid is known as the lift coefficient C_l, while the dragging coefficient relates the drag force to the density of the fluid. Lift and drag forces are usually expressed as

$$L = \frac{1}{2}\rho S V_0^2 C_l, \tag{4.14}$$

$$D = \frac{1}{2}\rho S V_0^2 C_d, \tag{4.15}$$

where S is the area of the airfoil (chord length c multiplied by the length of the blade), V_0 is the velocity of the free stream, and ρ is the density of the fluid.

For small values of the attack angle, the relation between the drag force and lift coefficient is linear; in Figure 4.8 a graph of force coefficients is presented. It can be seen that the characterized airfoil is symmetrical, because the starting lift force is zero; for an unsymmetrical shape the lift coefficient will start at a little offset. There is an ideal angle of attack so that the lifting force is maximum while the dragging is minimum.

Up to now we have seen the airfoil as a fixed structure. Nevertheless, the blades of the turbine are rotating; this rotation gives another component to the original velocity of the wind, and this component is due to the angular velocity and the distance from the center of the section of the blade. It is convenient to establish a new velocity known as the relative velocity of the airfoil. $V_{rel} = \sqrt{V^2 + (\omega r)^2}$, where V is the velocity of the air and ωr is the tangential velocity of the blade.

In Figure 4.9 we notice the effect of the tangential velocity on the relative velocity, as seen by an observer located in the airfoil. γ is the angle of the relative velocity, while ϕ is the orientation of the chord, also known as pitch angle. As mentioned previously, there is an attack angle α that gives the best performance of the turbine for maintaining it constant with respect to the

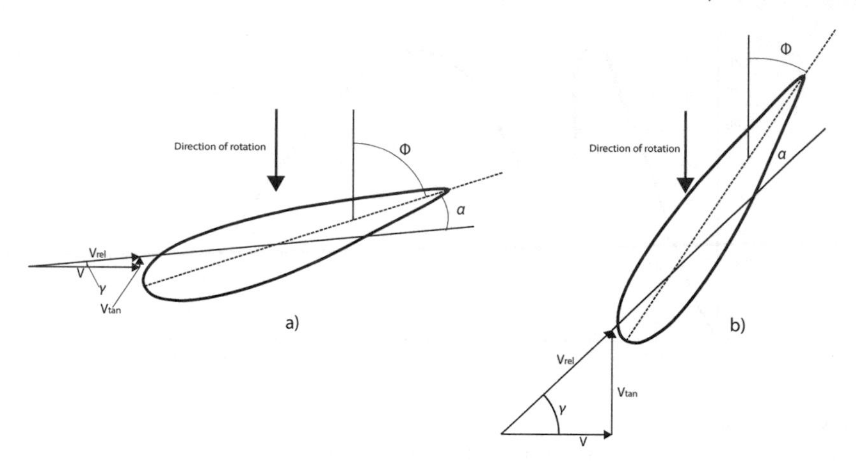

FIGURE 4.9
Velocity of the blade relative to the blade at two different locations from the center

relative velocity; there is the necessity of twisting the turbine, which is the reason for the form of large blades.

4.5 Constant-speed versus variable-speed turbines

As shown in the previous section, the power available in the wind is

$$P = \frac{1}{2} C_p \rho A V^3, \tag{4.16}$$

where C_P is known as the power coefficient (efficiency), ρ is the density of the fluid, A is the cross-sectional area, and V is the speed of the wind. The power coefficient C_P represents the efficiency of the turbine and is a function of the tip speed ratio (λ) and pitch angle ϕ. The tip speed ratio is defined as follows,

$$\lambda = \frac{\omega R}{V}, \tag{4.17}$$

where ω is the rotational speed of the blades, R is the rotor's radium, and v is the normal component of the wind speed. Figure 4.10 shows the $C_P - \lambda$ characteristics of an ideal airfoil. It can be seen that there is a specific λ where the turbine extracts the maximum energy from wind, so it is important to maintain the factor ω/V constant. The angle β is modified in some fixed-speed wind turbines in order to maintain the energy production below the maximum power capacity of the system in case of high wind periods, otherwise this will result in damage to the electric generator.

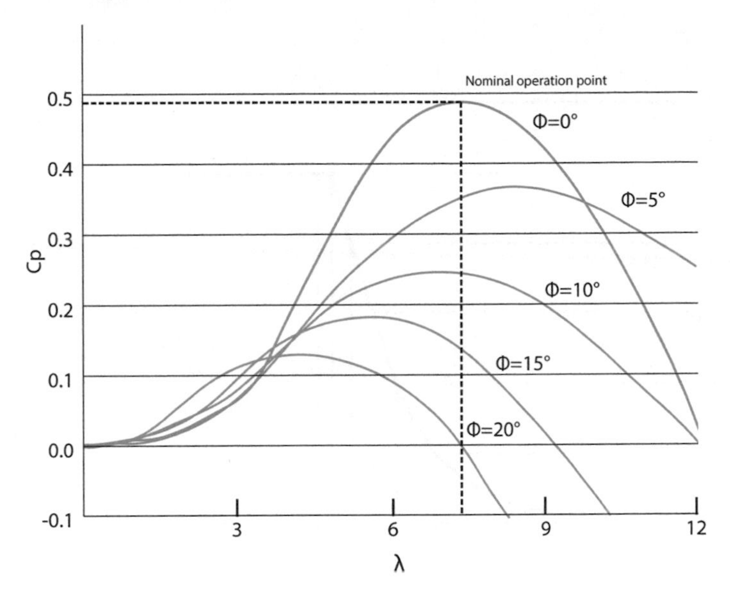

FIGURE 4.10
$C_p - \lambda$ characteristic for different values of pitch angle

Variable-speed turbines can provide greater efficiency that fixed ones, and in a large period of time, it is economically viable to implement a variable speed control for obtaining maximum wind energy. The speed control strategy consists of obtaining the maximum power point of the $C_p - \lambda$ curve at any time, as shown in Figure 4.11.

4.6 Wind energy resource

The first step in a wind energy project is to evaluate the potential of the zone in which it is proposed to locate the eolic park, the life-cycle costs, the reduction in greenhouse gases emission, and the proper wind turbine design (large turbines, small-scale turbines with diesel hybrid systems, etc.) The most important factor for evaluating a project is the average wind speed, because the available energy varies as the cubed wind-speed, and many studies have been made in order to characterize the wind. The variability in time of the wind can be divided into three scales according to [19], which are: large time scales (variation of the wind speed in periods of years), medium time scale (variation of the speed in periods up to a year), and the short time scale (variation of the

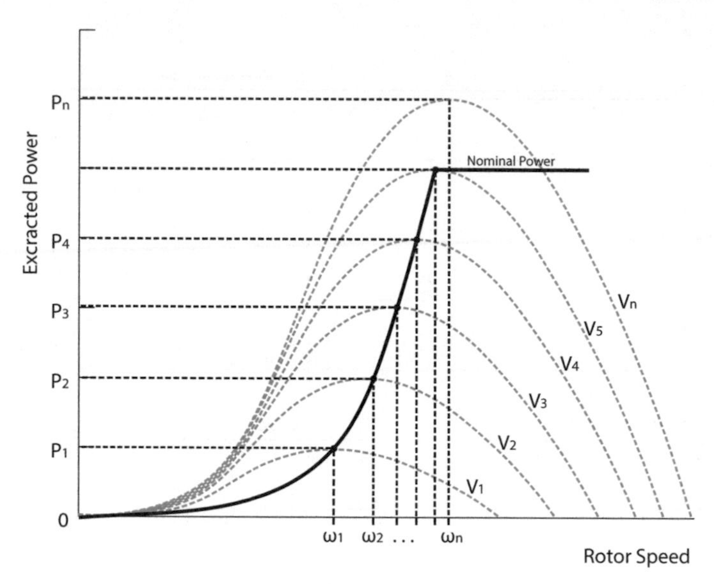

FIGURE 4.11
Wind energy conversion system maximum power extraction

wind speed during the day). The short time scale is very difficult to predict due to the high variability of the resource. On the other hand, medium and large scale are much more predictable; therefore, the wind generation projects are studied usually in terms of monthly variations covering one year. This variation is typically characterized in terms of statistical analysis; one of the most common distributions employed is the Weibull distribution [19]. This probability function expresses the probability of having certain average wind speed v_i during a year,

$$p(v_i) = \left(\frac{k}{c}\right)\left(\frac{v_i}{c}\right)^{k-1} exp\left[-\left(\frac{v_i}{c}\right)^{k}\right], \tag{4.18}$$

where c is the Weibull scale parameter and has the units of the wind speed, k is the unitless shape parameter, and v_i is a particular wind speed. The cumulative distribution function is expressed as follows:

$$p(v < v_i) = \left\{1 - exp\left[-\left(\frac{v_i}{c}\right)^{x}\right]\right\}. \tag{4.19}$$

The Weibull distribution parameters and the average wind speed are related with the complete gamma function

$$\overline{v} = c \cdot \Gamma\left(1 + \frac{1}{k}\right), \tag{4.20}$$

where $\overline{v}$ is the average wind speed, c is again the scale parameter, and k is the shape parameter. When $k = 2$, the Weibull distribution becomes a special case known as Rayleigh distribution, and the gamma function has a value of $\sqrt{\frac{\pi}{2}} \approx 0.8862$.

The variation of the hourly mean speed around the annual mean speed is small, as the shape parameter k is higher while the scale factor shows how windy (how high the average annual wind speed) a location is. An optimal scenario for an eolic park location is a place having high scale factor and a reduced shape factor. The two parameters can be estimated by the maximum likelihood method [12],

$$k = \left(\frac{\sum_{i=1}^{n} v_i^k ln(v_i)}{\sum_{i=1}^{n} v_i^k} - \frac{\sum_{i=1}^{n} ln(v_i)}{n} \right)^{-1}, \tag{4.21}$$

$$c = \left(\frac{1}{n} \sum_{i=1}^{n} v_i^k \right)^{\frac{1}{k}}, \tag{4.22}$$

where n is the number of non-zero data points, v_i is the wind speed in the time i. Equations 4.21 and 4.22 must be solved using an iterative procedure; some authors [56] recommend starting with the guess of $k = 2$. Another important definition that sometimes is employed is the wind power density (WPD), a quantitative indicator of the wind energy potential of a particular zone. This value is a combination of the effects of wind speeds, wind speed distribution, and the density of the air.

The WPD is defined as

$$WPD = \sum_{v_i=0}^{v_i=25} 0.5\rho\overline{v}_i^3, \tag{4.23}$$

where ρ is the air density, $P(v_i)$ is the probability to have a wind speed v_i during the year.

If we substitute the expression of the mean wind speed $\overline{v} = \sum_{v_i=0}^{v_i=25} v_i P(v_i)$;

$$WPD = 0.5\rho\overline{v}_i^3. \tag{4.24}$$

Figure 4.12 shows a mean wind speed map obtained from statistical studies [43].

4.7 Power generation system

The electrical power generated needs to be connected to the electrical grid. The electrical system has a fixed frequency and voltage; therefore, for attaching new power sources, some requirements must be covered. Those requirements are:

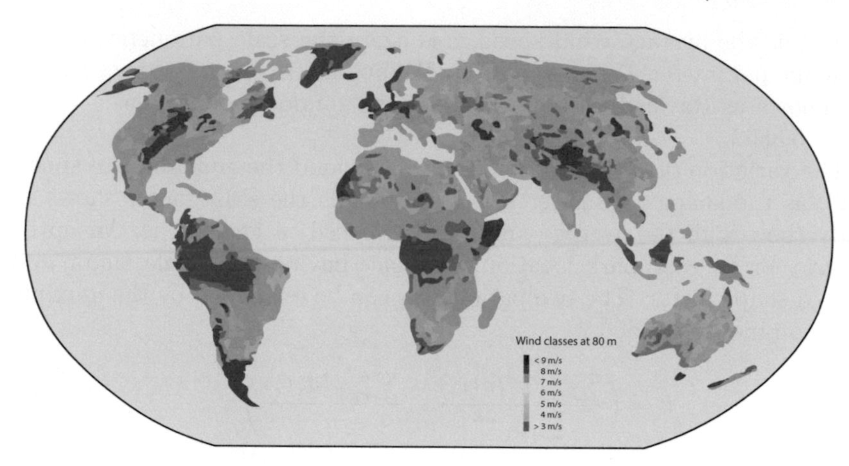

FIGURE 4.12
Global wind speed at 80m

- Equal sequence: The sequence of an electric three-phase machine is defined by the internal configuration of the windings and the rotating direction of the shaft It is very important to establish a sequence convention for coupling the generator in the same sequence of the grid.
- Equal voltage: The potential difference magnitude between the generated power and the electrical grid has to be close to zero in order to avoid damage by over-current when attachment is made.
- Equal phase shift: Not only the magnitude has to be equal but also the displacement. In other words, the instantaneous value of voltage has to be equal.

In the next subsection we review the most common configurations of wind turbines. First the fixed-speed turbines are explained, and then the variable-speed wind turbines are presented.

4.7.1 Fixed-speed wind turbines

As the electrical grid needs a fixed frequency and voltage magnitude, the easiest way to satisfy those requirements is by mounting an electric generator rotating at fixed speed. The first designs of wind turbines were made for being connected directly to the electrical grid like a motor through a soft starter, usually a variable step resistor. In Figure 4.13 is shown the general structure of a fixed-speed generator; a capacitor bank is used to compensate the inductive nature of the electric machine. This configuration was developed and used in Denmark [41], and because of that is known as the "Danish concept."

The induction generator used for this configuration is the squirrel cage,

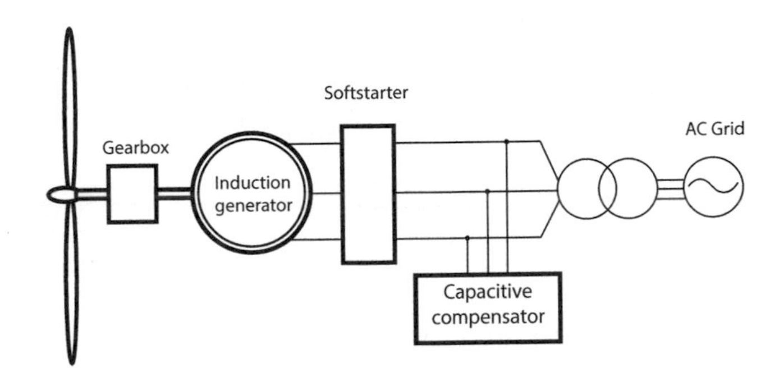

FIGURE 4.13
General structure of a fixed-speed wind turbine

because it is the roughest machine and also the most economical. After the turbine is rotating at the synchronous speed, the blades are positioned at their maximum efficiency attack angle and are left for extracting some mechanical power from the wind stream. When the induction machine is working like a generator, the rotational speed exceeds the synchronous speed and the electromagnetic torque is negative, inverting the current flux. The mechanical speed is very close to the synchronous speed, and the wind induces only small variations on it.

As shown in Figure 4.11, the aerodynamic efficiency has only a maximum at certain wind speeds, for the remaining wind speeds, the machine will operate at a point corresponding to the synchronous speed on the x axis of the curve. Another disadvantage is the incapability of the system to deal with the mechanical stress resulting from large wind gusts.

On the other hand, fixed-speed wind turbines are mechanically simple and have a low maintenance cost; there are no power electronic devices and their efficiency is high despite the fixed speed.

In order to increase the speed range of the generator, old designs are used to employ two winding sets: one for low wind speed, typically made of eight poles, and the second one for medium and large wind speeds (typically six or four poles). As seen in Chapter 2, the synchronous speed is inversely related to the number of poles. This design made the stator very big and expensive, therefore the wound rotor induction generator (WRIG) with access to the rotor windings represented the evolution of the multi-pole induction generators. The rotor of the WRIG was controlled by a variable resistance, controlled in turn by power electronics, depending on the measured speed of the wind, changing the torque-speed characteristic curve of the machine, and modifying the slip for certain values of extracted torque. The general scheme of this configuration is shown in Figure 4.14.

Obviously, the range of controllability remains very limited: about 10%

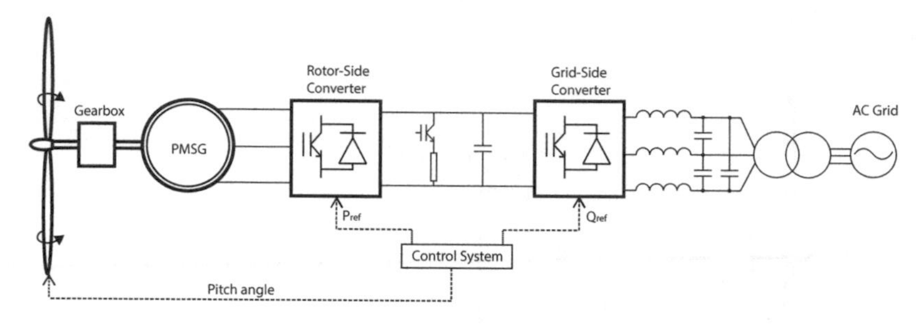

FIGURE 4.14
General structure of a full-converter variable-speed wind turbine

of the synchronous speed. The WRIG and the multi-pole design were the first attempts at variable speed wind turbines. The next subsection presents the modern techniques employed for increasing the speed range of the wind generators.

4.7.2 Variable-speed wind turbines

Nowadays, variable-speed wind turbines are the most used and manufactured wind energy systems. The variable operation requires power converters that partially or totally decouple the connection of the electric machine with the electrical grid. There are principally two types of configurations: the first one requires a complete power conversion from AC variable frequency and voltage, to a DC link that maintains a constant voltage, and back to AC at fixed voltage and frequency. The second one is the doubly-fed induction generator (DFIG), which decouples only the rotor part of the electric machine, keeping the stator fixed to the electrical grid.

Full-power converter schemes are very flexible in terms of type of generation implemented. In Figure 4.14 an induction generator is shown; nevertheless, synchronous generators, either wound rotor or permanent magnet, are equally valid. The permanent magnet and induction squirrel cage generators are the most popular in the industry, due to the absence of brushes and the disadvantages that this entails.

The rotor side inverter ensures the rotational speed of the electric machine. The principal advantage of controlling the speed is to adjust the speed according to the measured wind speed, in order to obtain at any point the maximum available power, as shown in Figure 4.11. On the other hand, the grid-side power converter attempts to transfer all the active power generated to the electrical grid while the reactive power is cancelled. For synchronous generators this is especially important, due to the inductive nature of this machine.

The DFIG is in reality a WRIG but with the rotor windings connected to a voltage converter, shown in Figure 4.15. The stator windings are con-

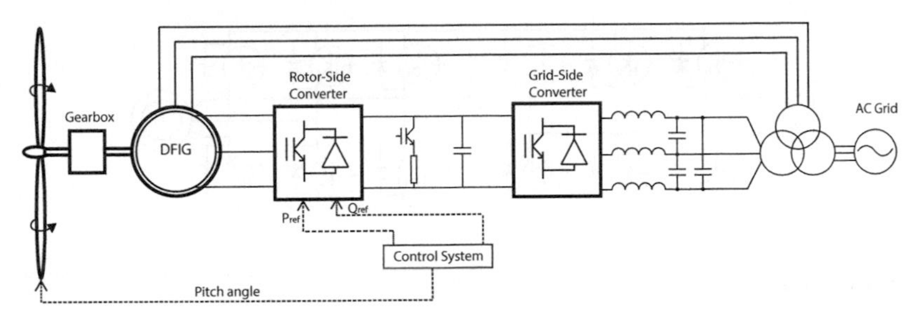

FIGURE 4.15
General structure of a DFIG-scheme wind turbine

nected directly to the electric grid; the advantage of doing this is that only a portion of the total power (about 30% according to [22]) flows through the power electronics devices. The DFIG allows variable-speed operation over a range typically about ±40% of the synchronous speed [13], and the size of the converter is directly related to the range of speed. The power electronics at the rotor side are responsible for controlling the active and reactive power entering or leaving the electric machine, while the converter connector to the grid side controls the DC-link voltage and ensures operation at large power factor.

The DFIG wind turbines are highly controllable and allow maximum power extraction at a wide range of wind velocity. In addition, the active and reactive power control is fully decoupled because of the independent control of the rotor current. When the turbine is operating at sub-synchronous speed, the current flows into the rotor, consuming power from the grid, and is turned back with the stator windings, while for super-synchronous speeds, both currents from the rotor and stator flow in the grid direction. The speed of the rotor influence, the fixed stator speed, and the rotational speed of the electric machine are shown with the following equation:

$$f_{\text{stator}} = \frac{n_{\text{rotor}} \times N_{\text{poles}}}{120} \pm f_{\text{rotor}}, \tag{4.25}$$

where n is the rpm speed of the rotor, N is the number of poles of the machine, and f_{rotor} is the electrical frequency of the current that flows through the rotor.

4.7.3 The back-to-back converters

Variable-speed operation is possible due to the power electronic converters' interface, allowing a partial decoupling from the grid. The back-to-back (B2B) converters consist simply of a force-commutated rectifier and a force-commutated inverter connected with a common DC-link, as shown in Figure 4.16.

The B2B converter has the following properties [21]:

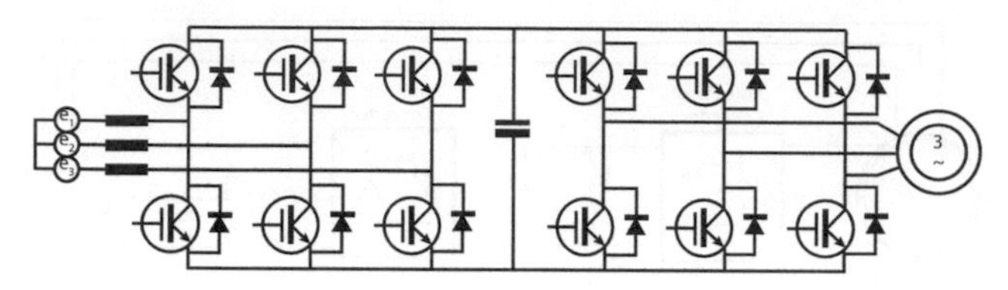

FIGURE 4.16
Back-to-back converter

- line-side converter (inverter) is controlled to produce near-sinusoidal output currents
- DC-link voltage is regulated by controlling the power flow to the AC grid
- inverter operates on the boosted DC-link, making it possible to increase the output power of a connected machine over its rated power

Particularly for the wind energy conversion system (WECS), one side of the converter is the rotor side, which ensures the rotational speed is adjusted within a large range, whereas the other side is the grid side, which transfers the active power to the grid and attempts to cancel the reactive power consumption.

Actually, power delivery can be controlled by focusing on the rotor's currents. If seen from a rotational d-q reference frame, modifying its d current would affect both active power (proportionally) and reactive power (inversely); however, its effect over the active power is much higher than that over the reactive power. On the other hand, varying the q current has a similar effect but is higher in magnitude. This analysis is shown in [73], which stands as a first step in a deeper analysis regarding how the behavior of some wind turbines can be manipulated, depending on power, speed, or torque demands.

4.8 Mathematical model of back-to-back converter

Consider the one phase B2B converter of Figure 4.17, where the back-to-back (B2B) converter is divided into 3 subsystems: the left side represents the first voltage source inverter, the right side represents the second voltage source inverter, and the central subsystem is the DC bus.

The equivalent circuits for each subsystem of Figure 4.17 are represented in Figure 4.18.

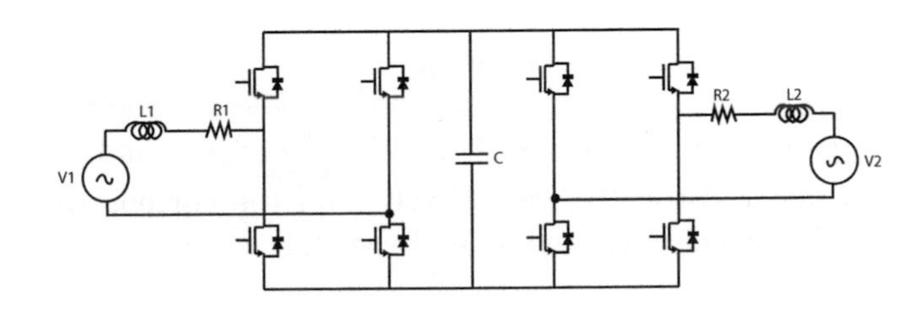

FIGURE 4.17
Back-to-back converter based on voltage source converters (VSC)

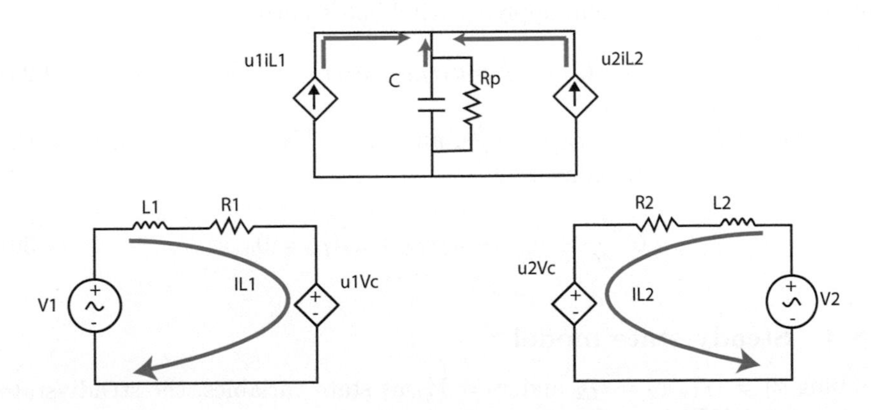

FIGURE 4.18
Equivalent circuits of back-to-back converter

4.8.1 Left side of back-to-back converter

Replacing the inverter by a controlled voltage source (see Figure 4.18), where $u_1 = 2u_{(VCS1)} - 1$ and $u_{(VCS1)} = [0, 1]$, the control voltage $u_1 = [-1, 1]$. Then, applying Kirchhoff's laws, the equation

$$v_1 = V_{L1} + V_{R1} + u_1 \tag{4.26}$$

is obtained. Substituting the voltages $V_{L1} = L_1 \left(\frac{d_{i_{L1}}}{dt} di_{L1} \right)$ and $V_{R1} = R_1 i_{R1}$ of inductor $L1$ and resistor $R1$, respectively, and the current $i_{R1} = i_{L1}$ in Equation 4.26, the model of the left side of the B2B converter is

$$v_1 = L_1 \frac{d_{i_{L1}}}{dt} + R_1 i_{L1} + u_1 V_c. \tag{4.27}$$

4.8.2 Right side of back-to-back converter

Using the procedure applied to get Equation 4.27, and considering the equivalent circuit of Figure 4.18, the model of the right side of the B2B converter is

$$v_2 = L_2 \frac{d_{i_{L2}}}{dt} + R_2 i_{L2} + u_2 V_c. \tag{4.28}$$

4.8.3 DC bus

The left and right converters are replaced by the voltage source 1 and voltage source 2, respectively. Then, applying Kirchhoff's current law,

$$i_c = i_{R_p} = u_1 i_{L1} + u_2 i_{L2}, \tag{4.29}$$

and replacing $i_c = C \frac{dV_c}{dt}$, $i_{R_p} = \frac{V_{R_p}}{R_p}$, and $V_C = V_{R_p}$, in Equation 4.28, the model of the DC bus remains as

$$C \frac{dV_c}{dt} + \frac{V_{R_p}}{R_p} - u_1 i_{L1} + u_2 i_{L2} = 0. \tag{4.30}$$

4.8.4 Steady-state model

Defining $x_1 = i_{L1}$, $x_2 = i_{L2}$ and $x_3 = V_C$ as state variables, the steady-state model of the B2B converter is

$$\begin{aligned} L_1 \dot{x}_1 + R_1 x_1 + u_1 x_3 &= v_1, \\ L_2 \dot{x}_2 + R_2 x_2 + u_2 x_3 &= v_2, \\ C \dot{x}_3 + \frac{1}{R_p} x_3 - u_1 x_1 - u_2 x_2 &= v_1. \end{aligned} \tag{4.31}$$

The model of the B2B converter in (4.31) is a discontinuous model since the

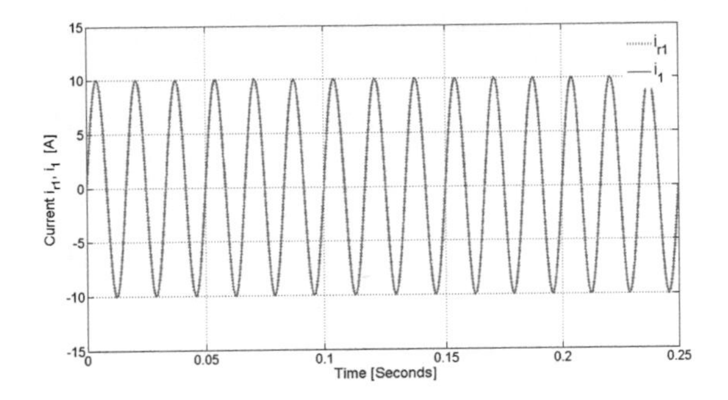

FIGURE 4.19
Simulation of tracking control of current i_1

inputs u_1 and u_2 are discontinuous. In order to analyze the B2B converter with tools for continuous systems, an average model is required. Average models of B2B converters are obtained by replacing the discontinuous controls u_1 and u_2 by pulse with modulated (PWM) signals.

The continuous model of the B2B converter is represented by

$$\begin{aligned} L_1\dot{x}_1 + R_1x_1 + \mu_1x_3 &= v_1, \\ L_2\dot{x}_2 + R_2x_2 + \mu_2x_3 &= v_2, \\ C\dot{x}_3 + \frac{1}{R_p}x_3 - \mu_1x_1 - \mu_2x_2 &= v_1, \end{aligned} \tag{4.32}$$

where μ_1 and μ_2 are continuous signals equivalent to the discrete signals u_1 and u_2.

4.8.5 Example back-to-back converter

Consider a B2B converter modelled by Equation 4.32, where the parameters are $L_1 = 13$ $[mH]$, $L_2 = 7$ $[mH]$, $R_1 = R_2 = 0.1$ $[\Omega]$, $C = 1500$ $[\mu F]$, $v_1 = 20sin2\pi(60)t$, $v_1 = 20sin2\pi(50)t$, and $R_p = 100$ $[\Omega]$.

In order to have the desired currents $i_{r1} = 10sin2\pi(60)t$ and $i_{r2} = 8sin2\pi(60)t$ in the left and right sides of the B2B converter, respectively, a simple feedback control is applied. Currents i_1 and i_2 are compared with the references i_{r1} and i_{r2} using feedback. Then, a control gain is designed for i_1 and i_2, track i_{r1}, and i_{r2}.

Simulations of the B2B converter are presented in Figures 4.19 through 4.21. It can be observed in Figure 4.19 that current $i_1(i_2)$ tracks $i_{r1}(i_{r2})$. Furthermore, voltage in the DC bus is maintained quasi-constant, as is shown in Figure 4.21.

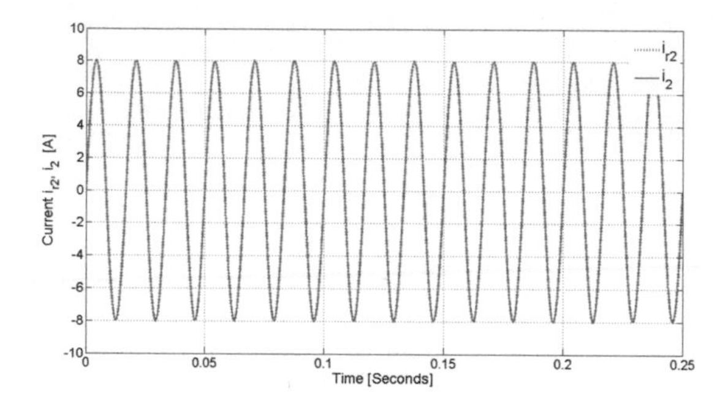

FIGURE 4.20
Simulation of tracking control for current i_2

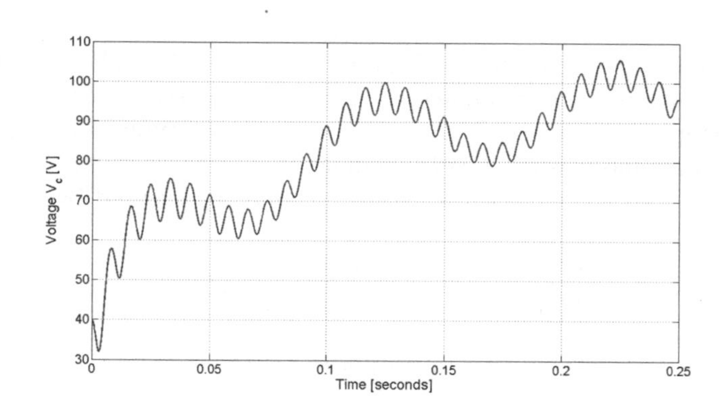

FIGURE 4.21
Simulation voltage in DC bus, V_c

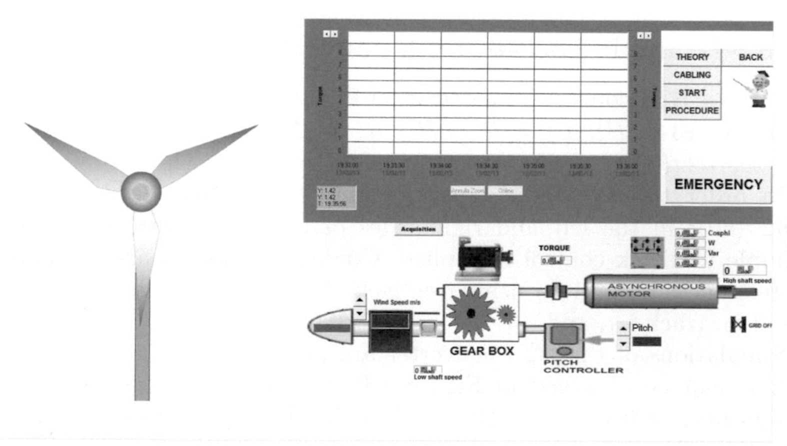

FIGURE 4.22
Pitch angle vs power extracted exercise

4.8.6 Exercises

The objective of this exercise is to obtain a pitch angle versus a power extracted characteristic curve for the simulated aero-generator and observe the power extraction depending on the wind speed.

Press ON the buttons "Torque and Enable" on the CH2 AD1 modulus. Press the Button START to simulate a force of wind of 18 [m/s]. Control the pitch of the blades to regulate the shaft speed to 1800 [rpm]. Observe the mechanical power with respect to the Pitch Angle.

Wind Speed [m/s]	**Pitch Angle** [°]	**Mechanical Power** [W]	**Torque** [Nm]	**High Speed Shaft** [rpm]	**Low Speed Shaft** [rpm]
18	0	0	0	0	0
18	20	4.4	0.3	154	1.5
18	22.5	18.4	0.6	443	4
18	27	63	0.6	1222	12
18	31.5	92.3	0.6	1483	15
18	40.5	92	0.6	1490	15

With the obtained data, graph the mechanical power versus pitch angle to obtain the optimal angle of the blades (see Graph 4.1). When the induction

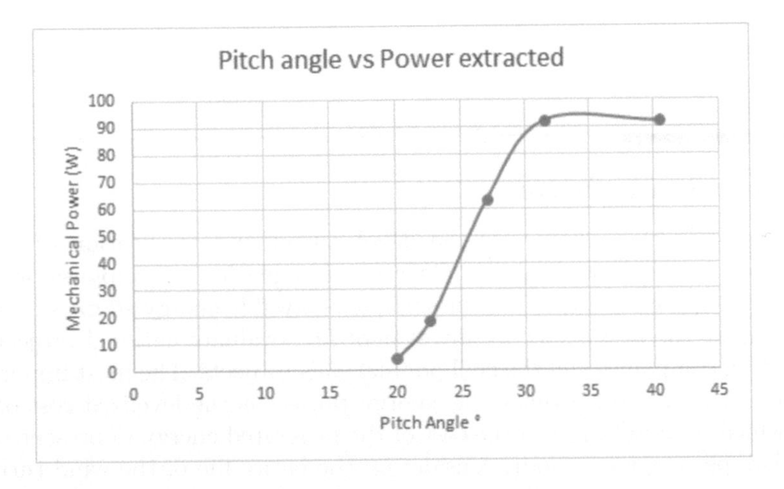

Graph 4.1
Pitch angle vs power extracted graph

machine is rotating at 1800 rpm, close the relay R2 for mounting the electric generator with the grid.

Once the electric generator and the grid are coupled, increment the speed

Wind Speed [m/s]	Mechanical Power [W]	Electrical Power [W]	Conversion Efficiency [%]
18	92.3	82	88.8
20	94.1	85	90.3
22	96.7	87	89.9
24	100.3	92	91.7

of the wind and measure the mechanical power extracted from the wind and the electrical power delivered to the grid. Keep the pitch angle at the optimal value (31.5°).

Use Diagram 4.1 for connections. Recommended equipment:

- Induction machine
- Brushless control
- Power circuit breaker
- Feeder manager relay
- Three-phase power meter
- Three-phase slip-ring asynchronous machine
- Three-phase supply unit

4.9 Economic factors

The most important factor for the development and application of a new technology is the economical one. Government agencies are employed for calculating energy cost of the most important renewable energy sources in order to determine the best choice for investment and evaluate national projects or supporting companies and the civil population's projects. The most important number for evaluating a renewable energy project is the levelized cost of energy, which is an indicator of the cost of the generated energy in present value in dollars per megawatt-hour, considering the entire life of the wind turbine, the costs of installation maintenance and operation of the wind project, and the rate of charge of the project. In this section, information extracted from [74] is presented; this information is based on installed land-based projects in the United States and estimated values for off-shore projects. The next table resumes the costs of installed capacity and energy production based on a 20-year operational life project. The conditions for calculating the total power produced were presented in Table 4.1.

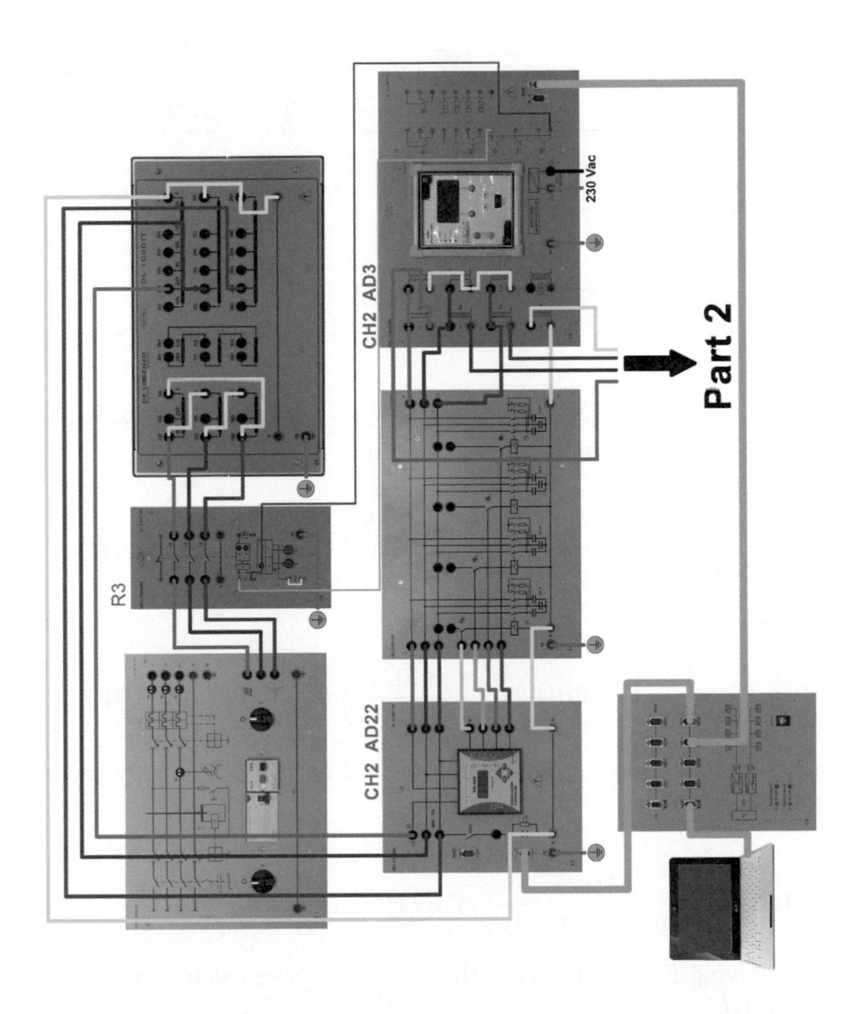

Diagram 4.1
Suggested electric connection for wind generation

TABLE 4.1
Wind properties

Wind parameter	Wind Land based	Offshore
Annual average speed at 80 [m]	7.75 [m/s]	8.9 [m/s]
Weibull K	2	2.1
Shear exponent	0.143	0.1
Air density	1.163 $[kg/m^3]$	1.225 $[kg/m^3]$
Turbine parameter	**Turbine Land based**	**Offshore**
Turbine hub height	80 [m]	80 [m]
Rotor net Cp	0.39	0.39
Rotor diameter	82.5 [m]	107 [m]
Turbine capacity	1.5 $[MW]$	3.6 $[MW]$

With this data, let's obtain the net energy production for year of each power conversion system, using Equation 4.16:

Land based:

$$P = \left(\frac{1}{2 \times 10^6}\right)(0.39)(1.165)\left(\frac{\pi(82.5)^2}{4}\right)(7.75)^3 = 0.564\ [MW].$$

Offshore:

$$P = \left(\frac{1}{2 \times 10^6}\right)(0.39)(1.225)\left(\frac{\pi(107)^2}{4}\right)(8.9)^3 = 1.513\ [MW].$$

The net annual energy production is defined as the total energy produced per installed capacity.

$$\text{AEP}_{\text{net}} = \frac{[MW]}{\text{Project capacity}}(365)(24) = \frac{[MW](8760)}{\text{Project capacity}} \tag{4.33}$$

Now, applying Equation 4.33 to the previous power calculation:

Land based:

$$\text{AEP}_{\text{net}} = \frac{(0.564)(8760)}{1.5} = 3293.8.$$

Offshore:

$$\text{AEP}_{\text{net}} = \frac{(1.513)(8760)}{3.6} = 3684.68.$$

On the other hand, the cost of installed projects is summed up in the following table. For land, project costs were calculated with data from an old project, observed in Figure 4.23. Considering a market adjustment for

TABLE 4.2
Installed capital costs for land and offshore wind projects

Element cost	Land-installation based on a 1.5 [MW] turbine Installed capacity cost [$/kW]	Energy production cost [$/MWh]	Off-shore based on a 3.6 [MW] turbine Installed capacity cost [$/kW]	Energy production cost [$/MWh]
Turbine cost				
Rotor	283.7	8.16	384.9	13.32
Drive train	699.7	20.13	706.3	24.48
Tower	302.6	8.71	697.71	24.18
Total	1286	37	1789	62
Installation cost				
Assembly, transport, & install	116.3478	5	1122.308	38.846
Electrical infrastructure	155.1304	2.5	562.1548	19.423
Support structure	77.565	4.5	1010.077	34.962
Development and project management	58.174	0.75	168.346	5.827
Others	38.783	0.25	56.115	1942
Total	446	13	2918	101
Soft costs				
Construction finance	57.333	1.667	167.437	5.812
Contingency	114.667	3.333	445.5	15.5
Insurance and surety bond	0	0	279.06	9.687
Total	172	5	893	31
Market price adjustment	195	6	0	0

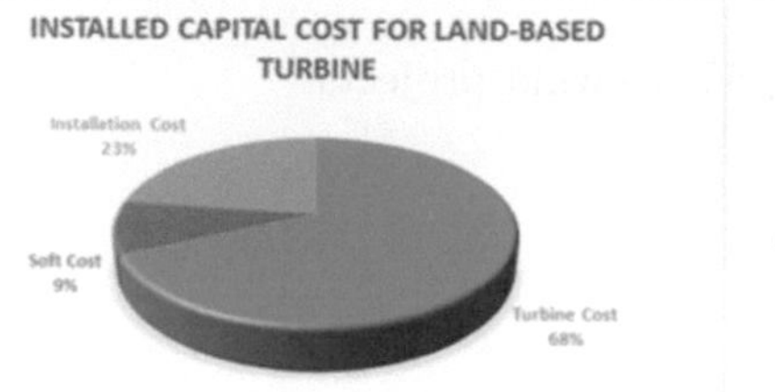

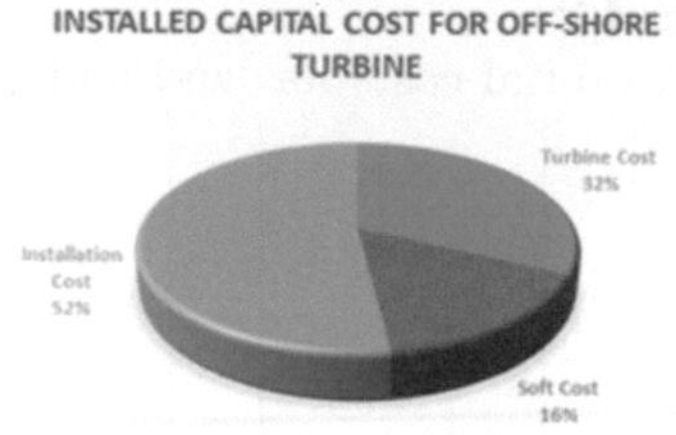

FIGURE 4.23
Installed capital cost for land-base and off-shore wind turbine projects

the price of typical payment in the 2011 market, the energy installation and production costs estimated were as shown in Table 4.2.

The total installed capacity cost considering the market price adjustment is:

Land based:

$$\text{ICC} = 1286 + 446 + 172 + 195 = 2099.$$

Offshore:

$$\text{ICC} = 1789 + 2918 + 893 = 5600.$$

For calculating the levelized cost of energy in [$/MWh], the formula

$$\text{LCOE} = \frac{(\text{ICC} \times \text{FCR}) + \text{AOE}}{\text{AEP}_{\text{ne}}/1000}$$

can be used, where

ICC Installed capacity cost [$/kW]

FCR Fixed charge rate [%]

AOE Annual operating expenses including taxes [$/kW/yr]

AEP Annual energy production [MWh/MW/yr]

The ICC and AEP_{net} have been calculated previously. The fixed charge rate (FCR) is defined by [74] as the amount of revenue required to pay the carrying charges (return on debt, return on equity, taxes, depreciation, and insurance) on an investment while capturing expected plant life, and is calculated by

$$\text{FCR} = \frac{d(1+d)^n}{(1+d)^n - 1} \times \frac{1 - (T \times \text{PVdep})}{(1 - T)}. \tag{4.34}$$

Table 4.3 shows an example of data related to some carrying charges.

Where d is the discount rate, n is the number of operational years, T is the tax rate, and PC_{dep} is the present value of depreciation.

TABLE 4.3
Data for FCR calculation

Data	Land based	Offshore
Discount rate (d)	8 %	10.5 %
Number of operational years (n)	20	20
Effective taxes (T)	38.9 %	38.9 %
Present value depreciation (PVdep)	81.1 %	77.8%

The FCR data used in the Tegen's report is:
Land based:

$$\text{FCR} = 9.5\%.$$

Offshore:

$$\text{FCR} = 11.8\%.$$

Finally, the annual operating cost is estimated from company's data and estimated for offshore projects, and presented in current time:
Land based:

$$\text{AOE} = 35 \left[\frac{\frac{\$}{kW}}{yr} \right].$$

Offshore:

$$\text{AOE} = 136 \left[\frac{\frac{\$}{kW}}{yr} \right].$$

Using Equation 4.34 and the data obtained, the levelized cost of energy for the wind projects is:
Land based:

$$\text{LCOE} = 72 \left[\frac{\$}{MWh} \right].$$

Offshore:

$$\text{LCOE} = 225 \left[\frac{\$}{MWh} \right].$$

According to [31], the expectancies for levelized cost to 2019 are as shown in Table 4.4.

4.10 Homework problems

Problem 4.1. *Broadly propose the parameters of a new eolic plant based on the graph below. Suppose the risk factor has been set as 50%, the head is* 90 $[m]$ *and there are 20% in losses.*

TABLE 4.4
U.S. levelized cost of energy expectancies for 2019

Plant type	Levelized capital cost [$/MWh]
Conventional coal	60
Conventional gas combined cycle	14.3
Conventional combustion turbine	27.3
Advanced nuclear	71.4
Geothermal	34.2
Biomass	47.3
Wind	64.1
Wind offshore	175.4
Solar thermal	195
Solar PV	114
Hydroelectrical	72

Solution 4.1. *From the kinetic energy equation:*

$$\dot{E} = \frac{1}{2}\rho A v^3 = \frac{1}{2}\left(1.225\ \left[\frac{kg}{m^3}\right]\right)\left(\pi\left(\frac{23\ [m]}{2}\right)^2\right)\left(6\ \left[\frac{m}{s}\right]\right)^3 = 54.97 \times 10^3\ [W].$$

However, only 59.3% of such power can be taken, as pointed out by physicists Albert Betz, so:

$$W = 32.6\ [kW].$$

Problem 4.2. *Suppose some given wind park is subject to a wind speed profile that rises as much as* 18 $[m/s]$ *and sinks as low as* 9 $[m/s]$ *(evenly) throughout a full operational day. If a mean power consumption of* 3 $[MW]$ *is desired, how many* 314 $[m^2]$ *ideal wind turbines are needed to reach the mean power requirement?*

Solution 4.2. *As both speeds are said to be distributed evenly, the mean wind speed is* 13.5 $[m/s]$. *By using the kinetic energy equation, it is possible to calculate the individual power of each turbine.*

$$\dot{E} = \frac{1}{2}\rho A v^3 = \frac{1}{2}\left(1.225\ \left[\frac{kg}{m^3}\right]\right)(314\ [m^2])\left(13.5\left[\frac{m}{s}\right]\right)^3 = 473.2 \times 10^3\ [W].$$

Due to maximum power extraction by Betz, each turbine can generate an ideal average of 280.61 $[kW]$.

So, there are 11 turbines required.

Problem 4.3. *Using the data from Problem 4.2 and Figure 4.8 (lift and drag coefficients graph), determine the optimal dimensions of each blade (3 blades) considering a* $20°$ *attack angle, an overall design limit of* 72.63 $[kNm]$ *torque, and* 457 $[N]$ *drag force.*

Solution 4.3. *The* $20°$ *attack angle exhibits a peak lift coefficient of about* 1.22*, while the drag coefficient is as small as* 0.08*. Both force equations can be written as follows (the maximum wind gusts rise up to* 18 $[m/s]$*):*

$$L = \frac{1}{2}\rho S V_0^2 C_l = \frac{1}{2}\left(1.2225\ \left[\frac{kg}{m^3}\right]\right) S \left(18\ \left[\frac{m}{s}\right]\right)^2 (1.22) = 241.11S,$$

$$D = \frac{1}{2}\rho S V_0^2 C_d = \frac{1}{2}\left(1.2225\left[\frac{kg}{m^3}\right]\right) S \left(18\left[\frac{m}{s}\right]\right)^2 (0.08) = 15.88S.$$

Knowing that $S = c \times l$*, and that the blades are 10 [m] long, both the above equations can be constrained to determine the chord length (all three blades contribute equally to such forces):*

$$\frac{72.63 \times 10^3\ [Nm]}{3\ [blade] \times 10\ [m]} = 241.11S = (242.11)(10\ [m])c \therefore c = 1\ [m],$$

$$\frac{457\ [N]}{3\ [blade]} = 15.88S = (15.88)(10\ [m])c \therefore c = 1\ [m].$$

Therefore, due to design constraints, the chord length must be 96 $[cm]$.

Problem 4.4. *For constant torque operation, an attack angle controller is installed on a wind turbine that operates under the same conditions as in Problem 4.2. Consider the airfoil area to be* $S = (10\ [m])(0.96\ [m])$*, the attack angle to be constrained between* $0°$ *and* $15°$*, and the lift coefficient to behave linearly between* $(0°, 0)$ *and* $(15°, 1)$*. Find the attack angle interval in which the generator must operate to compensate wind variations.*

Solution 4.4.

$$L(V_0, C_l) = \frac{1}{2}\rho S V_0^2 C_l = \frac{1}{2}\left(1.225\ \left[\frac{kg}{m^3}\right]\right)(9.6\ [m^2])V_0^2 C_l = 5.88\ \left[\frac{kg}{m}\right] V_0^2 C_l.$$

This leads to the maximum (18 $[m/s]$) *and minimum* (9 $[m/s]$) *operating points:*

$$L_{max} = 1.91 \times 10^3 C_l \therefore T_{max} = 57.3 \times 10^3 C_l,$$

$$L_{min} = 476.3C_l \therefore T_{min} = 14.29 \times 10^3 C_l.$$

The behavior of C_l *is also known:* $C_l(d) = \frac{d}{15}$*. So, the highest attack angle must compensate for slow wind as follows:*

$$T_{min} = 14.29 \times 10^3 \left(\frac{d}{15}\right)\bigg|_{d \to 15°} = 14.29\ [kNm]\ @15°.$$

This forces the angle to move at fast winds operation:

$$T_{max} = 57.3 \times 10^3 \left(\frac{d}{15}\right) = 14.29\ [kNm]\ \textit{leading to}\ d = 3.74°.$$

Problem 4.5. *Measure the wind power density (WPD) for a place where the wind speed follows the Weibull distribution with parameters $k = 2$ and $c = 9$.*

Solution 4.5. *The Weibull distribution can be solved for these parameters as follows:*

$$P(v_i) = \left(\frac{k}{c}\right)\left(\frac{v_i}{c}\right)^{k-1} e^{-\left(\frac{v_i}{c}\right)^k} = (0.22)\left(\frac{v_i}{9}\right) e^{-\left(\frac{v_i}{9}\right)^2}.$$

This can be used on the WPD equation:

$$WPD = \sum_{v_i=0}^{v_i=25} \left(\frac{1}{2}\right) \rho v_i^3 P(v_i) = 589.64 \left[\frac{W}{m^2}\right].$$

4.11 Simulation

As wind exhibits different speed-rates, the incorporation of an asynchronous machine is quite transparent, as shown in Figure 4.24 . Its rotor is not necessarily moving at the synchronous speed, so every speed surplus can be taken as active power generation. However, in the case of a squirrel-cage asynchronous generator, the magnetization field is taken from the grid, so the machine will always require reactive power, making necessary the incorporation of parallel capacitors to compensate for its inductive nature. In addition, speed variations could lead to an improper power profile, which cannot be amended if the access to the rotor's currents is restricted.

This has led to more complex asynchronous machine designs (together with increasingly complex power electronics boards), such as the wound rotor double fed induction machine, WR DFIG, which is actually leading the wind generation industry.

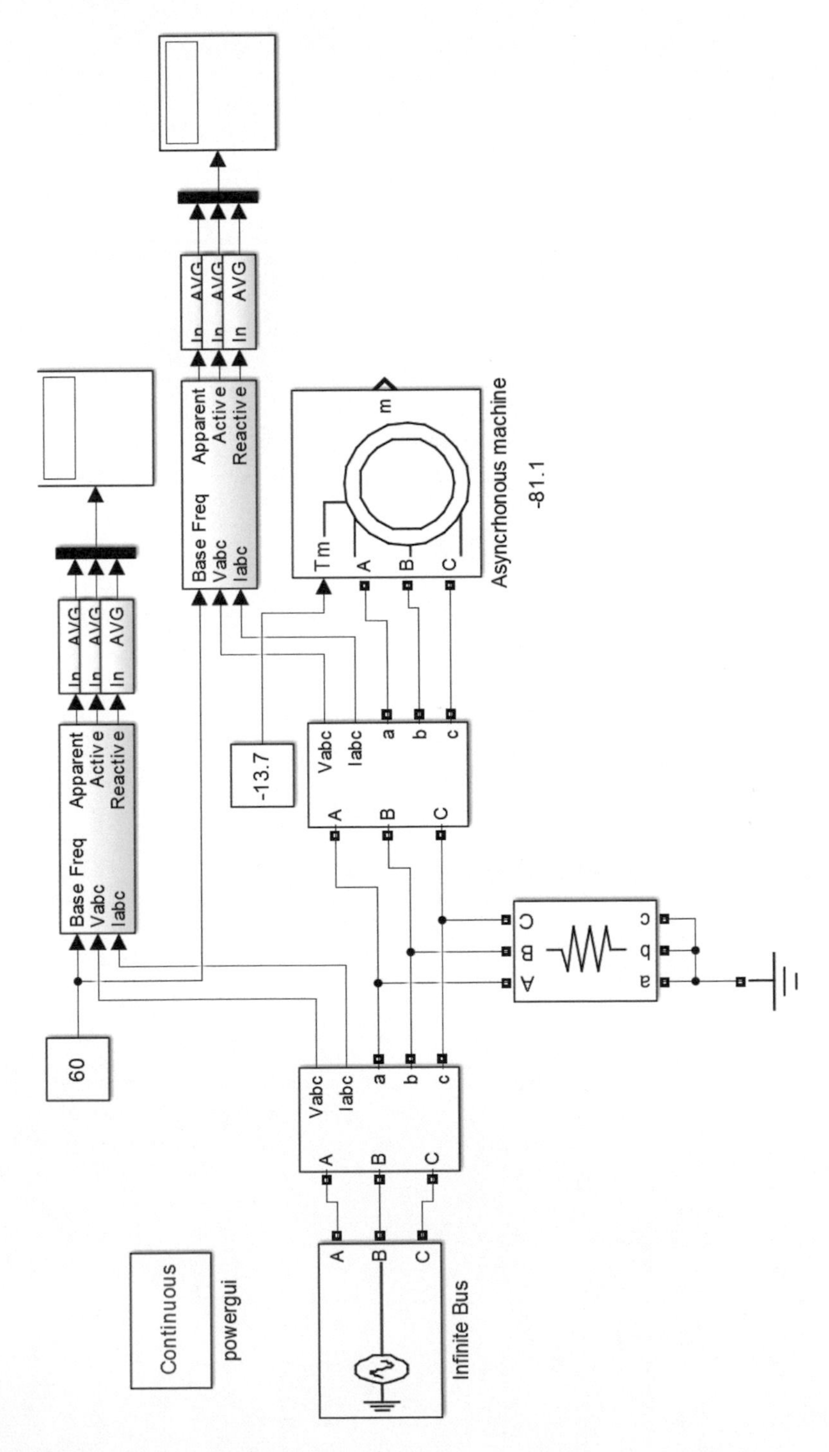

FIGURE 4.24
Asynchronous machine integration to the bus simulation

5

Solar Energy

The sun is a huge source of energy that keeps most environmental processes running. Its light and heat are also paramount for different human activities, even economic ones like agriculture. However, most of the available energy delivered by the sun (about 700×10^6 [TWh] daily) over the Earth's surface is not purposely transformed into useful forms of energy, so its enormous potential is hindered by the inability to capture it.

Besides, it would be impossible to actually take most of the sun's energy. Yet there are lots of opportunities to take advantage of it. Its heat can be directly used in water heaters, and its light can be used thanks to the photovoltaic effect of some semiconductors. This has led to a whole industry that develops technology regarding the different ways in which the sun's energy can be converted, mostly to electrical energy.

There are two main ways to use solar energy. Firstly, the sun's light can be redirected to a single spot, greatly increasing the amount of energy per unit area. This can boil or pre-heat water if the light is directed to a metallic container. Secondly, and perhaps more importantly, the light can be captured through photo-voltaic cells, directly transforming light's energy into electricity. Both techniques are referred to as *thermal* and *photovoltaic* solar energy sources, respectively [76].

Although the sun can provide huge amounts of usable energy, there are still some issues to address regarding its availability. The sun's power is only available during the daytime, and can only be used fully for a limited part of the day and depending on the weather. This makes solar energy subject to sometimes unknown factors, leading to an intermittent performance, which make such systems depend on energy storage technologies.

Some of these issues are partially solved by forecasting the amount of energy on a specific place on earth. Hence, solar power plants can be installed depending on the potential $[\frac{kWh/m^2}{\text{year}}]$ of that location, and assuming a 10% overall efficiency [33, p. 300]. It is also noteworthy that solar panels' efficiency will drop to their 80% to 88% nominal performance in 25 years.

It is clear that besides the amount of available energy the sun provides, it is nonsense to estimate its thorough usage as a justification for solar energy's incorporation to daily life. The sun provides (some hundreds of thousands of times) more energy than what is needed by humans, but such energy is not readily available, it is intermittent, and the means used to "catch" it are expensive, polluting, and mostly inefficient. Thus, solar energy should be seen not as a flawless solution but a promising area in which all kinds of resources must be invested so that solar power is taken advantage of effectively.

In the same spirit, this chapter shows data regarding solar energy availability, and deals specifically with photo-voltaic solar panels to introduce their behavior, modeling techniques, and the areas in which they can be improved.

5.1 Capturing solar energy

Independent from the technique used to extract energy from the sun, there are some natural effects that must be acknowledged before any consideration can be made regarding solar power. *Solar radiation* should be understood as the part of the sunlight that reaches earth's atmosphere, carrying most of the sun's available energy. However, different particles like carbon dioxide, methane, ozone, and water vapor can be hit by such rays before they get to the earth's surface [76, p. 211].

Once particles are hit, the light can be "filtered" (only some specific wavelengths of the light can go through the particle) or re-directed. The aforementioned effects are called *dispersion* and *reflection*, respectively. In addition, particles are also capable of transforming sun's radiation into heat; hence, part of the available energy is *absorbed* by those particles [76, pp. 211–212].

The above principles explain how solar radiation can warm up the air and how it can be reflected inside earth's atmosphere. Overall, 30% of such radiation is reflected back to space. The remaining energy is responsible for the water cycle, ocean currents, and wind courses. Alternatively, radiation can also be absorbed by the earth's surface as heat, or it can be converted into electricity.

As introduced above, there are two different ways in which solar energy can be used. If photo-voltaic (photo = light) devices are used, the sun's light energy is directly converted into electricity; otherwise, it is the solar heat (thermal radiation) that is taken into account [33, p. 302]. The basic principle regarding solar heat redirection is shown in Figure 5.1.

Photo-voltaic cells and panels operate quite differently. Broadly, the photovoltaic effect refers to the phenomenon of electron emission from a material after it has been hit by light. The light's energy ejects the electrons off the material, so the energy conversion is straightforward from light to electricity. Such electromotive force (voltage) will depend on different factors regarding the material's properties. Only some kinds of materials can produce enough electromotive force to be useful in the process of energy transfer.

It is clear that the performance of any device intended to capture solar power will depend on how effectively the light or heat is acquired. The direction of the sunlight and its power due to weather conditions (and the device's cleanliness) are paramount operating factors. The sun changes its position as

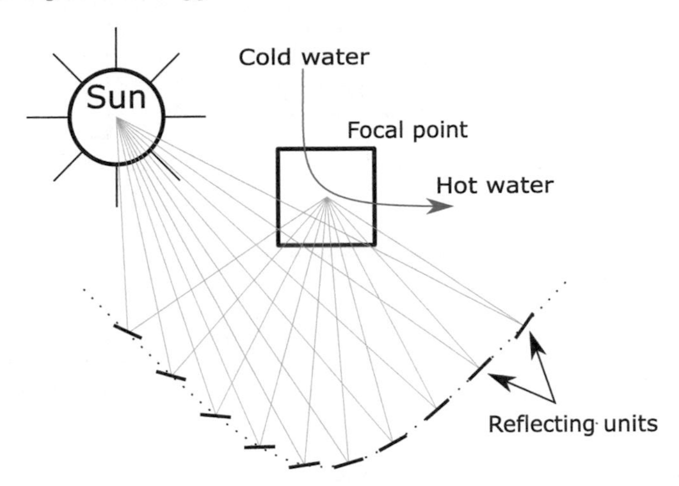

FIGURE 5.1
Solar reflectors as heaters — basic principle

the day goes by, so the incidence of the light rays is not always the same, and to guarantee maximum conversion effectiveness, the reflectors or panels should move optimally. In addition, the unpredictable conditions of the useful rays make any control system devoted to solar power extraction operate over always-changing operating points.

However, most solar panels' installations are fixed (they do not orient to face the sun directly), and clouds, leaves, dirt, etc., can shade them (even partially). This implies that a solar power plant depends on many conditions that make the system not as reliable as other traditional power sources. So, their application is sometimes better as complementary systems to help the energy transfer process of another (more dependable) plant.

For instance, solar thermal facilities are commonly used as a pre-heater for the incoming water to a thermal power plant. Any amount of energy taken from the sun will reduce the required combustible material needed to boil the water. This process is not intended to work on its own as it depends on the same solar power's intermittent nature, which, anyway, reduces the required fuel and pollutant gases.

The above description corresponds to that of many applications of alternative power sources. Some geothermal units and photo-voltaic systems are also used in that way. The main benefit of such a cooperative interconnection is that the required electrical reliability and dependability are still obtained, and the intermittent drawbacks of the aforementioned technologies are eliminated.

A different way to build reliable solar power plants is to specifically forecast "ideal" zones where the light incidence can be guaranteed for a specific amount of minimum energy capacity. Installation of power plants in deserts (e.g., the Atacama desert at Chile) and off-shore locations make a successful integration

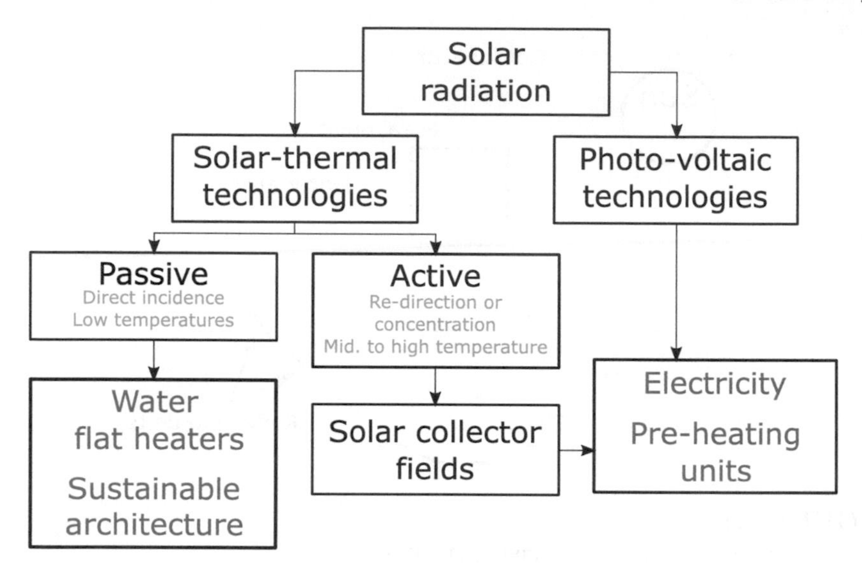

FIGURE 5.2
Different solar power technologies [76, pp. 238–240]

possible. A brief schematic about different solar power technologies is shown in Figure 5.2.

Solar irradiation can be estimated statistically, or directly measured at different locations to obtain an overall [kWh/m^2] metric. It is known that there are two latitudinal high irradiation bands throughout the earth over the tropics, at about 23° north and south, respectively. Such zones can reach up to a 2800 [kWh/m^2] accumulated energy for a year. Figure 5.3 shows the world zones where sunlight energy is higher than 2500 [kWh/m^2/yr].

5.2 Introduction to photo-voltaics

As introduced above, the photo-voltaic effect is not exclusive of semiconductors; however, these materials permit sufficient electromotive force to enable power transfer to a load. Electric current will appear whenever sunlight photons hit the material strongly enough to separate electrons and holes, both with negative and positive charges, respectively. Such interactions enable charged particles to flow, as shown in Figure 5.4.

However, de-attached electrons cannot flow just as they "jump" off the material because recombination occurs. A p-n junction is needed so the hole-electron separation becomes possible and recombination is prevented. The

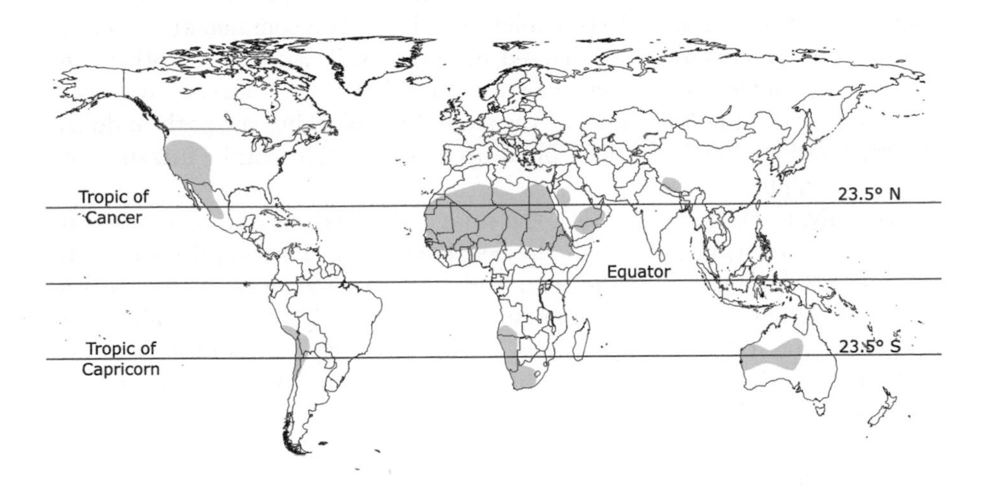

FIGURE 5.3
World zones where sun iradiance is higher than 2500 [kWh/m^2/yr]

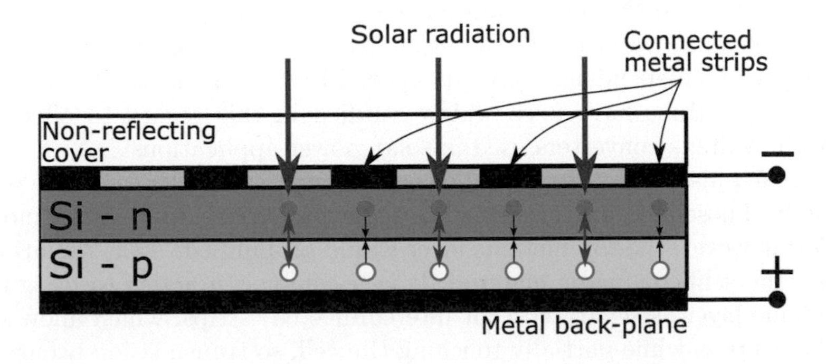

FIGURE 5.4
Schematic representation of a solar cell [51, pp. 65–69]

voltage present at the terminals of the solar cell is actually a consequence of the p-n junction band [51, p. 66].

Figure 5.4 shows a schematic representation of both aforementioned electronic interactions. Firstly, the p-n junction behaves as a common diode, which implies that the electronically charged materials will attract each other, developing a *forbidden* band where electrons and holes are "clogged" together: both try to reach the other side, at the same time blocking the path to do so. The aggregation of such particles build up a voltage that can be measured at the cell's terminals.

Secondly, as the photons touch the solar cell, electrons and holes are separated and their transit through the cell is enabled. This implies there are two currents interacting in the solar cell: the *photo-current* I_L due to the photo-voltaic effect, and the diode current I_D related to the recombination process [51, pp. 66–67]. Those effects behave inversely, as more light power would increase I_L but would also reduce the forbidden band. Hence, the voltage dependent on such a band, necessary for power transfer, will be reduced whenever I_L increases.

The voltage-current relation of a solar cell (which will be modeled later) can be foreseen from the aforementioned observations; consider a constant light power to be delivered to the cell. If no load is connected to the cell, the terminal voltage will be that of the p-n junction. Such voltage will never be higher if a load is connected, as the current flow will diminish the forbiden band. If too much current is required, then the voltage will drop to zero as the junction potential will decay accordingly.

It is important to notice that the electrical behavior of the solar cell is dependent on light power but also on the connected load. This means that even having great solar incidence, the cell could deliver small amounts of energy if the optimal V-I operating point is disregarded. In fact, there is only one equivalent load for which the power transfer is maximum, so power electronics are needed to be attached to the cell's output to match such optimal impedance, achieving load independence. For instance, MPPT (maximum power point tracking) algorithms have been widely studied in this regard to effectively control the voltage converters used in solar power applications.

Figure 5.4 also shows the metallic conductors attached to each side of the cell's body. Those metallic terminals are needed to catch and transmit moving particles; however, the top metallic layer would also impede solar radiation in reaching the semiconductor material. It is a common practice to design the top metallic layer as separated (but interconnected) strips, which allow solar radiation to pass while partially touching the cell, so transmission occurs. On the contrary, the backplane metal layer can be placed continuously.

Solar cells are built from different semiconductor materials. Silicon (Si), galium arsenide (GaAs), and germanium (Ge) are three popular components of solar cells. Each material exhibits a distinctive characteristic regarding junction potentials and how much power they can absorb depending on the light's

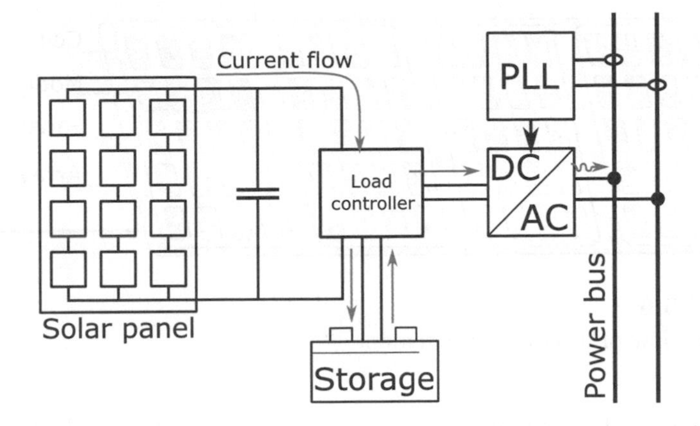

FIGURE 5.5
Simple setting of solar photovoltaic panel

wavelength components. A typical crystalline Si cell presents a 0.5 1.0 [V] potential and current density of 20–40 [mA/cm^2] [65].

As can be easily foreseen, the only possible way of obtaining higher voltages from solar power is to combine more than one cell in series. On the other hand, if the current is desired to be increased, more than one cell must be arranged in parallel. Such distribution commonly holds up to 60 or 72 cells, forming a single solar *panel* [65, p. 6].

Solar panels deliver DC current, which can be directly used by very specific devices. However, most electric appliances take AC inputs, so a voltage conversion (currently available at up to 95% efficiency) is needed even for local (or stand-alone) applications. In addition, due to intermittent power availability, a storage system is required. MPPT, storage management, and output voltage regulation are tasks covered by a *load controller* (see Figure 5.5).

Load controllers are power electronics devices comprised of different DC-DC converters, capable of managing a regulated output partially independent from the panel's variability. Such independence is guaranteed by a storage system that is controlled so that it takes additional power when available and delivers power if required. Constant operation will be guaranteed if the average solar power can cover load requirements.

DC-DC converters are made of semiconductors together with passive devices intended to "convert" from a voltage input magnitude to another at high efficiency (about 96%). Such conversion is achieved by switching one or a set of transistors, taking portions of the input power depending on the desired output. Passive elements attached to such a switching device are commonly used as filters to provide a smooth, regulated output. Traditional converters include the *Buck*, intended to step down the input voltage, the *Boost*, used if the voltage is required to rise above the input magnitude, and the *Buck-Boost*, which can perform both operations. Such devices can be voltage or

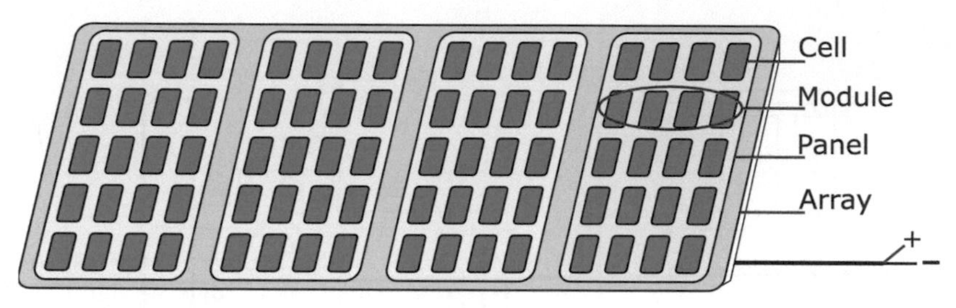

FIGURE 5.6
Schematic view of a panel arrangement

current controlled, enabling the load controller operation and allowing different sources to be merged on the same DC bus almost regardless of the original electrical conditions of each device.

Once the output has been regulated, there are two options for available power usage. For *stand-alone* operation, the solar panel acts like the only power provider so that different loads can be connected directly to the load controller if their required DC input matches that of the controller. If AC is required, an inverter is needed. Stand-alone operation can be seen in Figure 5.5 if the PLL block is disregarded, and the power bus is considered to be only locally available.

The second option is called *grid-connected*, and implies that the solar power is directly injected into the infinite bus. As the public electric grid is AC driven, the AC inverter is required. In addition, the electric grid operates at fixed-frequency and fixed-amplitude voltage, so a *phase-locked loop* (PLL) is required to drive the inverter. This means the inverter's generated AC voltage must match the voltage already present in the power bus, so current injection can occur. Figure 5.5 shows this operation if the power bus is considered to be the public electric grid.

5.2.1 Solar arrangements

For the sake of clarity, a fully functional solar array is shown in Figure 5.6. This figure tries to clarify some of the most used names for different hierarchical components of a solar generator. Each solar array is comprised of a number of solar panels, each made with a certain amount of modules that incorporate different solar cells. Cells are clustered in modules because the electrical output of the panel is desired to have specific voltage-current ratings. Modules are also arranged so the desired electrical output is furnished to meet required characteristics.

On ideal operation, a module behaves (electrically) as a cell, with increased voltage or current ratings. Likewise, the panel itself can be modeled as a cell, having the combined effect of all connected modules. Solar arrangements can be clustered in series or parallel to increase their output voltage or current,

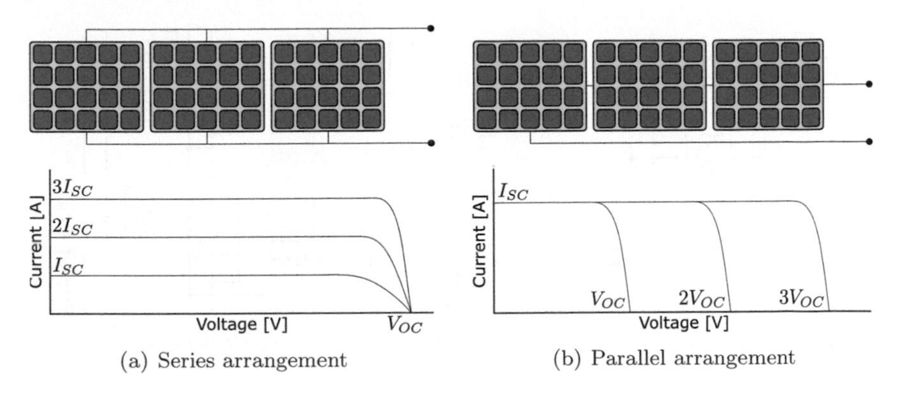

(a) Series arrangement

(b) Parallel arrangement

FIGURE 5.7
Solar cells/panels arrangements — ideal behavior

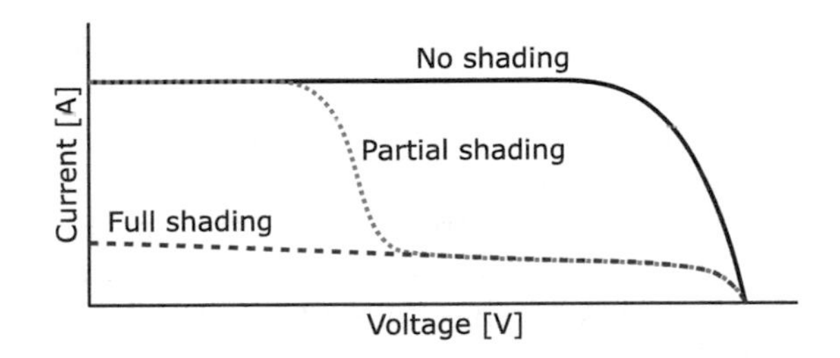

FIGURE 5.8
Partial shading effects on electric output

respectively. A hybrid connection is common among cells within the same panel, while a solar array is often comprised of parallel connected panels. These relations are graphically shown in Figure 5.7.

However, the actual electric behavior of interconnected cells is not the same as that of a single cell. Individual cells' output depends on irradiation and temperature; hence, a cell's arrangement will depend on the individual sunlight and heat conditions. This implies that if one only cell is shaded, the module will exhibit a different voltage-current as a whole.

Figure 5.8 shows how the individual interaction of each cell makes the overall operation of a bigger arrangement exhibit an unwanted behavior. The only way in which a bigger arrangement could operate as a single cell is if the light and temperature conditions are the same for the whole arrangement. There are many issues around the *partial shading* effect, mostly due to the difficulties in finding the optimal operating point regarding power extraction.

Solar arrangements are partially shaded by clouds, birds, leaves, sand, dust, dry salt, etc., depending on their location. It's hard to consider an ever-lasting operational solar arrangement operating ideally. Hence, maintenance is normally low, but required. The association of solar cells commonly requires

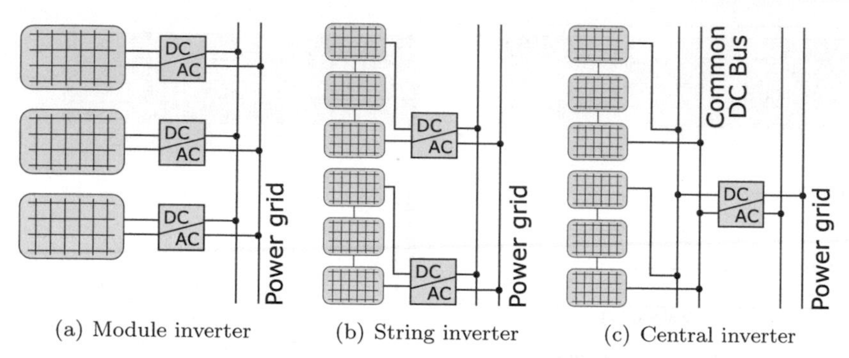

(a) Module inverter (b) String inverter (c) Central inverter

FIGURE 5.9
Different topologies regarding inverters' connections

the use of additional diodes to avoid panel damage; this is further explained in Section 5.3.1.

In addition, solar arrangements intended for AC use require inverter modules to effectively provide or interact with an AC bus. There are different topologies in which these modules are added to the panels' arrangements. Figure 5.9 shows three ways of connecting solar panels to an AC bus.

Notice that the different AC connections represented in Figure 5.9 mainly differ in the amount of power the inverter module manipulates. Figure 5.9(a) shows individual inverters associated to a single panel each. This implies that the power through the inverter is minimal but the synchronization with the power grid must be performed individually. Figure 5.9(b) presents a "string" topology that takes series-connected panels into a single inverter module. In this case, the inverter modules need to manipulate higher power but the synchronization to the power grid is performed considering "blocks" of solar panels. Finally, Figure 5.9(c) shows a single inverter which, regardless of the panels' arrangement behind it, performs a full-power conversion. Evidently, the inverter required for "central" operation is the one that processes the maximum amount of power among topologies; however, it is the only module to be synchronized to the power grid.

5.2.2 Solar cell types

The most general classification for solar cells is performed in terms of the nature of the material used in their construction. This classification specifically separates organic and inorganic cells depending on the materials used to take advantage of the photoelectric effect. Broadly speaking, inorganic cells use semiconductor materials like those introduced above, while organic cells use organic polymers or molecules.

Organic cells work under a similar concept regarding charges separation due to sunlight power. However, instead of having impure semiconductors,

TABLE 5.1
Some characteristics of different cell types, adapted from [72]

Cell type	**Required space** [m^2/kW]	**Efficiency** [%]
Monocrystalline	7 to 9	17–22
Polycrystalline	9 to 11	13–18
Amorphous	16 to 20	7–13

electron donor and acceptor polymers are built and placed together so the current flux is possible. Organic cells can be further classified depending on the disposition of both different polymers and other characteristics that can be parallel furnished to their finishing, such as transparency, light weight, flexibility, etc.

Besides, organic cells are a promising technique, mostly due to their low fabrication costs and cleaner deposition, but they also exhibit low efficiency (about 1/3 toward silicon ones) and chemical degradation [52]. This has hindered their usage in applications where inorganic cells are commonly used.

Organic cells are the subject of current research and are undoubtedly a topic that will be developed in the years to come; however, inorganic cells have also made themselves a place in power generation, so a more specific classification around them is needed as they dominate the current solar power panorama.

Inorganic solar cells are classified broadly as *crystalline* and *thin-film*. Crystalline cells are made from semiconductor "patches" sliced from bigger blocks. Such blocks are normally required to be thick (hundreds of [μm]) due to manufacturing obstacles. Thin-film cells are made through a chemical process that ends with a semiconductor under 10 [μm] [72, p. 377].

The above classification clearly points to the thickness of the final semiconductor block to be used. However, additional features must be considered as part of such differentiation. For instance, silicon can be arranged to exhibit a *mono-crystalline* structure (continuous crystal lattice structure throughout the block) only for crystalline cell types, while thin-film can only incorporate amorphous arrangements. Some electrical characteristics in this regard are shown in Table 5.1.

Table 5.1 and the above classification clearly show the main trade-off related to solar cell selection. A heavier, more expensive crystalline cell will exhibit better electric characteristics, while a lighter, cheaper one will make a poorer use of light incidence. It is worth mentioning that organic cells can be found to exhibit efficiencies around 9%; however, they still cannot replace thin-film cells due to some existing issues regarding their operational stability.

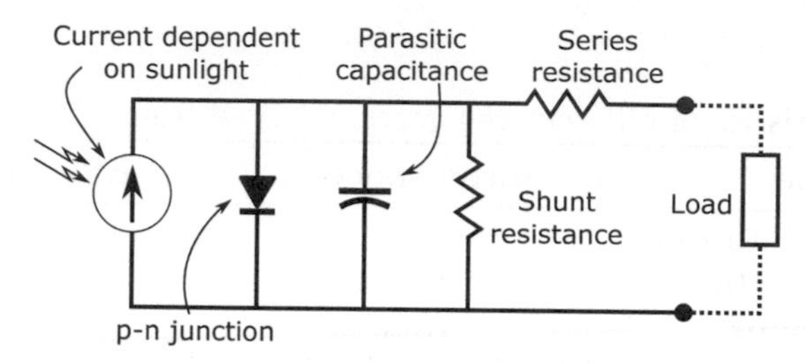

FIGURE 5.10
Solar cell equivalent circuit

5.3 Photo-voltaic panel modeling

As introduced in Section 5.2, the large-scale behavior of a solar cell involves a typical p-n junction as a diode. Sunlight will make electrons and holes get separated, making a certain amount of current available to flow through the load. The diode will not allow any current to pass through it unless its terminals' voltage goes beyond the forbidden band potential, so the generated current will be made available to the load unless the voltage exceeds the p-n junction limit. This interaction will generate a voltage-current characteristic like those shown in Figures 5.7 and 5.8.

There are additional non-idealistic effects inside the solar cell, which can be foreseen directly from Figure 5.4. The particles moving throughout the cell's body will not flow freely, and the movement through the terminals will also find electronic interferences. Both aforementioned effects can be accounted for with two equivalent resistors, one in series and one parallel to the current flow. Also, the particles' location and their clear block separation resembles a parallel capacitance. These interactions can be depicted through an equivalent circuit as the one shown in Figure 5.10.

For the sake of simplicity, the parasitic capacitance and the shunt resistor are commonly disregarded. Such simplification is still useful for many big-scale analyses and provides an easy way to fulfill the cell's model. If the series resistor R_s and a resistive R_L are considered together with the p-n junction and the sunlight current, the circuit can be easily solved by using Kirchoff's law:

$$I_L = I_G - I_D, \tag{5.1}$$

where I_L is the load current, I_G is the generated photo-current, and I_D is the current flowing through the diode. The diode's current can be easily known by using Shockely's equation, shown in (5.2),

$$I_D = I_0 \left(\exp\left[\frac{V_D}{nV_T}\right] - 1 \right), \tag{5.2}$$

where I_0 is the reverse bias saturation current, V_D is the diode's voltage, n is the ideality factor, and V_T is the "thermal voltage," defined as:

$$V_T = \frac{kT}{q}, \tag{5.3}$$

where k is the Boltzmann's constant, T the absolute temperature in [K], and q is the charge of a single electron.

It is known that

$$k = 1.38064852 \times 10^{-23} \; [m^2 \; kg \; s^{-2} \; K^{-1}],$$
$$q = 1.60217662 \times 10^{-19} \; [C],$$

so $V_T \approx 26 \; [mV]$ at 300 [K].

The above equations can be put together after substituting the diode's terminals' voltage $V_D = V_{\text{out}} + I_L R_s$ as follows:

$$I_L = I_G - I_0 \left(\exp \left[\frac{q}{AKT} (V_{\text{out}} + I_L R_s) \right] - 1 \right). \tag{5.4}$$

This simplified model allows the designer to easily know some voltage-current parameters. For instance, the open circuit voltage V_{OC} can be computed if $I_L = 0$, so

$$\frac{I_G + I_0}{I_0} = \exp \left[\frac{qV_{\text{out}}}{AKT} \right],$$
$$V_{\text{out}} = \ln \left(\frac{I_G + I_0}{I_0} \right) \frac{AKT}{q}. \tag{5.5}$$

Similarly, short-circuit conditions imply $V_{\text{out}} = 0$, so the equation is reduced to simply

$$I_{\text{SC}} = I_G. \tag{5.6}$$

Equation (5.4) is a simplified form for computing the electric characteristics of a solar cell, and it is normally accepted as a good approximation of solar cells' operation. However, if a more precise model is desired, the leakage current through the shunt resistance can be easily accounted for as:

$$I_L = I_G - I_0 \left(\exp \left[\frac{q}{AKT} (V_{\text{out}} + I_L R_s) \right] - 1 \right) - \frac{V_{\text{out}} + I_L R_s}{R_p}. \tag{5.7}$$

The generated current I_G can be calculated depending on manufacturer's test results over the cell. Such parameters are referred to as *standard test conditions* (STC) and account for the cell's behavior at $G^* = 1000 \; [W/m^2]$ light power and $T^* = 25° \; C$ temperature. Manufacturer's tests result in a standard generated current I_G^* and a temperature coefficient μ_{SC} in [A/°C]. Such parameters allow the user to calculate

$$I_G = \frac{G}{G^*} \left(I_G^* + \mu_{\text{SC}} \Delta T \right), \tag{5.8}$$

where G and G^* stand for present and STC irradiance, respectively, and $\Delta T = T - T^*$. If (5.8) is used together with (5.4) or (5.7), the behavior of a solar cell can be fully computed for any given electric condition.

It is noteworthy that the previously mentioned ideality factor accounts for manufacturing issues regarding the p-n junction or materials variability. This variable is $A = 1$ for an ideal junction and can go as high as 2 or even more depending on the manufacturer's specifications. It can be seen mathematically that a higher A would result in a "slower" rise of the diode's current due to an identical voltage. The solar cell's model will be clearly affected by this variable as the current will drop faster for higher As.

Another commonly neglected issue is related to the scale current I_0, sometimes called "reverse bias saturation current." Any p-n junction will exhibit a negative current flow if inversely biased. Alternatively, the exponential behavior presented in (5.2) uses $-I_0$ to effectively account for such a negative current, making the exponential equation reach $-I_0$ instead of 0 at $-\infty$ (that is why it is also referred to as "scale" current). It is important to mention that a p-n junction cannot hold too long before breaking if inversely biased; hence, there is a negative threshold voltage after which (5.2) cannot hold anymore.

Solar cells (panels) commonly incorporate a blocking diode at their positive terminal. This diode prevents the cell for being inversely biased, so the junction will be protected and I_0 will be a mere scale factor. Parallel diodes can also be used to bypass solar arrangements subjected to partial or full shading, allowing other connected solar cells/panels to deliver power continuously.

Finally, it is important to notice a feature present in Equations (5.4) and (5.7). The load current I_L can be found on both sides of the equation, making it hard to fulfill a voltage-current separation. This difficulty is due to the exponential nature of the equation, which hinders analytical solving possibilities. For instance, the cell model is usually computed by taking $I_L(t^-)$ as the current inside the exponential part of the equation, using software capable of variables feedback.

It is possible to find numerical results for the above-presented models by using a successive approximation method. Such a method can be as simple as equation balancing (substituting I_L until both sides become equal), or it can be automated through a numeric solver. An analytic alternative is also possible if the Lambert function $W(z)$, $z \in \mathbb{C}$ (5.9) is used. In this case the exponential can be directly solved and the voltage and current decoupled; however, a way to calculate $W(z)$ must be available.

$$z = W(z)e^{W(z)}. \tag{5.9}$$

5.3.1 Modeling cells arrangements

In the preceding sections, the series and parallel arrangements of solar cells have been discussed. It is clear that a sum is enough to account for a collective behavior of solar cells, depending on their actual connection (see Figure 5.7).

In this way, for series connected cells,

$$V_{\text{out}} = \sum_{i=1}^{N} V_{\text{out}}^{(i)} \tag{5.10}$$

will hold, having a total amount of N series connected cells. On the other hand,

$$I_L = \sum_{i=1}^{N} I_L^{(i)} \tag{5.11}$$

will exhibit the behavior of N parallel connected cells.

The above equations are valid if all conforming cells are assumed to be equal, and to operate under equal conditions. This makes it possible to simply multiply the current or voltage by N for parallel or series operation, respectively. In this case (the ideal case), Equation (5.4) can be modified for both cases as in (5.12) and (5.13).

$$I_L = I_G - I_0\left(\exp\left[\frac{q}{AKT}\left(\frac{V_{\text{out}}}{N} + I_L R_s\right)\right] - 1\right) \qquad \text{series,} \tag{5.12}$$

$$I_L = NI_G - NI_0\left(\exp\left[\frac{q}{AKT}\left(V_{\text{out}} + \frac{I_L R_s}{N}\right)\right] - 1\right) \qquad \text{parallel.} \tag{5.13}$$

However, cells are not ideal and partial shading may occur in cell arrangements. This forces such solar modules to commonly incorporate bypass diodes, as shown in Figure 5.11. Suppose the third cell is shaded as shown, so the current flowing through it would be much less than the current the remaining cells could provide. This behavior will make the shaded cell to be seen as a load, actually exhibiting an inverse polarization. Whether or not a diode is placed in parallel, the current is free to flow avoiding the shaded panel.

If no bypass diodes are used in series arrangements, a problem commonly referred to as a *hot spot* can occur [51, p. 110]. The shaded cell will heat as a result of its load behavior (power absorption). Panel manufacturers commonly provide intermediate connections at the output to allow the addition of bypass diodes for sets of series connected cells. Such sets of cells are commonly rated under 24 [V] as a rule of thumb [51, p. 113].

As can be foreseen from the above comments, a similar issue regarding parallel connection will also exist. If two cells or panels are connected in parallel and their behavior is different, it is possible to have one of the modules absorb the energy the other one generates. Figure 5.12 shows this situation assuming the second module has been somehow partially shaded.

The voltage-current behavior of both the aforementioned modules is also shown in Figure 5.12. Notice that the difference between both modules is not large; however, in open circuit operation, the parallel output voltage would make one of the modules operate at $I^{(1)}$ and the other at $I^{(2)}$. This clearly shows that the second module will be taking $I^{(2)} = I^{(1)}$. A *blocking diode* can

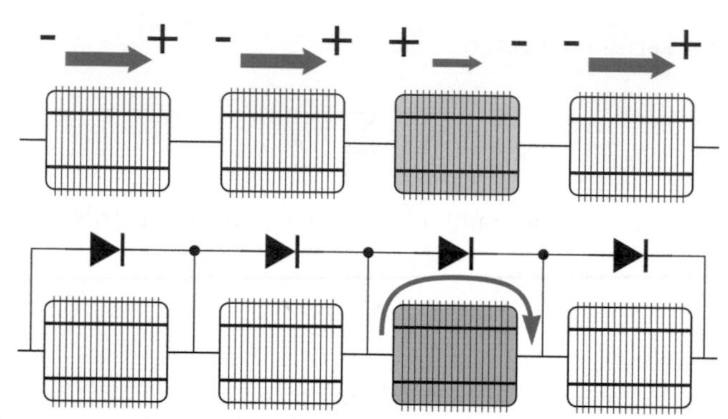

FIGURE 5.11
Bypass diode in series cell arrangements

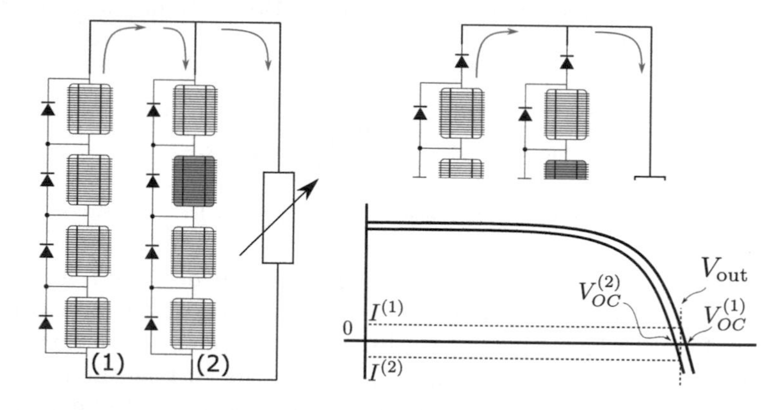

FIGURE 5.12
Blocking diodes used for parallel connection

be used to prevent the undesired incoming current flow, as is also shown in Figure 5.12.

The graph in Figure 5.12 also shows that such behavior is unlikely. Actually, blocking diodes are mostly disregarded in everyday operation and become necessary only if, as a rule of thumb, the connection implies a common bus higher than 120 [V] [51, p. 114].

5.3.2 Maximum power

Solar cells and panels do not operate optimally in terms of power extraction. Actually, the highest efficiency will depend on the load connected at the output. In fact, if a load is directly connected to a solar cell, the voltage at its terminals and the resulting current flowing through the load will match a single point in the voltage-current curve discussed so far.

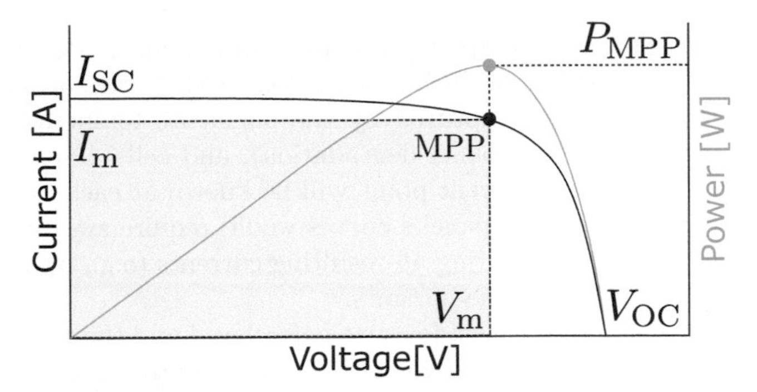

(a) Graphical view of maximum power point of a solar cell

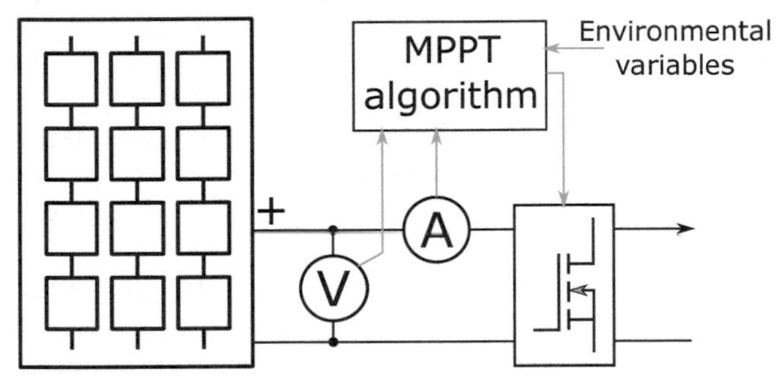

(b) Schematic view of MPPT alternative

FIGURE 5.13
Maximum power extraction concept

Such a point depends on the cell's characteristics and can be estimated by using its equivalent circuit (see Figure 5.10). This implies that the power obtained from the cell, $P = VI$, will be dominated by the cell's parameters, and subjected to the specific connected load. In most applications, the solar generators are expected to be load independent, and actually, to be connected to the power grid, which can ideally take any amount of current.

It is then important to guarantee that the power extracted from sunlight will be "maximum," i.e., at the cell's highest efficiency. This issue was introduced in Section 5.2 where the load controller was told to enable MPPT (maximum power point tracking) operation. In addition, Figure 5.5 shows a schematic regarding solar panel usage considering load independence.

MPPT implies that a DC-DC converter can behave as a varying load, controlled by a supervisory algorithm. Such an algorithm seeks for an equivalent impedance (seen from the panel as a connected load) that fulfills maximum power extraction. The voltage-power relation is shown in Figure 5.13(a) where the aforementioned point is marked as MPP. The MPPT schematic view is shown in Figure 5.13(b).

There is a common misunderstanding regarding MPPT that must be elucidated promptly. The voltage-current and voltage-power curves shown throughout this chapter are not known during operation. Such curves depend on the actual sunlight power being captured by the panel, its temperature, possible non-idealities (shading or panel degradation), and cells' interconnection. Moreover, only one voltage-current point will be known at each time instant. This implies that deriving the panel's curves would require sweeping all possible output voltages and capturing all resulting currents (e.g., by varying the connected load) at different time-instants.

As the curves cannot be known for sure beforehand and the voltage sweeping is operationally unavailable, it would be impossible to know if the panel's operation is optimal, having only one measured operating point. MPPT algorithms offer an alternative to effectively seeking for P_{MPP}, by "intelligently" moving through the curve. This indeed can take advantage of knowing the panel's parameters and can be seen as a sweeping in itself, but it is an efficient one. The design of better MPPT algorithms is a research endeavor of different groups around the world, who are trying to face fast-changing conditions, partial shading (as the curve could be swept incorrectly), and power electronics efficiency.

As a final comment, the gray rectangle shown in Figure 5.13(a) with base V_m and height I_m has an area equivalent to the delivered power. This means that finding the maximum power point is equivalent to finding the circumscribed rectangle with the largest area. In this spirit, a parameter commonly referred to as *fill factor*, referring to the quality of the solar panel, measures how completely the aforementioned rectangle "fills" the voltage-current curve, as shown in (5.14).

$$\text{FF} = \frac{V_m I_m}{V_{\text{OC}} I_{\text{SC}}} \tag{5.14}$$

5.4 A brief example: Solar umbrella monitoring with LabVIEW

The goal of this project was to study the behavior of electric consumption of small electronic devices, as well as the characteristics of solar generated energy at common spaces inside universities. The prototype consisted of three flexible solar panels, mounted over an outdoor umbrella, with a battery-based storage system, a load manager to power small electronic devices through an inverter, a weather station, current sensors, LED lights for illumination, and a communication interface, as shown in Figure 5.14.

The current sensors detect the current generated by the solar panel as well as the current delivered by the inverter. The weather station is a set of sensors of humidity, temperature, and light, to identify the exact conditions at which the solar energy is generated. A strip of LEDs was mounted in the inside of

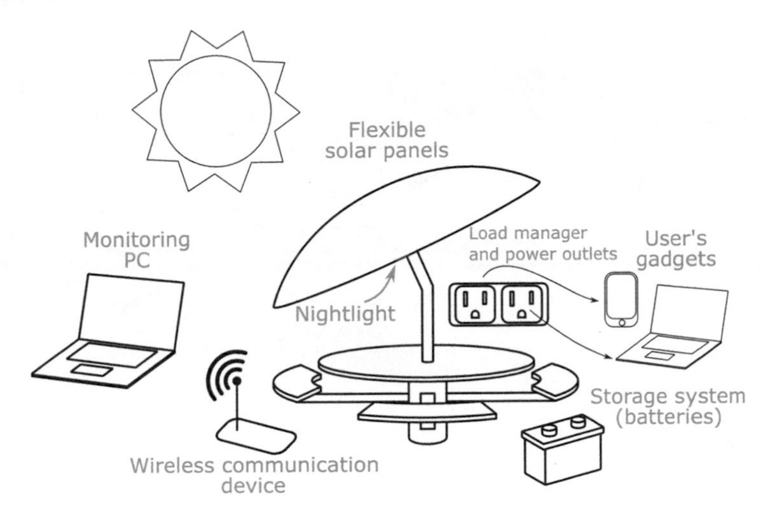

FIGURE 5.14
Block diagram of the solar umbrella monitoring system

the umbrella, so the intensity of the LEDs was dependent on the amount of ambient light detected. Finally, all the information collected was sent to a remote computer wirelessly via X-Bee protocol. The remote computer uses LabVIEW to show the information and analyze it, as shown in Figure 5.15.

This allowed the tracking of the electrical consumption of devices connected to the inverter in real time. This information was used for didactic purposes in courses of engineering. Also, for research purposes, information about the efficiency of the solar panel in different weather conditions was gathered.

5.5 Exercises

Assume that there is a factory connected to a three-phase generator, and the possibility to add a source of solar energy to reduce the consumption of energy from the generator. The solar panel will be tilted to recreate the movement of the sun during the day to observe the differences in power.

Recommended equipment and connections:

- Lamps for photovoltaic trainer
- Photovoltaic inclinable module
- Inverter grid
- Electrical power digital measuring unit

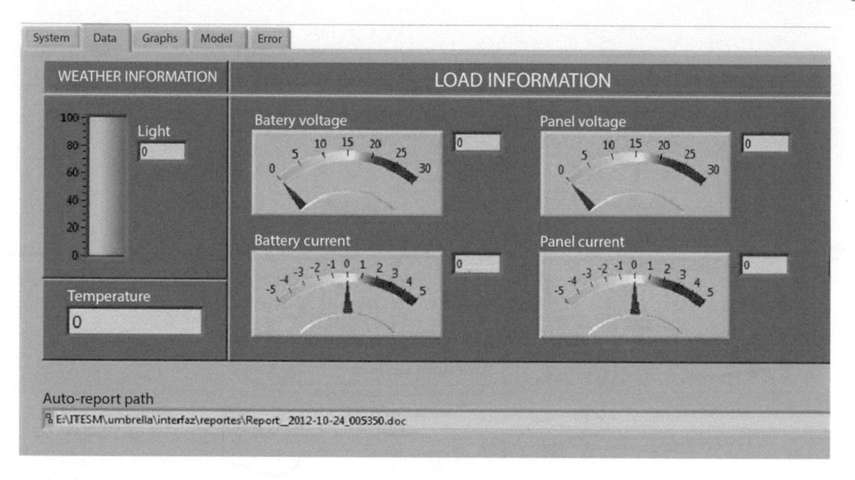

FIGURE 5.15
Information on remote computer

- Maximum demand meter
- Three-phase supply unit
- Resistive load

Connect the lamps and point the light towards the photovoltaic panel (90°); the lamps act as the light source. The photovoltaic panel is then connected to the inverter grid, and this module is then connected in series to the thermo-magnetic differential switch to activate or deactivate the load. The electric power digital meter module is connected to the switch in order to measure the output power coming from the inverter.

Finally, connect the three-phase supply unit, assuming it is the power generator of the factory, to the maximum demand meter and then to the Y-connected load (see Diagram 5.1).

1. Set the three resistive loads to the 2nd value
2. Set the lamps to 100% of light intensity
3. Put the solar panel in front of the light at 90° degrees. Then vary the angle by 10° degrees until the contribution of the solar panel is 0 [W]. For every position of the solar panel, write down the voltage and current generated and calculate the power; this can be obtained in the screen of the electrical power digital measuring unit. Results should be similar to those shown in Table 5.2.
4. Measure with the maximum demand meter the active power provided by each of the lines in the source for each of the previous positions of the solar panel. Results should look like those preseted in Table 5.3.

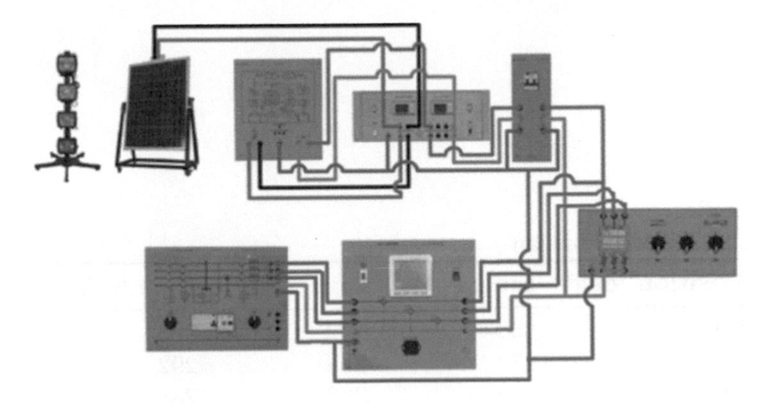

Diagram 5.1
Solar energy exercise

TABLE 5.2
Panel's dependency on incidence angle

Position [deg]	Voltage [V]	Current [A]	Power [W]
90°	18.4	1.06	19.504
80°	15.3	1.08	16.524
70°	14	0.9	12.6
60°	15.3	0.86	13.158
50°	17.9	0.728	13.0312
40°	17.9	0.656	11.7424
30°	15.8	0.55	8.69
20°	15.1	0.358	5.4058

5. Repeat steps 3 and 4 but modifying the light intensity of the lamps to 75%. Write down the results as shown in Table 5.4.

A solar panel is exposed directly to the sunlight to obtain the most energy from the sun and convert it to electricity, but there is a downside for this. As the solar panel is in direct contact with the sunlight, it begins to heat up and its efficiency starts to degenerate. In this exercise the lamps will act as the light source, and at the same time, they will heat the panels. Measurements will be taken every five minutes to observe the efficiency.

Recommended equipment and connections:

- Lamps for photovoltaic trainer
- Photovoltaic inclinable module
- Inverter grid

TABLE 5.3
One-phase solar contribution

Position [deg]	Line 1 [W]	Line 2 [W]	Line 3 [W]
90°	268	292	292
80°	232	288	292
70°	240	288	288
60°	244	288	292
50°	248	288	288
40°	256	292	288
30°	260	288	288
20°	272	288	288

TABLE 5.4
Results considering 75% of original light power

Position [deg]	Voltage [V]	Current [A]	Power [W]	Line 1 [W]	Line 2 [W]	Line 3 [W]
90°	16.9	0.552	9.3288	264	288	288
80°	17.8	0.544	9.6832	264	292	288
70°	16.8	0.57	9.576	264	288	288
60°	14.4	0.528	7.6032	268	288	288
50°	13.3	0.52	6.916	268	288	288
40°	14.8	0.344	5.0912	276	288	288
30°	14.1	0.304	4.2864	276	288	288
20°	14.7	<0.300	-	284	288	288

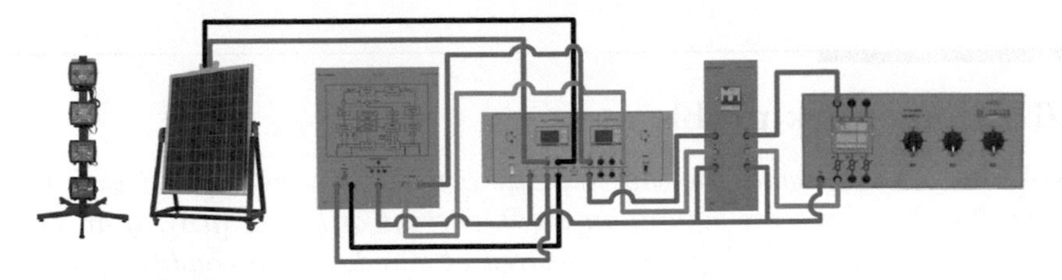

Diagram 5.2
Solar energy exercise connection diagram

TABLE 5.5
Panel's power dependency on temperature

Time [min]	Voltage [V]	Current [A]	Power [W]	Temp. [C]
0	18.4	1.134	20.8656	35°
5	17.5	1.168	20.44	62.6°
10	17	1.168	19.856	75.5°
15	16.8	1.167	19.6056	79.4°

- Electrical power digital measuring unit
- Resistive load
- Laser thermometer

Connect the lamps and point the light towards the photovoltaic panel (90°); the lamps act as the light source. The photovoltaic panel is then connected to the inverter grid module; this module is then connected in series to the thermo-magnetic differential switch to activate or deactivate the load. The electric power digital meter is connected to the switch in order to measure the output power coming from the inverter (as seen in Diagram 5.2).

1. Set the resistive load to the 2nd value
2. Set the lamps to 100% of light intensity
3. Put the solar panel in front of the light at 90° degrees. Take 4 measurements every 5 minutes of the voltage and current generated by the solar panel, and take the temperature of the panel with the thermometer. Write down the results as shown in Table 5.5

5.6 Homework problems

Problem 5.1. *Estimate the short-circuit current of a solar panel with form factor $A = 2$ and reverse saturation diode current of $I_0 = 1$ $[\mu A]$, if an open-circuit voltage $V_{OC} = 0.81$ $[V]$ is measured at standard test conditions.*

Solution 5.1. *The open-circuit equation can be used:*

$$V_{OC} = 0.81 = \frac{AKT}{q} \ln\left(\frac{I_L - I_0}{I_0}\right) = \frac{2\left(1.38 \times 10^{-24} \left[\frac{J}{K}\right]\right)(298\ [K])}{1.6022 \times 10^{-19}[C]} \ln\left(\frac{I_L - 1 \times 10^{-6}\ [A]}{1 \times 10^{-6}\ [A]}\right).$$

So, a direct evaluation can lead to $I_L = 7.07$ [A].

Problem 5.2. *Estimate the series resistance R_s parameter of a solar cell model with $A = 3$ and $I_0 = 1$ $[\mu A]$, if it is known to exhibit the following pairs at the same irradiance: (10 [V],3 [A]) and (20 [V],1.5 [A]).*

Solution 5.2. *Substituting in the panel equation:*

$$I = I_L - I_0 \left(\exp\left[\left(\frac{q}{AKT}\right)(V + IR_s)\right] - 1\right).$$

$$I_i = I_L - 1 \times 10^{-6} \left(\exp\left[\left(\frac{q}{3K(298)}\right)(V_i + I_i R_s)\right] - 1\right).$$

Thus, for both pairs:

$$I_L = 3 + 1 \times 10^{-6} \left(\exp\left[(12.986)(10 + 3R_s)\right] - 1\right),$$
$$I_L = 1.5 + 1 \times 10^{-6} \left(\exp\left[(12.986)(20 + 1.5R_s)\right] - 1\right).$$

Solving the above equations is not easy, analytically speaking; however, a numerical algorithm like the "bisection" can be used to identify R_s as precisely as required. Some iterations of a basic search procedure are shown in the table below. So, $R_s = 6.6665039$ $[\Omega]$.

R_s	***Eq1***	***Eq2***
6.0000000	8.18E+151	3.57E+157
7.0000000	6.80E+168	1.03E+166
6.5000000	2.36E+160	6.06E+161
6.7500000	4.00E+164	7.90E+163
6.6250000	3.07E+162	6.92E+162
6.6875000	3.51E+163	2.34E+163
6.6562500	1.04E+163	1.27E+163
6.6718750	1.91E+163	1.72E+163
6.6640625	1.41E+163	1.48E+163
6.6679688	1.64E+163	1.60E+163
6.6660156	1.52E+163	1.54E+163
6.6669922	1.58E+163	1.57E+163
6.6665039	1.55E+163	1.55E+163

Problem 5.3. *Draw the V-I curve of a panel composed of 20 series cells with the following parameters (each) and at standard test conditions:* $A = 3$, $R_s = 0.06$ $[\Omega]$, $I_sc = 3$ $[A]$, $I_0 = 1$ $[\mu A]$.

Solution 5.3. *The cell panel equation can be better processed (after substituting) in the following form:*

$$i = 3 - 1 \times 10^{-6} \exp\left(778.83 \times 10^{-3} i + 12.98 \frac{v}{20}\right).$$

Notice that the voltage has been divided by the number of series cells to depict series cells connection.

This can be solved through the Lambert equation. However, it can also be solved numerically for different voltages. The table below shows some iterations of a simple search method.

Voltage [V]	***Calculated Current [A]***	***Test Current [A]***
0	2.99999	2.99999
5	2.99973	2.99973
10	2.99322	2.99322
14	2.91453	2.91453
15	2.84506	2.84506
18	2.29342	2.29343
20	1.54992	1.54993
22.98015	0.00002	0

The V-I curve can be then plotted if multiple voltage values are solved as shown in the above table.

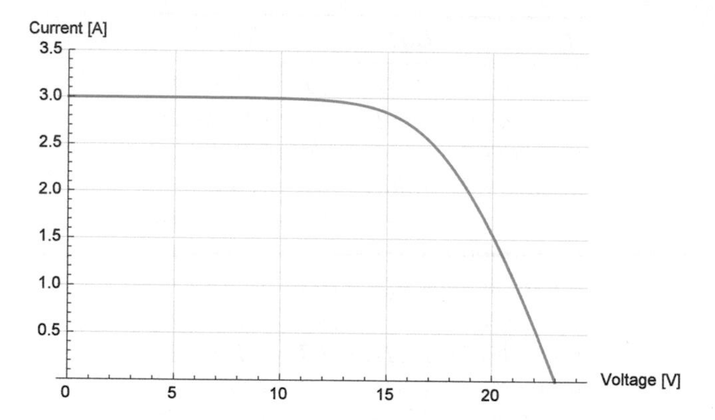

Problem 5.4. *Using the same data from Problem 5.3, plot the effect of changing the form factor as follows:* $A = [1, 5] \in \mathbb{Z}$.

Solution 5.4. *As the form factor intervenes in the weighing of both voltage and current inside the diode equation, the limit at which the equivalent diode will start to take current is modified.*

The same considerations for the solution of Problem 5.3 hold. However, it will be necessary to solve numerically for each form factor. Results can be seen in the graph below.

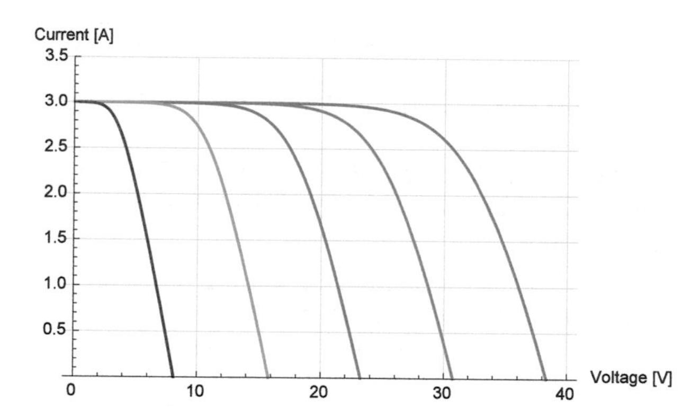

Problem 5.5. *Find the optimal operating point (maximum power point) of the panel described in Problem 5.3.*

Solution 5.5. *The panel equation can be substituted to derive*

$$i = 3 - 1 \times 10^{-6} \exp\left(778.83 \times 10^{-3} i + 12.98 \frac{v}{20}\right).$$

Power can be calculated directly as $P = vi$, *so the preceding equation leads to*

$$P = 3v - v(1 \times 10^{-6}) \exp\left(778.83 \times 10^{-3} \frac{P}{v} + 12.98 \frac{v}{20}\right).$$

Again, this equation could be solved directly by using the Lambert equation; however, a numerical approximation is also possible, as shown in the table below. Hence, the maximum power point can be found at about 16.25 [V] and 2.69 [A]. An analytical solution leads to $v = 16.29275$ *[V] and* $i = 2.684$ *[A].*

Voltage [V]	*Calculated Power [W]*	*Test Power [W]*	*Voltage [V]*	*Calculated Power [W]*	*Test Power [W]*
5	*14.99867*	*14.99867*	*15*	*42.67585*	*42.67585*
8	*23.98514*	*23.98513*	*16*	*43.6656*	*43.66561*
11	*32.85804*	*32.85803*	*17*	*43.33859*	*43.33859*
14	*40.80342*	*40.80341*	*18*	*41.28162*	*41.28162*
17	*43.33859*	*43.33859*	*19*	*37.20969*	*37.20968*
21	*22.64282*	*22.64282*			

Voltage [V]	*Calculated Power [W]*	*Test Power [W]*	*Voltage [V]*	*Calculated Power [W]*	*Test Power [W]*
15.5	*43.30722*	*43.30722*	*16.75*	*43.56792*	*43.56792*
16	*43.6656*	*43.66561*	*16.5*	*43.69448*	*43.69448*
16.5	*43.69448*	*43.69447*	*16.25*	*43.72476*	*43.72475*

5.7 Simulation

Solar panels can be seen as a direct current source, similar to a battery. This implies they require an AC inverter so they can be integrated to an electric network. In addition, their electric output varies depending on temperature and light irradiance, so their power profile must be manipulated by voltage converter circuits to guarantee the maximum power extraction and the proper synchronization to the grid.

Even though automatic control and power electronics are currently integrated in commercial solutions to install solar panels domestically, it is important to understand the effects of power tracking, cooling, and solar panel variations with respect to reported nominal characteristics.

The block diagram shown in Figure 5.16 exhibits a solar panel as a simple block. This model could be used to sweep the desired environmental parameters to test panel's performance.

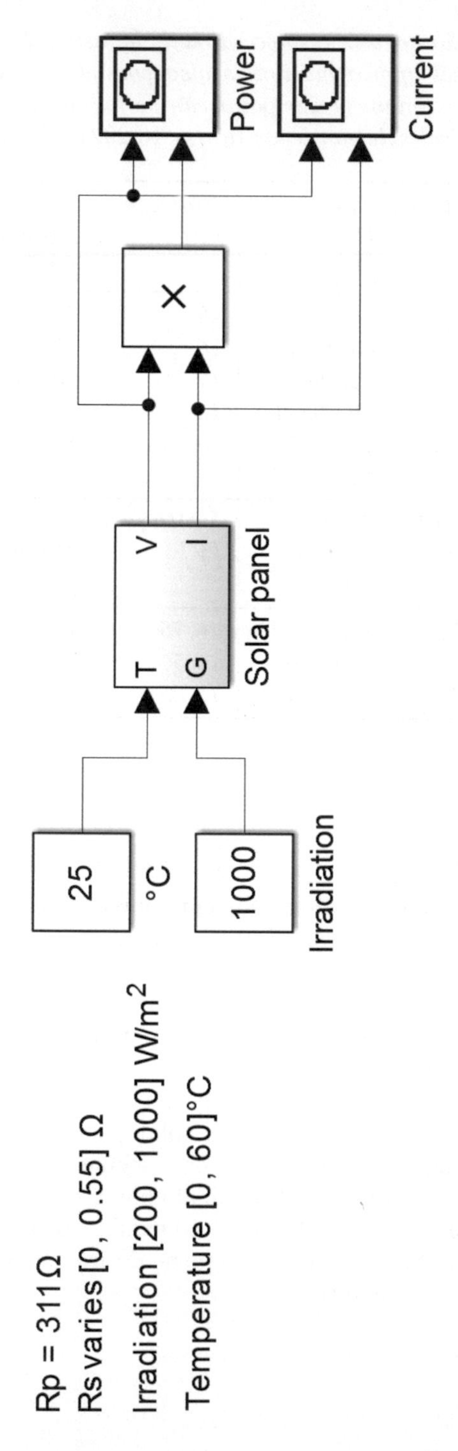

FIGURE 5.16
Simulation of a solar panel

6

Electric Power Transmission

Transmission of electric power is the key for the development of the modern electric system in the world. The necessity of carrying power at high voltage through tall steel towers that cross countries in all directions has enhanced the analysis of the transmission lines with equivalent circuits (inductances, resistance, and capacitance). There are underground transmission lines and overhead ones; in this chapter, the latter will be analyzed due to their greater popularity.

A network of transmission lines is referred as "power grids." Traditional power grids transmit three-phase alternating current at voltages from 69 $[kV]$ to 765 kV from the generators to the load. Less impressive but not less important are the city's and railway's power lines, which transmit power generated to the loads.

In this chapter, series resistance, series inductance, and shunt capacitance for three-phase power lines will be analyzed, after which, the ABCD model of transmission lines is presented and its π equivalent circuit.

6.1 Generic structure of a transmission tower

The transmission system consists mainly of aluminum or copper conductors, isolation strings, a tower for supporting the components, and shield or overhead ground wires. Figure 6.1 shows a waist-type tower for high voltage transmission. Shield cables aim to protect the system from lightning and the tower is made of a conductor material (steel) for discharging electricity to the ground.

Traditional conductors are arranged in strands because they are easy to manufacture by adding layers for a larger conductor area, and are flexible and easy to handle compared with an equivalent-area solid conductor.

Although copper is a better conductor, the abundant and less expensive aluminum has replaced it in most overhead transmission lines. For giving strength to aluminum conductors, strands of steel are commonly added; these conductors are known as aluminum-conductor steel-reinforced cable (ACSR). Other common conductors are all-aluminum conductor (AAC), all aluminum-alloy conductor (AAAC), and aluminum conductor alloy reinforced (ACAR).

Often, high voltage lines have many conductors per phase, or a bundle.

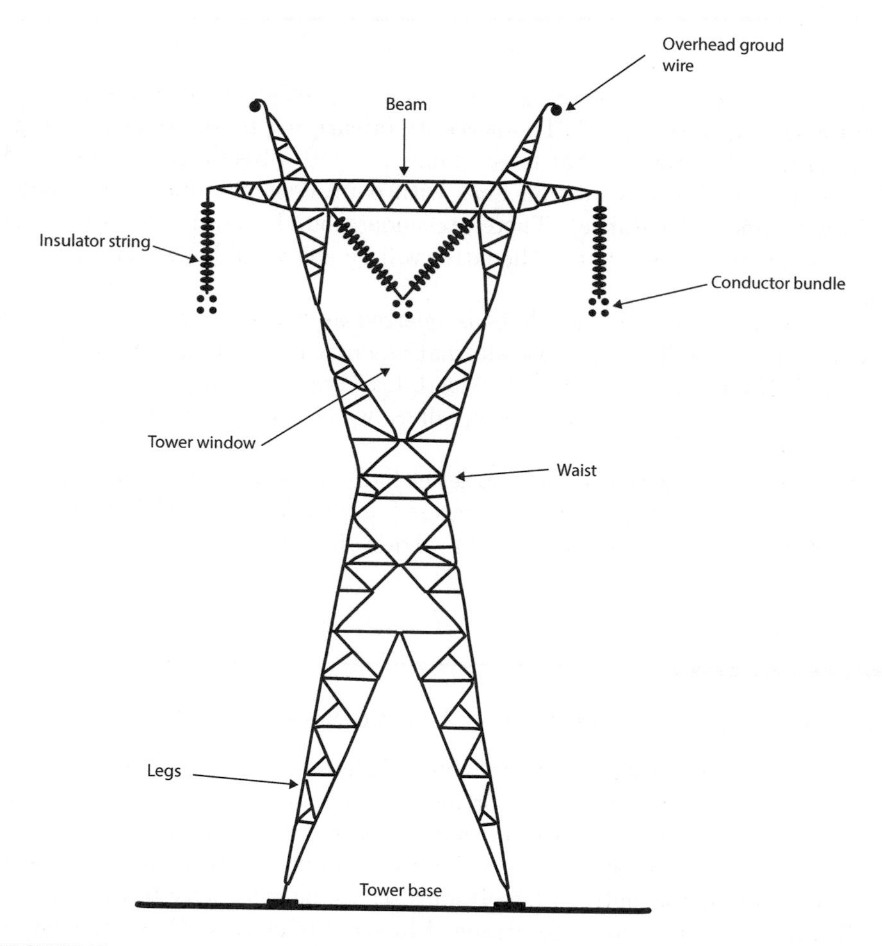

FIGURE 6.1
Waist-type power tower components

Figure 6.1 shows a transmission line of four conductors per phase, decreasing the strength of the electric fiend at the surface in order to reduce the losses due to the corona effect.

Insulator strings of discs, made mainly of porcelain, are used to suspend transmission lines and separate them from the tower. Figure 6.1 shows a V-shaped arrangement in the middle bundle for avoiding swings due to air currents. The porcelain discs are designed to support mechanical loads of about 7500 $[kgf]$.

The transmission tower is the supporting unit for overhead transmission lines carrying the transmission cables at a safe height from the ground. Many designs are implemented according to economic factors, the ambient conditions, and mechanical loads (wind loads, weight of cables, etc.).

Although high voltage towers are made of steel, the transmission lines of cities can be made of wood and concrete because they are not designed for discharging energy to the ground.

6.2 Transmission line parameters

For modeling transmission lines, it is necessary to establish physics that give them their electrical characteristics, so a precise model can be developed. In this section, resistance, inductance, and capacitance for a solid non-magnetic conductor are discussed.

6.2.1 Electric resistance

The resistance model presented follows these assumptions:

- Electric resistance varies linearly with the temperature
- No presence of "skin effect"
- Spiraling of the conductors does not affect its resistance

Electric resistance is typically presented as a DC value dependent on the temperature, directly proportional to the conductor length, and inversely proportional to the cross-sectional area of the conductor,

$$R_{dc} = \frac{\rho_T \times l}{A}, \tag{6.1}$$

where ρ_T is the conductor resistivity dependent on the temperature, l is the length, and A is the cross-sectional area.

The resistivity of the conductor varies linearly over traditional operation ranges according to

$$\rho_T T2 = \rho_T T1 \left(\frac{T_2 + T}{T_1 + T} \right), \tag{6.2}$$

TABLE 6.1
Resistivity and temperature constant of some conductor metals

Material	Resistivity at $20°C$ $[\Omega m] \times 10^8$	Temperature constant $(°C)$
Annealed copper	1.72	234.5
Hard-drawn copper	1.78	241.5
Aluminum	2.85	228.3
Silver	1.59	243
Brass	6–9	480
Iron	10	180
Steel	12–88	180–980

where ρ_{T2} is the resistivity at temperature 2 (T2). ρ_{T1} is the resistivity at temperature 1 (T1). T is the temperature constant of the material. T2 and T1 have the same units as constant T.

In Table 6.1, some metals' temperature constant and resistivity at $20°C$ are presented.

Electric resistance increases due to the "skin effect." On an alternating current, the distribution of charge is not uniform, tending to crowd toward the conductor surface, with smaller current density at the conductor center [39]. At a higher frequency, the skin effect increases; for nominal frequency of power systems (60 $[Hz]$ or 50 $[Hz]$), the conductors' manufacturers provide the percentage of electric resistance due to this phenomenon.

6.2.2 Inductance

To model the three-phase inductance of a transmission line, let's start with the solid cylindrical conductor inductance model. Inductance is present due to the permeability μ of the medium. The permeability of the air is very close to the vacuum's permeability, $\mu_0 = 4\pi \times 10^{-7}$ $[H/m]$, and also, the permeability of non-ferromagnetic materials is very close to μ_0, 0.999 and 1.000022 for copper and aluminum, respectively.

The steps followed for determining the inductance due to both the internal and external magnetic fluxes are:

1. Obtain the magnetic field intensity due to the current flowing through the conductor (Ampere's law)

$$\oint H \, dl = I_{\text{enclosed}}. \tag{6.3}$$

2. Obtain the magnetic flux density B

$$B = \mu H. \tag{6.4}$$

3. Obtain the magnetic flux

$$\phi = BA. \tag{6.5}$$

4. Obtain magnetic linkage flux

$$\lambda = \text{fraction of the flowing current} \times \phi. \tag{6.6}$$

5. Obtain the inductance from linkage flux

$$L = \lambda / I. \tag{6.7}$$

Assumptions:

- Current distributes uniformly through the cross-sectional area of the conductor (no skin effect)
- Conductor material is non-magnetic($\mu = \mu_0$)
- Electric conductors are made of a single solid bar (or bundle of bars)

Consider the electric conductor shown in Figure 6.2. The current is flowing out to the page, r is the radius of the cross sectional area, and the lined internal circle is the path of integration for obtaining the magnetic field. Also consider that the conductor has a length l.

6.2.2.1 Inductance due to internal flux linkage

The first step is to obtain the magnetic field H implementing Ampere's law, Equation (6.3). The path of integration is the circumference of radius $x < r$,

$$\oint H_x \, dl = H_x(2\pi x) = I_{\text{enclosed}}. \tag{6.8}$$

As mentioned in the assumptions, the current I_{enclosed} is distributed uniformly,

$$I_{\text{enclosed}} = \frac{\pi x^2}{\pi r^2} I = \left(\frac{x}{r}\right)^2 I. \tag{6.9}$$

The magnetic field is equal to

$$H_x = \frac{xl}{2\pi r^2}. \tag{6.10}$$

The next step is to obtain the magnetic flux density

$$B_x = \mu_0 H_x = \frac{\mu_0 I}{2\pi x} = 2 \times 10^{-7} \frac{I}{x}. \tag{6.11}$$

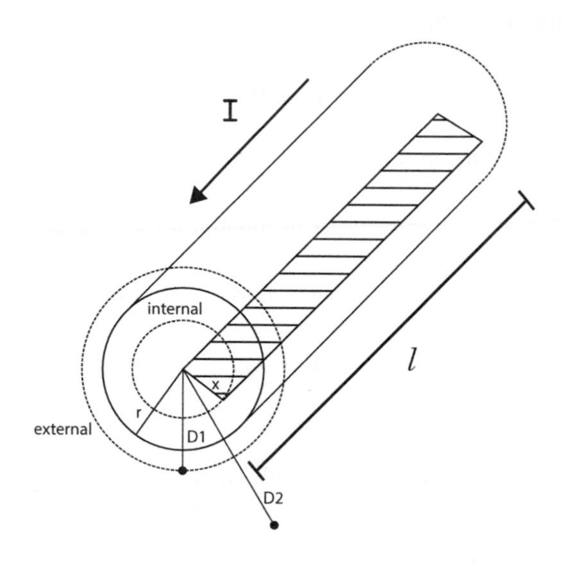

FIGURE 6.2
Solid cylindrical conductor and paths of integration

Once we have obtained the magnetic flux density, let's consider the lined area shown in Figure 6.2 for obtaining the differential magnetic flux $d\phi$:

$$d\phi_x = B_x l dx = \frac{\mu_0 x l}{2\pi r^2} l dx. \tag{6.12}$$

For obtaining the differential flux linkage $d\lambda$ it is important to remember that only the enclosed fraction of the current proportional to $(x/r)^2$ is linked. So, the actual differential internal linkage flux is

$$d\lambda_{\text{int}} = \left(\frac{x}{r}\right)^2 d\phi_x = \frac{\mu_0 x^3 I}{2\pi r^4} l dx. \tag{6.13}$$

Integrating the differential linkage flux from 0 to r,

$$\lambda_{\text{int}} = \frac{\mu_0 I}{2\pi r^4} l \int_0^r x^3 dx = \frac{\mu_0 I}{8\pi} l. \tag{6.14}$$

Finally, the inductance is calculated as

$$L_{\text{int}} = \frac{\lambda_{\text{int}}}{I} = \frac{\mu_0 l}{8\pi} = \frac{1 \times 10^{-7}}{2} l. \tag{6.15}$$

6.2.2.2 Inductance due to external flux linkage

Similarly, consider the external path of integration of Figure 6.2, that is, $x > r$. Applying the Ampere's law again,

$$H_x = \frac{1}{2\pi x}. \tag{6.16}$$

The magnetic permeability of air is equal to $\mu_0 = 4\pi \times 10^{-7}$ $[H/m]$, so the magnetic flux density is computed as

$$B_x = \mu_0 H_x = \frac{\mu_0 I}{2\pi x} = 2 \times 10^{-7} \frac{I}{x}. \tag{6.17}$$

For the external flux, the total current is linked, so the flux is equal to the linkage flux, calculating the differential linkage flux:

$$d\phi = d\lambda_{\text{ext}} = B_x l dx = 2 \times 10^{-7} \frac{I}{x} l dx. \tag{6.18}$$

Observe points D1 and D2 in Figure 6.2. For calculating the actual linkage flux, it is necessary to integrate Equation (6.18) from D1 to D2,

$$\lambda_{\text{ext}} = \int_{D_1}^{D_2} 2 \times 10^{-7} \frac{I}{x} l dx = 2 \times 10^{-7} I \ln\left(\frac{D_2}{D_1}\right) l. \tag{6.19}$$

Then, the inductance of external flux is

$$L_{\text{ext}} = \frac{\lambda_{\text{ext}}}{I} = \frac{\mu_0 l}{8\pi} = 2 \times 10^{-7} \ln\left(\frac{D_2}{D_1}\right) l. \tag{6.20}$$

6.2.2.3 Total inductance

The total linkage flux from the center of the conductor to an arbitrary point of

$$\lambda_{\text{total}} = \lambda_{\text{int}} + \lambda_{\text{ext}} = 2 \times 10^{-7} Il \left[\frac{1}{4} + \ln\left(\frac{D_2}{D_1}\right)\right] =$$

$$2 \times 10^{-7} Il \left[\ln e^{\frac{1}{4}} + \ln\left(\frac{D}{r}\right)\right].$$

$$\lambda_{\text{total}} = 2 \times 10^{-7} Il \ln\left(\frac{D}{re^{-\frac{1}{4}}}\right) = 2 \times 10^{-7} Il \ln\left(\frac{D}{r'}\right). \tag{6.21}$$

$$r' = re^{-\frac{1}{4}}. \tag{6.22}$$

$$L_{\text{total}} = 2 \times 10^{-7} l \ln\left(\frac{D}{r'}\right). \tag{6.23}$$

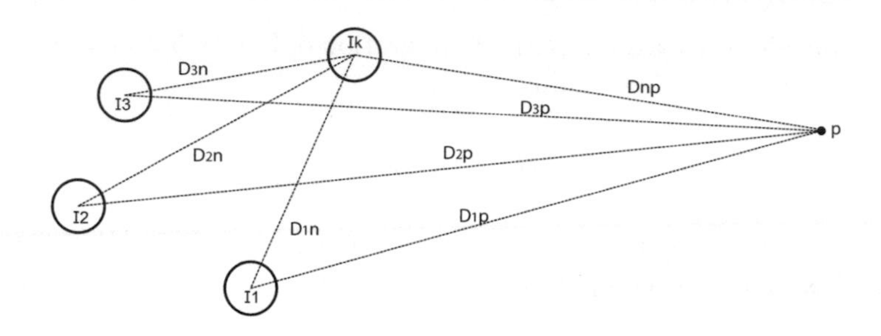

FIGURE 6.3
Array of N conductors

6.2.2.4 Inductance in a point P due to an N conductors system

Once the linkage flux of a conductor is known for some given point in space, its flux linkage can be known relative to other conductors. The flux linking conductor n to point P, due to its own current n (see Figure 6.3), is

$$\lambda_{\mathrm{nP}} = 2 \times 10^{-7} Il \ln \left(\frac{D_n}{r'_n} \right). \tag{6.24}$$

The inductance of a conductor is due to its own current but also to the current of adjacent conductors. So the total inductance of a conductor n to a point in the space P due to all the currents k is

$$\lambda_{\mathrm{nP}} = 2 \times 10^{-7} l \sum_{k=1}^{K} I_k \ln \left(\frac{D_{Pk}}{D_n k} \right). \tag{6.25}$$

In Equation (6.25), when $n = k$, $D_{nn} = r'_n = re^{-\frac{1}{4}}$.

6.2.2.5 Inductance in the infinite

Now let's consider an infinite distance between the electric conductors and the point P. Also consider that the total current of the electric system is zero $\sum_{k=1}^{K} I_k = 0$.

If we algebraically manipulate Equation (6.25), representing it as a subtraction of natural logarithms, we obtain

$$\lambda_{nP} = 2 \times 10^{-7} l \left(\sum_{k=1}^{K} I_k \ln D_{Pk} - \sum_{k=1}^{K} I_k \ln D_{nk} \right). \tag{6.26}$$

Then we separate the last element of the positive part:

$$\lambda_{nP} = 2 \times 10^{-7} l \left(I_K \ln D_{PK} + \sum_{k=1}^{K-1} I_k \ln D_{Pk} - \sum_{k=1}^{K} I_k \ln D_{nk} \right). \tag{6.27}$$

On the other hand, if the sum of currents equals zero,

$$I_K = -\sum_{k=1}^{K-1} I_k. \tag{6.28}$$

If we substitute Equation (6.28) into Equation (6.27), we obtain

$$\lambda_{nP} = 2\times 10^{-7} l \left(\sum_{k=1}^{K-1} I_k \ln D_{PK} - \sum_{k=1}^{K-1} I_k \ln D_{Pk} - \sum_{k=1}^{K} I_k \ln D_{nk} \right),$$

$$\lambda_{nP} = 2\times 10^{-7} l \left(\sum_{k=1}^{K-1} I_k \ln \frac{D_{Pk}}{D_{PK}} + \sum_{k-1}^{K} I_k \ln \frac{1}{D_{nk}} \right). \tag{6.29}$$

Now consider an infinite length of point P, then $D_{Pk} = D_{PK}$ and $\ln 1 = 0$.

Finally, the inductance of a conductor referred to the infinite is only the last part of Equation (6.29),

$$\lambda_n = 2\times 10^{-7} l \sum_{k=1}^{K} I_k \ln \frac{1}{D_{nk}}. \tag{6.30}$$

6.2.2.6 Inductance of a monophasic system

Applying Equation (6.30) to a single phase system, considering conductor 1 is positive ($I_1 = I$) and conductor 2 is negative ($I_2 = -I$),

$$\lambda_1 = 2\times 10^{-7} l \left(I \ln \frac{1}{r_1'} - I \ln \frac{1}{D_{12}} \right) = 2\times 10^{-7} l I \ln \frac{D_{12}}{r_1'}, \tag{6.31}$$

$$\lambda_2 = 2\times 10^{-7} l \left(I \ln \frac{1}{D_{21}} - I \ln \frac{1}{r_2'} \right) = -2\times 10^{-7} l I \ln \frac{D_{21}}{r_2'}, \tag{6.32}$$

and the total inductance is

$$L_1 = \frac{\lambda_1}{I_1} = 2\times 10^{-7} l \ln \frac{D_{12}}{r_1'}, \tag{6.33}$$

$$L_2 = \frac{\lambda_2}{I_2} = 2\times 10^{-7} l \ln \frac{D_{21}}{r_2'}, \tag{6.34}$$

$$L_{\text{total}} = L_1 + L_2 = 2\times 10^{-7} l \left(\ln \frac{D_{12}}{r_1'} + \ln \frac{D_{21}}{r_2'} \right),$$

$$= 2\times 10^{-7} l \frac{D_{12}}{\sqrt{r_1' + r_2'}}. \tag{6.35}$$

6.2.2.7 Inductance of a three-phase system

Assumptions:

- The system is totally transposed

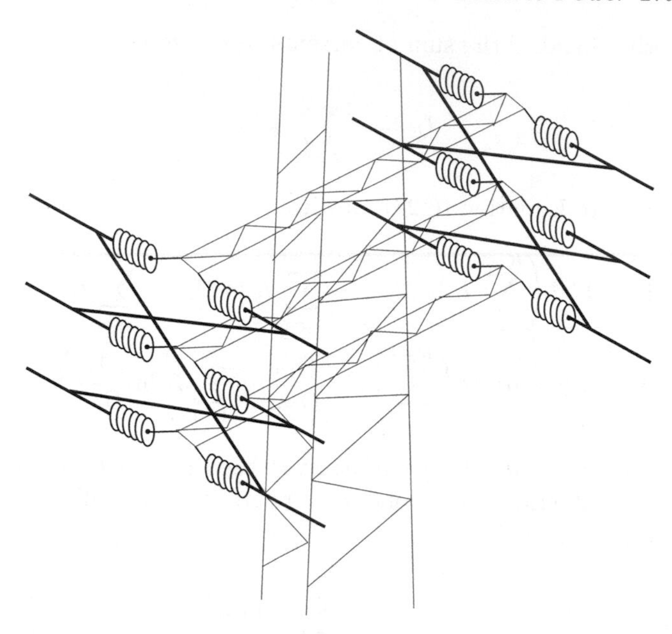

FIGURE 6.4
Transposition tower

- The conductors are identical (r' is equal for the three phases)

Let's consider only phase "a" of the system: in three-phase balanced systems usually analysis is reduced and represented as a single-phase-per-unit system,

$$\lambda_a = 2 \times 10^{-7} l \left(I_a \ln \frac{1}{D_{aa}} + I_b \ln \frac{1}{D_{ab}} + I_c \ln \frac{1}{D_{ac}} \right). \tag{6.36}$$

If distances between conductors are not equal, inductances per phase are different and as this is not convenient for a power system, a technique called transposition is implemented. Transposition consists of interchanging positions of the phases in order to maintain an average constant distance between them.

In Figure 6.4 with a scheme of a transposition tower is shown, a double-circuit line, which conductors of the same phase usually place opposite each other, on the left side. Previous to the transposition the phases from top to bottom could be A-B-C, while on the right, C'-B'-A'. Then, phases of the left side are transposed to C-A-B, while the right side changes to B'-A'-C'. The opposite phase issue is made in order to reduce the inductance because the fluxes counteract each other.

Assume a complete transposed line. From Equation (6.36) let's obtain the

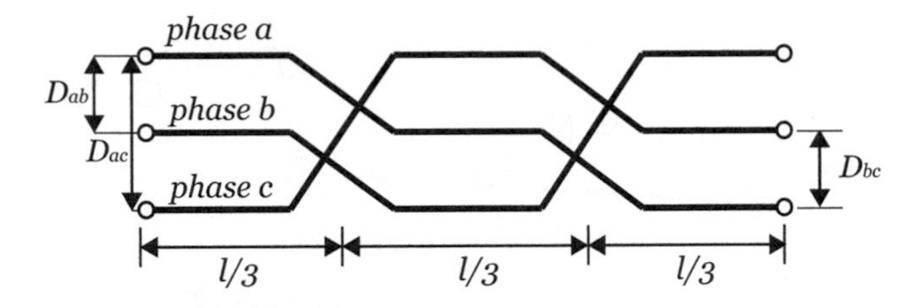

FIGURE 6.5
Transposed three-phase line

average linkage flux taking phase a of Figure 6.5,

$$\lambda_{a1} = 2\times 10^{-7} l \left(I_a \ln \frac{1}{r'_a} + I_b \ln \frac{1}{D_{ab}} + I_c \ln \frac{1}{D_{ac}} \right), \tag{6.37}$$

$$\lambda_{a2} = 2\times 10^{-7} l \left(I_a \ln \frac{1}{r'_a} + I_b \ln \frac{1}{D_{bc}} + I_c \ln \frac{1}{D_{ab}} \right), \tag{6.38}$$

$$\lambda_{a3} = 2\times 10^{-7} l \left(I_a \ln \frac{1}{r'_a} + I_b \ln \frac{1}{D_{ac}} + I_c \ln \frac{1}{D_{bc}} \right). \tag{6.39}$$

Obtaining the average:

$$\lambda_{\text{average}} = \frac{\lambda_{a1} + \lambda_{a2} + \lambda_{a3}}{3} =$$

$$\frac{2\times 10^{-7} l}{3} \left(3I_a \ln \frac{1}{r'_a} + I_b \ln \frac{1}{D_{ab} D_{bc} D_{ac}} + I_c \ln \frac{1}{D_{ab} D_{bc} D_{ac}} \right). \tag{6.40}$$

Total current is zero, so $-I_a = I_b + I_c$,

$$\lambda_{\text{average}} = \frac{2\times 10^{-7} l}{3} \left(3I_a \ln \frac{1}{r'_a} - I_a \ln \frac{1}{D_{ab} D_{bc} D_{ac}} \right),$$

$$\lambda_{\text{average}} = 2\times 10^{-7} l \left(I_a \ln \frac{\sqrt[3]{D_{ab} D_{bc} D_{ac}}}{r'_a} \right). \tag{6.41}$$

As all the phases have the same average distance, inductance of any of them will be

$$L = 2\times 10^{-7} l \left(\ln \frac{\sqrt[3]{D_{ab} D_{bc} D_{ac}}}{r'_a} \right). \tag{6.42}$$

6.2.3 Geometric mean radius and geometric mean distance

Up to now, we have considered conductors as solid cylinders. Nevertheless, real transmission lines are typically formed by a bunch of conductors distributed in strands so that inductance decreases. Hence the concepts of geometric mean distance and geometric mean radius are very useful for simplifying a power

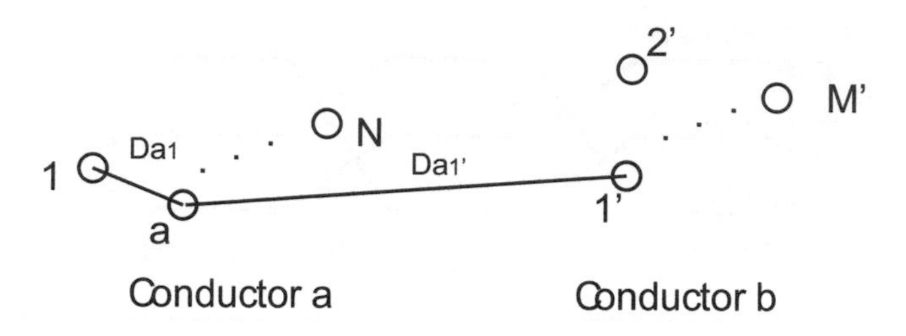

FIGURE 6.6
Single-phase system with composite conductors

transmission line into an equivalent single-conductor-per-phase system. For standard stranded conductors, manufacturers provide the geometric mean radius, which is equivalent to r' of a solid conductor.

Consider a single-phase system conformed by N identical sub-conductors for conductor a and M identical sub-conductors for conductor b.

Assumptions:

- Sub-conductors are identical and carry the same amount of current

Now, if we calculate the total flux ϕ_a of sub-conductor "a" shown in Figure 6.6, we obtain

$$\phi_x = 2 \times 10^{-7} l \left[\frac{1}{N} \sum_{n=1}^{N} \ln \frac{1}{D_{xn}} - \frac{1}{M} \sum_{m=1}^{M} \ln \frac{1}{D_{xm}} \right]. \tag{6.43}$$

In this case, the linking flux λ_a is not equal to the total flux, because only the fraction of current $1/N$ is linked by the magnetic flux, so the actual linkage flux is

$$\lambda_x = 2 \times 10^{-7} l \frac{1}{N} \left[\frac{1}{N} \sum_{n=1}^{N} \ln \frac{1}{D_{xn}} - \frac{1}{M} \sum_{m=1}^{M} \ln \frac{1}{D_{xm}} \right] =$$
$$2 \times 10^{-7} lI \left[\frac{1}{N^2} \sum_{n=1}^{N} \ln \frac{1}{D_{xn}} - \frac{1}{MN} \sum_{m=1}^{M} \ln \frac{1}{D_{xm}} \right]. \tag{6.44}$$

Finally, by logarithms laws we obtain:

$$\lambda_x = 2 \times 10^{-7} lI \ln \frac{\sqrt[MN]{\prod_{x=1}^{N} \prod_{m=1}^{M} D_{xm}}}{\sqrt[N^2]{\prod_{x=1}^{N} \prod_{n=1}^{N} D_{xn}}}. \tag{6.45}$$

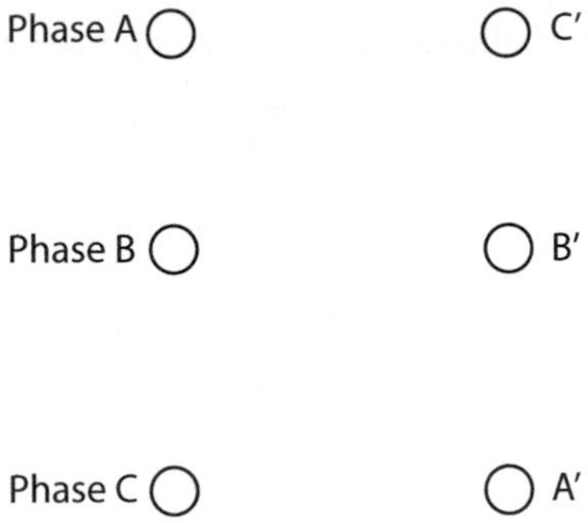

Phase A C′

Phase B B′

Phase C A′

FIGURE 6.7
Double circuit line

If

$$\text{GMD} = \sqrt[MN]{\prod_{x=1}^{N}\prod_{m=1}^{M} D_{xm}}, \tag{6.46}$$

$$\text{GMR} = \sqrt[N^2]{\prod_{x=1}^{N}\prod_{n=1}^{N} D_{xn}}, \tag{6.47}$$

then

$$L_x = 2 \times 10^{-7} l \ln \frac{\text{GMD}}{\text{GMR}}. \tag{6.48}$$

The same procedure can be made for a three-phase system. From the previous section we know the inductance of a three-phase system. Replacing r'_a by GMR in Equation (6.42) for a complete solution, we get:

$$L = 2 \times 10^{-7} l \left(\ln \frac{\sqrt[3]{D_{ab} D_{bc} D_{ac}}}{\text{GMR}} \right). \tag{6.49}$$

6.2.4 Inductance of three-phase double circuit lines

As shown in Figure 6.4, some transmission lines are formed by two identical three-phase circuits, in parallel; however, some lines are placed opposite each other as shown in Figure 6.7. In this configuration the distance between phases is not obvious, so we need to calculate it. Again consider that lines are completely transposed so the average inductance is balanced.

It is convenient to obtain equivalent distances between phases as follows:

$$D_{AB} = \sqrt[4]{D_{ab} D_{ab'} D_{a'b} D_{a'b'}}, \tag{6.50}$$

$$D_{BC} = \sqrt[4]{D_{bc} D_{bc'} D_{b'c} D_{b'c'}}, \tag{6.51}$$

$$D_{AC} = \sqrt[4]{D_{ac} D_{ac'} D_{a'c} D_{a'c'}}. \tag{6.52}$$

Then, the equivalent geometric mean distance per phase is

$$\text{GMD} = \sqrt[3]{D_{AB}D_{BC}D_{AC}}. \tag{6.53}$$

Similarly, we can obtain the equivalent radius of each phase for the GMR:

$$\text{GMR}_A = \sqrt[4]{(r'_a D_{aa'})^2} = \sqrt{r'_a D_a a'}, \tag{6.54}$$

$$\text{GMR}_B = \sqrt{r'_b D_{bb'}}, \tag{6.55}$$

$$\text{GMR}_C = \sqrt{r'_c D_{cc'}}. \tag{6.56}$$

where r'_a is $re^{-\frac{1}{4}}$ for a solid conductor and the geometric mean radius of a stranded conductor.

For a complete transposed line, the average GMR is

$$\text{GMR}_{\text{average}} = \sqrt[3]{\text{GMR}_\text{A}\text{GMR}_\text{B}\text{GMR}_\text{C}}. \tag{6.57}$$

Finally, the inductance per phase to neutral is

$$L = 2 \times 10^{-7} l \left(\ln \frac{\text{GMD}}{\text{GMR}_{\text{average}}}\right) v(t) = V_m \cos{(wt + \theta)}. \tag{6.58}$$

6.2.5 Capacitance of transmission lines

Capacitance of transmission lines comes from the electric field resulting from a charging charge flowing through conductors. An electric field describes a resulting force per unit charge and has units of Newton per Coulomb, which is equivalent to volts per meter. The potential difference between conductors and the relationship between the charge and the potential difference define the capacitance of the conductors. The ground also plays an important role because capacitance also comes from different potential between phases and ground. The units of the capacitance are the Faradays, represented by [F], which is defined as the amount of capacitance present when a coulomb produces a potential difference of one volt [23].

The steps followed for determining the capacitance due to the charge of conductors are:

1. Obtain the electric field strength from Gauss's law:

$$\oiint_S E \cdot dA = \frac{Q}{\epsilon}. \tag{6.59}$$

2. Obtain the voltage between conductors (or ground) from the electric field:

$$V = \int_{D_1}^{D_2} E_x dx. \tag{6.60}$$

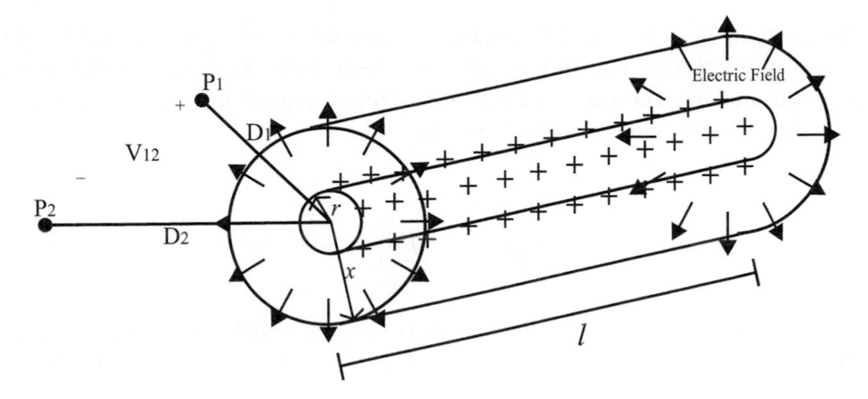

FIGURE 6.8
Solid cylindrical conductor with uniform charge distribution and Gaussian surface of integration

3. Obtain the capacitance:

$$C = \frac{q}{V}. \tag{6.61}$$

Assumptions:

- Conductor is uniformly charged, solid, and cylindrical
- Conductor is very long and end effects are negligible
- Resistivity of the conductor is zero, so there is no electrical field inside it
- Permittivity of the air is $\epsilon_0 = 8.854 \times 10^{-12}$ $[F/m]$

Consider the electric conductor shown in Figure 6.8. The electric charge is distributed uniformly thought the conductor, r is the radius of the cross sectional area of the conductor, and the external circle is the Gaussian surface of integration for obtaining the electric field. Consider also that the conductor has a length l, and the electric field lines exit radially outward from the conductor in all directions.

The first step is to calculate the electric field resulting from the electric charge q of the cylindrical conductor. The Gauss law states that the electric field crossing the closed surface is equal to the electric flux. Electric flux is equal to the total charge divided by the permittivity of the free space,

$$\oiint_S E \cdot dA = \frac{Q}{\epsilon}. \tag{6.62}$$

The electric field inside the conductors equals zero, because we have established that we are analyzing a perfect conductor. The electric field outside can be determined by selecting a Gaussian surface with radius $x > r$ and length equal to the conductor length l. As all the electric field lines are parallel to the area of the surface, the cosine of the dot product equals one and the integral of the electric field equation (6.62) yields

$$E_x 2\pi x l = \frac{q}{\epsilon_0},$$
$$E_x = \frac{q}{\epsilon_0 2\pi x l}. \tag{6.63}$$

Once the electric field has been calculated, the potential difference between concentric cylinders at distances D_1 and D_2 from the center of the conductor is

$$V_{12} = \int_{D_1}^{D_2} E_x dx = \int_{D_1}^{D_2} \frac{q}{\epsilon_0 2\pi x l} dx = \frac{q}{2\pi\epsilon_0 l} \ln \frac{D_2}{D_1}. \tag{6.64}$$

Finally, the capacitance of the conductor between point 1 and 2 is

$$C_{12} = \frac{q}{V_{12}} = \frac{2\pi\epsilon l}{\ln \frac{D_1}{D_2}} v(t) = V_m \cos(wt + \theta). \tag{6.65}$$

6.2.5.1 Capacitance of a single-phase line

Let's start with a single, phase-single, conductor power system. The charge of positive conductors is uniform q and the negative conductors, by conservation of charge, will have an equal but negative charge $-q$. From the previous section we know how to compute the voltage:

$$V_{xy} = \frac{q}{2\pi\epsilon_o l}\left[\ln \frac{D_{yx}}{D_{xx}} - \ln \frac{D_{yy}}{D_{xy}}\right] = \frac{q}{2\pi\epsilon_0 l}\left[\ln \frac{D_{yx} D_{xy}}{D_{xx} D_{yy}}\right]. \tag{6.66}$$

As $D_{xy} = D_{yx}$ and $D_{xx} = r_x$, $D_{yy} = r_y$, Equation (6.66) yields:

$$V_{xy} = \frac{q}{2\pi\epsilon_o l}\left[\ln \frac{D_{xy}^2}{r_x r_y}\right] = \frac{q}{\pi\epsilon_0 l}\left[\ln \frac{D_{xy}}{\sqrt{r_x r_y}}\right]. \tag{6.67}$$

The capacitance of the single line is:

$$C_{xy} = \frac{\pi\epsilon_0 l}{\ln \frac{D}{\sqrt{r_x r_y}}}. \tag{6.68}$$

6.2.5.2 Capacitance of a three-phase line

Now consider a three-phase system neglecting the effect of neutral conductors and earth. Again, by conservation of charge, consider $q_a + q_b + q_c = 0$. The voltage from phase "a" to a point in space P is

$$V_{aP} = \frac{1}{2\pi\epsilon_0 l}\left[q_a \ln \frac{D_{aP}}{r_a} + q_b \ln \frac{D_{bP}}{D_{ab}} + q_c \ln \frac{D_{cP}}{D_{ac}}\right]. \tag{6.69}$$

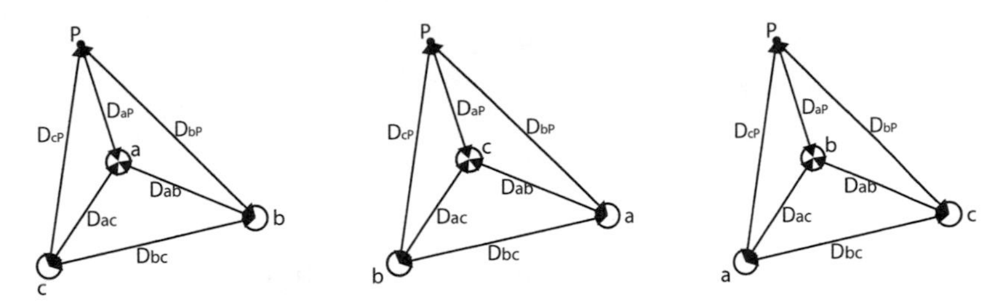

FIGURE 6.9
Unsymmetrically spaced, complete transposed 3-phase transmission line

Considering a complete transposed line as in Figure 6.9 to calculate the average voltage between conductor "a" and point "P" due to all the charges,

$$V_{aP_1} = \frac{1}{2\pi\epsilon_0 l}\left[q_a \ln\frac{D_{aP}}{r_a} + q_b \ln\frac{D_{bP}}{D_{ab}} + q_c \ln\frac{D_{cP}}{D_{ac}}\right],$$
$$V_{aP_2} = \frac{1}{2\pi\epsilon_0 l}\left[q_a \ln\frac{D_{bP}}{r_a} + q_b \ln\frac{D_{cP}}{D_{bc}} + q_c \ln\frac{D_{aP}}{D_{ab}}\right],$$
$$V_{aP_3} = \frac{1}{2\pi\epsilon_0 l}\left[q_a \ln\frac{D_{cP}}{r_a} + q_b \ln\frac{D_{aP}}{D_{ac}} + q_c \ln\frac{D_{bP}}{D_{bc}}\right].$$

So, the average is

$$V_{aP\text{ average}} = \frac{V_{ab1}V_{ab2}V_{ab3}}{3} =$$
$$\frac{1}{6\pi\epsilon_0 l}\left[q_a \ln\frac{D_{aP}D_{bP}D_{cP}}{r_a^3} + q_b \ln\frac{D_{aP}D_{bP}D_{cP}}{D_{ab}D_{bc}D_{ac}} + q_c \ln\frac{D_{aP}D_{bP}D_{cP}}{D_{ab}D_{bc}D_{ac}}\right]. \tag{6.70}$$

Substituting $-q_a = q_b + q_c$ in Equation (6.70),

$$V_{aP\text{ average}} = \frac{1}{6\pi\epsilon_0 l}\left[q_a \ln\frac{D_{aP}D_{bP}D_{cP}}{r_a^3} - q_a \ln\frac{D_{aP}D_{bP}D_{cP}}{D_{ab}D_{bc}D_{ac}}\right]$$
$$= \frac{q_a}{6\pi\epsilon_0 l}\left[\ln\frac{D_{aP}D_{bP}D_{cP}}{r_a^3}\right],$$

$$V_{aP\text{ average}} = \frac{q_a}{2\pi\epsilon_0 l}\ln\frac{\sqrt[3]{D_{ab}D_{bc}D_{ac}}}{r_a} = V_a. \tag{6.71}$$

The capacitance to neutral is

$$C_{an} = \frac{q_a}{V_a} = \frac{2\pi\epsilon_0 l}{\ln\frac{\sqrt[3]{D_{ab}D_{bc}D_{ac}}}{r_a}}. \tag{6.72}$$

Similarly to the inductance section, if we have bundled conductors with no

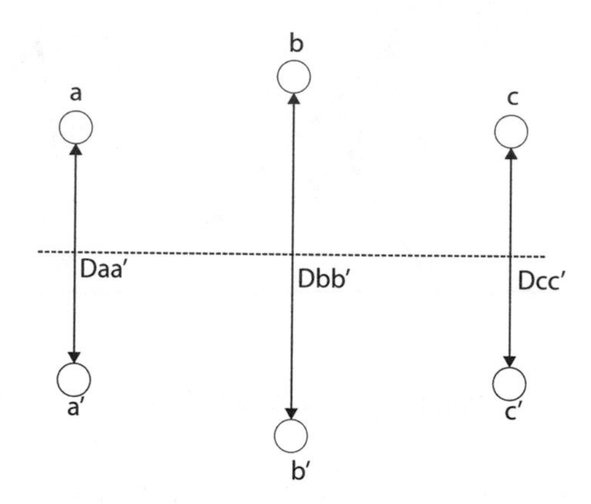

FIGURE 6.10
Image method for a three phase transmission line

uniform distribution, the geometric mean distance and geometric mean radius remain valid.

So, for bundled conductors with well-defined distance between phases, the equation

$$C_{an} = \frac{q_a}{V_a} = \frac{2\pi\epsilon_0 l}{\ln \frac{\sqrt[3]{D_{ab}D_{bc}D_{ac}}}{\text{GMR}}} \tag{6.73}$$

is a good approximation, while the exact model used for not-well-defined distance between phases, as that of double-line transmission lines, is

$$C_{an} = \frac{q_a}{V_a} = \frac{2\pi\epsilon_0 l}{\ln \frac{\text{GMD}}{\text{GMR}_{\text{average}}}}. \tag{6.74}$$

Equations (6.46), (6.47) and from (6.50) to (6.57) of the inductance section are valid for obtaining the GMD and GMR in Equation (6.74). The only difference is that the distance of a conductor with respect to itself is the radius ($D_{xx} = r_x$) instead of $r'_x = re^{(-1/4)}$. This comes from the assumption that the electric field within the conductor is zero, while for the magnetic field this is not valid because all the conductors have a magnetic permeability.

6.2.5.3 Effect of earth on the capacitance of conductors

The electric flux lines of a conductor have an end placed on an equipotential conductor, such that, according to the images method (see Figure 6.10) which models the earth plane as a perfect conductor, a charge at certain height above the plane will induce an opposite charge inside the earth at a distance of twice the height from the original conductor.

Again, consider a point P far from the system for calculating the voltage from phase "a" to this point,

$$V_{aP} = \frac{1}{2\pi\epsilon_0 l}\left[q_a\left(\ln\frac{D_{aP}}{r_a} - \ln\frac{D_{a'P}}{D_{aa'}}\right) + q_b\left(\ln\frac{D_{bP}}{D_{ab}} - \ln\frac{D_{b'P}}{D_{ab'}}\right) + q_c\left(\ln\frac{D_{cP}}{D_{ac}} - \ln\frac{D_{c'P}}{D_{ac'}}\right)\right]. \quad (6.75)$$

Considering an infinite distance of point P, it is valid to assume:

$$D_{a'P} = D_{aP},\ D_{b'P} = D_{bP},\ D_{c'P} = D_{cP}.$$

And equation (6.75) yields:

$$V_{aP} = \frac{1}{2\pi\epsilon_0 l}\left[q_a\left(\ln\frac{D_{aa'}}{r_a}\right) + q_b\left(\ln\frac{D_{ab'}}{D_{ab}}\right) + q_c\left(\ln\frac{D_{ac'}}{D_{ac}}\right)\right]. \quad (6.76)$$

For a complete transposed line:

$$V_{aP_1} = \frac{1}{2\pi\epsilon_0 l}\left[q_a\left(\ln\frac{D_{aa'}}{r_a}\right) + q_b\left(\ln\frac{D_{ab'}}{D_{ab}}\right) + q_c\left(\ln\frac{D_{ac'}}{D_{ac}}\right)\right],$$

$$V_{aP_2} = \frac{1}{2\pi\epsilon_0 l}\left[q_a\left(\ln\frac{D_{bb'}}{r_a}\right) + q_b\left(\ln\frac{D_{bc'}}{D_{bc}}\right) + q_c\left(\ln\frac{D_{ba'}}{D_{ba}}\right)\right],$$

$$V_{aP_3} = \frac{1}{2\pi\epsilon_0 l}\left[q_a\left(\ln\frac{D_{cc'}}{r_a}\right) + q_b\left(\ln\frac{D_{ca'}}{D_{ca}}\right) + q_c\left(\ln\frac{D_{cb'}}{D_{cb}}\right)\right].$$

The average voltage is

$$V_a = \frac{1}{6\pi\epsilon_0 l}\left[q_a \ln\frac{D_{aa'}D_{bb'}D_{cc'}}{r_a^3} + q_b \ln\frac{D_{ab'}D_{bc'}D_{ca'}}{D_{ab}D_{bc}D_{ca}} + q_c \ln\frac{D_{ac'}D_{ba'}D_{cb'}}{D_{ab}D_{bc}D_{ca}}\right]. \quad (6.77)$$

Since $D_{ab'} = D_{ba'};\ D_{bc'} = D_{cb'};\ D_{ca'} = D_{ac'}$,

$$V_a = \frac{1}{6\pi\epsilon_0 l}\left[q_a \ln\frac{D_{aa'}D_{bb'}D_{cc'}}{r_a^3} - q_a \ln\frac{D_{ab'}D_{bc'}D_{ca'}}{D_{ab}D_{bc}D_{ca}}\right],$$

$$V_a = \frac{q_a}{6\pi\epsilon_0 l}\ln\left[\frac{D_{ab}D_{bc}D_{ca}}{r_a^3}\cdot\frac{D_{aa'}D_{bb'}D_{cc'}}{D_{ab'}D_{bc'}D_{ca'}}\right]. \quad (6.78)$$

Finally, the capacitance is equal to:

$$C = \frac{q_a}{V_a} = \frac{2\pi\epsilon_0 l}{\ln\left[\frac{\sqrt[3]{D_{ab}D_{bc}D_{ca}}}{r_a}\cdot\sqrt[3]{\frac{D_{aa'}D_{bb'}D_{cc'}}{D_{ab'}D_{bc'}D_{ca'}}}\right]}. \quad (6.79)$$

6.2.6 Exercises

Calculate the following transmission line parameters for the transmission line modules.

Consider a 360 $[km]$ transmission formed by bundles of conductors. The bundles have the following parameters:

Resistance $[\Omega/km]$	Equivalent Diameter $[m]$	Geometric Mean Radius $[m]$
0.03611	0.1160	0.2769

The transmission line is completely transposed and has the configuration shown in Figure 6.11

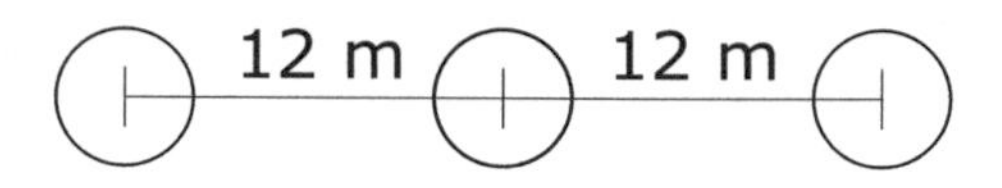

FIGURE 6.11
Length of transmission lines

Obtain the resistance in $[\Omega]$, the inductance in $[mH]$, and the capacitance in $[\mu F]$.

Obtain the π equivalent circuit placing half of the total capacitance at each end of the transmission line and the total impedance and resistance in the middle.

Insert all bridging plugins to the capacitance of the transmission line model, and with an impedances meter, obtain the capacitance of one end of the transmission line with respect to neutral and compare it with the calculated value.

Also obtain the inductance of one line and compare your result with the inductance calculated.

Calculate the transmission line parameters for the second module with the configuration shown in Figure 6.12 and the following parameters:

Resistance $[\Omega/km]$	Equivalent Diameter $[m]$	Geometric Mean Radius $[m]$
0.033	0.1148	0.1960

Do the same procedure as that employed for calculating the parameters of the first module.

Connect the equipment as shown in the connection Diagram 6.1.

With an LCR meter, measure three values of capacitance and inductance from capacitive loads and inductive loads, respectively, and find their impedance (assume a 60 $[Hz]$ frequency operation).

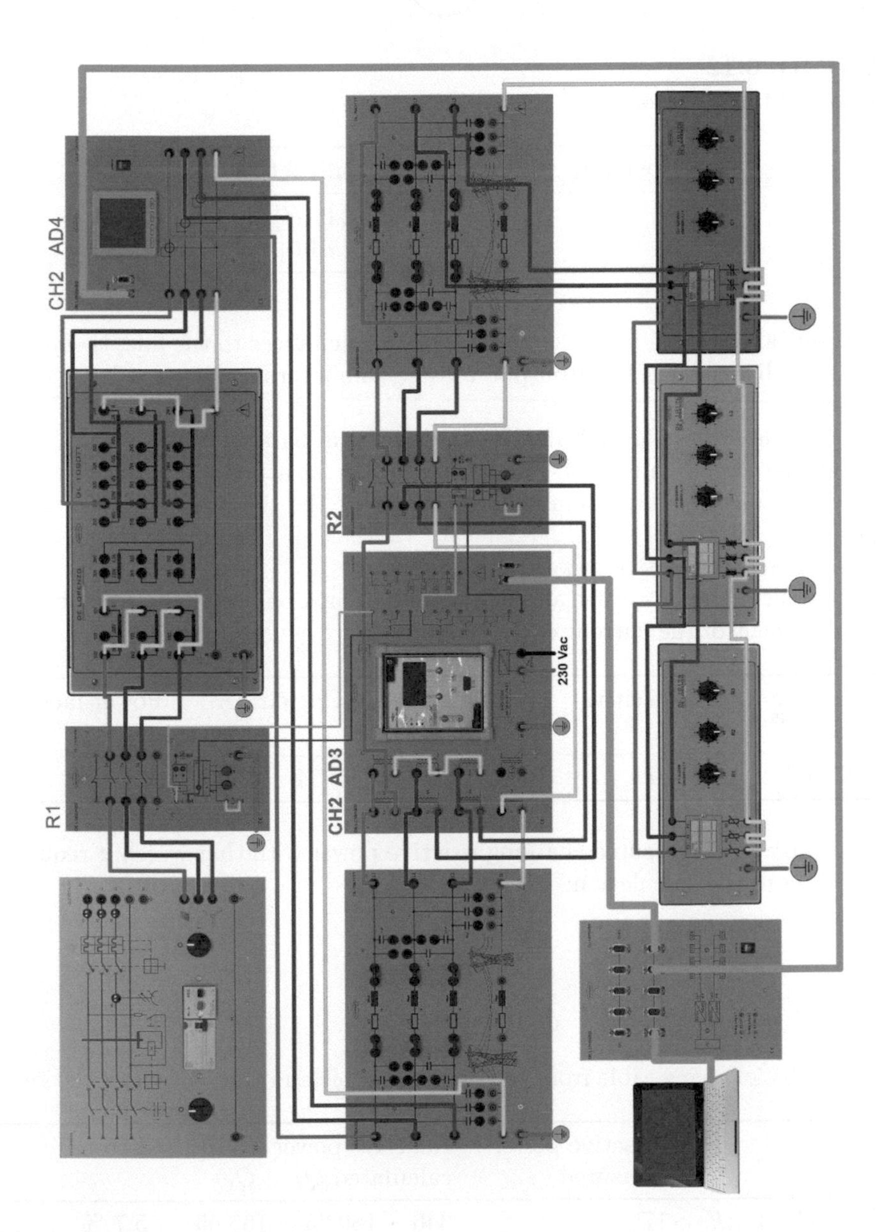

Diagram 6.1
Connections for different loads

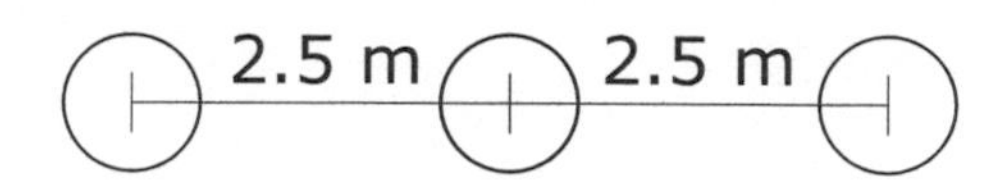

FIGURE 6.12
Length of transmission lines

Position	Capacitance	Inductance
1	1.5 $[\mu F]$	2.6 $[H]$
2	2.5 $[\mu F]$	1.814 $[H]$

Insert all bridging plugs connecting the capacitances to the line, then close the Relay R1 and check the output voltage coming from the secondary transformer.

Close the Relay R2 and monitor the electrical parameters.

Set the load R and L, C to zero and take measurements of the power consumed with the maximum demand meter by the shunt capacitance of the line.

Measure the reactive power consumed.

Increment the capacitive and inductive load in a balanced way and observe the increment of the current consumed and the reactive power.

Inductive load	Capacitive load	Reactive power [VAr]	Current [A]	Power factor
1.27 $[H]$	3 $[\mu F]$	175	0.34	0.7

Compare the measured charging reactive power with that which it requires according to the calculation:

$$Q_C = \omega C_{an} V_{LL}^2,$$
$$Q_L = \omega L_L I_L^2,$$
$$Q_{\text{Load}} = Q_L - Q_C.$$

Fill the following table from the previous table and the above calculations.

Load	Reactive power measured	Reactive power calculated $Q_L - Q_C$	Error %
1.27 $[H]$, 3 $[\mu F]$	175	$346 - 180.95 = 165.05$	5.7 %

Recommended equipment and connections:

- Resistive load
- Inductive load

- Capacitive load
- Overhead line model — Long
- Overhead line model — Medium
- Power circuit breaker
- Three-phase power meter
- Three-phase transformer
- Three-phase supply unit
- Feeder manager relay

6.3 Transmission line mathematical model

In the previous section we obtained the parameters for transmission lines. Nevertheless, it is important to know that every single tranche of the transmission line can be modeled as a series impedance with its correspondent shunt capacitance. In this section, the equivalent ABCD model, which relates the voltage and current of one side of the transmission line with the voltage and current of the other side, is presented. The ABCD model is convenient for presenting the exact model of a transmission line because the ABCD matrix can relate non-linear relationships between inputs and outputs.

Some simplified models just ignore the differential equations that represent the exact model of the transmission line and approximate the model as a T circuit, assuming that all the capacitance is located at half the transmission line distance, or as a pi circuit that takes half of the total capacitance at each end of the transmission line. This is valid for medium length (more than 80 $[km]$ and less than 250 $[km]$), while for short length lines (less than 80 $[km]$), according to [39] even the shunt capacitance can be neglected.

For medium and long transmission lines, the π circuit is widely used to analyze power flow because many computer programs are designed for a circuit representation of components, and the ABCD parameter of the network is a matrix operation. In this section, the exact ABCD parameter is presented first (see Figure 6.14), and then the equivalent pi circuit is presented (see Figure 6.15). This last form will be used in the following section for power flow analysis.

6.3.1 Exact ABCD model of a transmission line

Consider a transmission line with the following series impedance conformed by a resistance and an inductance, and a shunt admittance formed by a ca-

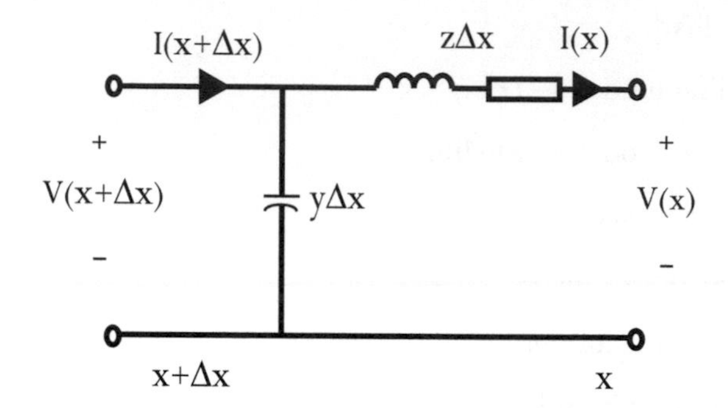

FIGURE 6.13
Differential section of a transmission line

pacitance:

$$z = R + j\omega L \; [\Omega/m],$$
$$y = j\omega C \; [S/m].$$

The voltage on the left side of Figure 6.13 is

$$V(x + \Delta x) = V(x) + (z\Delta x)I(x). \tag{6.80}$$

Rearranging:

$$zI(x) = \frac{V(x + \Delta x) - V(x)}{\Delta x}. \tag{6.81}$$

If $\Delta x \to 0$,

$$zI(x) = \frac{dv(x)}{dx}. \tag{6.82}$$

On the other hand, for the current, the KCL equation gives

$$I(x + \Delta x) = I(x) + (y\Delta x) \times V(x + \Delta x). \tag{6.83}$$

Rearranging and assuming $\Delta x \to 0$,

$$yV(x) = \frac{I(x + \Delta x) - V(x)}{\Delta x} = \frac{dI(x)}{dx}. \tag{6.84}$$

Differentiating and substituting the current derivative,

$$yV(x) = \frac{1}{z}\frac{d^2V(x)}{dx^2}, \tag{6.85}$$

which can be expressed as

$$\frac{d^2V(x)}{dx^2} - yzV(x) = 0. \tag{6.86}$$

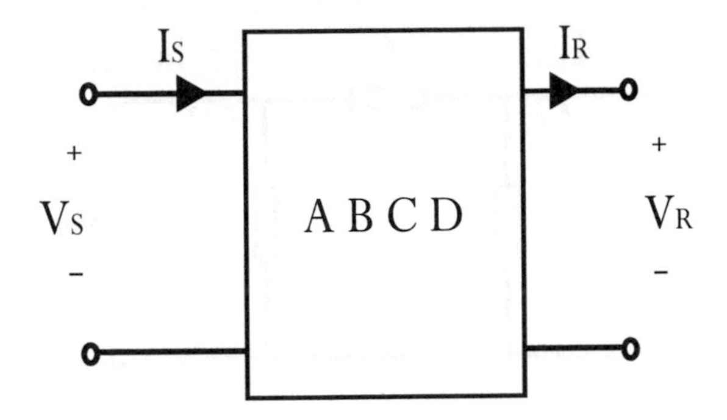

FIGURE 6.14
ABCD representation of a transmission line

If the differential equation is solved, the solution has the form

$$V(x) = A_1 e^{\gamma x} + A_2 e^{-\gamma x}, \tag{6.87}$$

where $\gamma = \sqrt{yz}$, also called propagation constant. On the other hand, substituting the solution (6.87) into (6.82),

$$I(x) = \frac{A_1 e^{\gamma x} + A_2 e^{-\gamma x}}{Z_C}, \tag{6.88}$$

where $Z_C = \sqrt{\frac{z}{y}}$ is called the characteristic impedance. Now, solve A_1 and A_2 with the following boundary conditions: $x = 0 \to V(0) = V_R = A_1 + A_2$, $I(0) = I_R = \frac{A_1 - A_2}{Z_C}$.

Substituting A_1 and A_2 leads to

$$V(x) = \frac{e^{\gamma x} + e^{-\gamma x}}{2} V_R + Z_C \left(\frac{e^{\gamma x} - e^{-\gamma x}}{2} \right) I_R, \tag{6.89}$$

$$I(x) = \frac{1}{Z_C} \left(\frac{e^{\gamma x} - e^{-\gamma x}}{2} \right) V_R + \left(\frac{e^{\gamma x} + e^{-\gamma x}}{2} \right) I_R. \tag{6.90}$$

Representing in matrix form, recognizing the hyperbolic function cosh and sinh, and considering a length l,

$$\begin{bmatrix} V_S \\ I_S \end{bmatrix} = \begin{bmatrix} A & B \\ C & D \end{bmatrix} \begin{bmatrix} V_R \\ I_R \end{bmatrix} = \begin{bmatrix} \cosh \gamma l & Z_C \sinh \gamma l \\ \frac{1}{Z_C} \sinh \gamma l & \cosh \gamma l \end{bmatrix} \begin{bmatrix} V_R \\ I_R \end{bmatrix}, \tag{6.91}$$

where

$$\gamma = \sqrt{yz}, \; Z_c = \sqrt{\frac{z}{y}}. \tag{6.92}$$

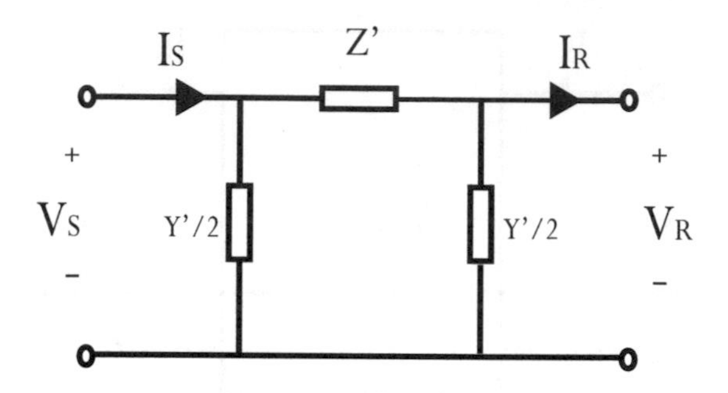

FIGURE 6.15
Transmission line equivalent pi circuit

6.3.2 Equivalent π circuit

Most power flow programs use the equivalent π circuit because they were created for representing power system elements as impedances connected together. So, an equivalent circuit was created identical in structure to a nominal π circuit (where the impedance is all in the center of the model and half of the admittance is situated at each end of the transmission line), but with the following corrections for obtaining the exact transmission line model. In [78, 47] the mathematical formulation of the π circuit can be found,

$$Z' = Z_C \sinh(\gamma l) = ZF_1 = Z\frac{\sinh(\gamma l)}{\gamma l}, \tag{6.93}$$

$$\frac{Y'}{2} = \frac{\tanh(\gamma l/2)}{Z_C} = \frac{Y}{2}F_2 = \frac{Y}{2}\frac{\tanh(\gamma l/2)}{\gamma l/2}. \tag{6.94}$$

Here we are just showing the resulting equations.

6.3.3 Exercises

Assemble Diagram 6.2 using the large transmission line model and connect all the plugins in the transmission line model. Set the primary side of the three-phase transformer in Δ connection 380 $[V]$ and set the secondary-side to Y UN – 5%.

Take measure of the voltage V, current I, and power consumed (both active P and reactive Q) of the entire power system (including transmission lines) and the voltage and current after the transmission lines. Fill the following table, where 1 is the input measure of the transmission line and 2 is the measure at the output of the transmission line.

Observe the increase in reactive power due to the increment of current flowing through the transmission line's inductor. Also, observe the voltage drop due to the transmission line impedance and how it increments with the current.

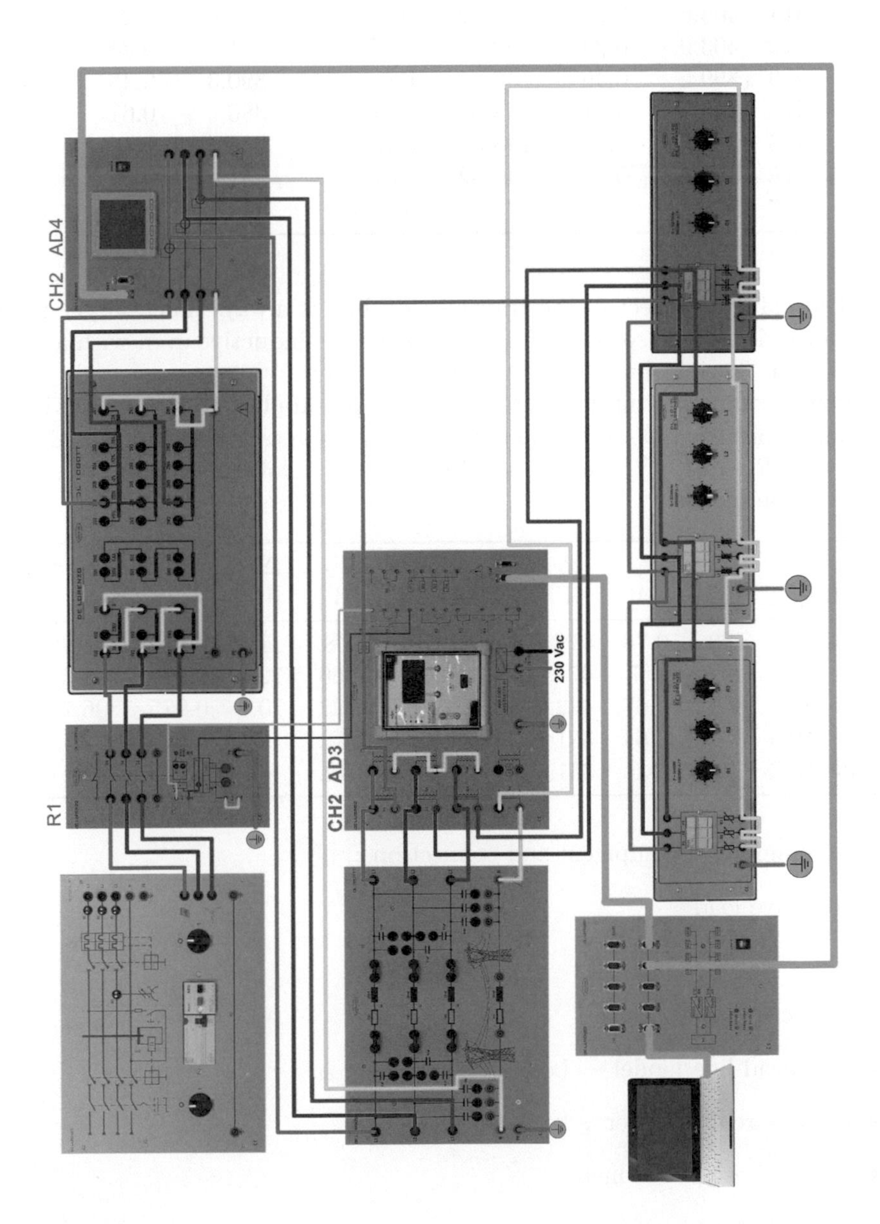

Diagram 6.2
Connections for matched-load practice

R	V1 [V]	I1 [A]	P1 [W]	Q1 [VAr]	V2 [V]	I2 [A]
R1	405.6	0.21	148.52	9	404.6	0.21
R2	403.9	0.29	148.52	7.32	401	0.29
R3	399.5	0.46	148.52	62.8	390.3	0.45
R4	395.5	0.61	148.52	135.58	380	0.61
R5	391.5	0.78	148.52	237.58	366.4	0.75
R6	388.2	0.97	148.52	375	351.4	0.96
R7	388	1.05	148.52	452.8	343.6	1.03

The inductive effect of the transmission line is always much larger than the capacitive effect, so the reactive power will be inductive even with a very low current flowing through the transmission line.

Power companies must maintain the voltage supplied to the users within a range typically ±5% of the nominal voltage, maintain the load impedance constant ($R_4 = 312$ [Ω]), and obtain the power and voltage drop of the transmission line model.

Supply voltage	V_1 [V]	I_1 [A]	P_1 [W]	V_2 [V]	I_2 [A]	P_2 [W]	ΔP [W]	η	**Voltage drop [%]**
UN+5%	409.4	0.73	519.5	398.2	0.73	508.1	11.4	0.978	97.264
UN	392.6	0.7	478.2	382.3	0.7	466	12.2	0.974	97.376
UN–5%	376.3	0.67	432.9	364	0.67	422	10.9	0.975	96.731
UN–10%	354.9	0.64	391.2	345.8	0.64	381.7	9.5	0.976	97.436
UN–15%	336.4	0.6	351.5	327.7	0.6	342.1	9.4	0.973	97.414

Recommended equipment and connections:

- Resistive load
- Inductive load
- Capacitive load
- Overhead line model — Long
- Power circuit breaker
- Three-phase power meter
- Three-phase transformer
- Three-phase supply unit
- Feeder manager relay

6.4 Power flow analysis

In power system analysis, it is very common to visualize the system as flow of power. The objective is to compute voltage and phase at each bus. One bus can be seen as a point of connection between elements (a node). As a result of the calculation the real and reactive power can be obtained, and you can visualize the total load flowing through a conductor. The objective of this calculation is to operate the system at its maximum efficiency point and maintain the voltage range of all the buses at an ideal range. The power flow is analyzed from the single-line diagram of a power system; the input data is the bus, transmission line, and transformers data.

For buses with loads and/or generators connected, the convention is that the power generated is positive while the power consumed is negative,

$$P_k = P_{G_k} - P_{L_k}, \tag{6.95}$$

$$Q_k = Q_{G_k} - Q_{L_k}, \tag{6.96}$$

where k is the number of the bus.

Buses can be classified into three types:

- Swing bus: Is set as the reference, and it is convenient to establish a value of $1.0\angle 0°$ per unit. The unknown is the power flow P_1 and Q_1.
- Load bus: The input data is P_k and Q_k, and load buses are the most frequent buses on power system. The unknown is the voltage of the bus and its angle $V_k\angle\delta_k^\circ$.
- Voltage controlled bus: These buses represent generator, switched shunt capacitor, or voltage regulated buses. The input data is P_k and V_k and the unknowns are the Q_k and δ_k.

Transmission lines and electric machines parameters are represented as admittances. An admittance matrix is calculated for convenience as

$$Y_{\text{bus}} = \begin{bmatrix} Y_{11} & \cdots & Y_{1n} \\ \vdots & \ddots & \vdots \\ Y_{n1} & \cdots & Y_{nn} \end{bmatrix},$$

where the diagonal elements Y_{kk} are the sum of admittances connected to the bus k. The off-diagonal elements Y_{kn}, where $k \neq n$, are the negative of the sum of the admittances connected between bus k and bus n.

The negative signs are present because for convention, it is assumed that all the currents are leaving the bus,

$$I = Y_{\text{bus}} V. \tag{6.97}$$

Using the Y_{bus} matrix, the currents can be calculated from

$$I_k = \sum_{n=1}^{N} Y : knV_n. \tag{6.98}$$

For each bus, the complex power flow of the bus k is

$$S_k = P_k + jQ_k = V_k I_k^*, \tag{6.99}$$

where I_k^* is the conjugate of the current of node k. If we substitute I_k^* with the current of each bus, we get

$$S_k = P_k + jQ_k = V_k \left[\sum_{n=1}^{N} T_{kn} V_n \right]^*. \tag{6.100}$$

By using polar notation, $V_n = V_n e^{j\delta n}$; $Y_{kn} = Y_{kn} e^{j\theta_{kn}}$, and separating real and imaginary parts, we get

$$P_k = V_k \sum_{n=1}^{N} Y_{kn} V_n \cos\left(\delta_k - \delta_n - \theta_{kn}\right), \tag{6.101}$$

$$Q_k = V_k \sum_{n=1}^{N} Y_{kn} V_n \sin\left(\delta_k - \delta_n - \theta_{kn}\right). \tag{6.102}$$

There are several methods for solving the power flow problem. One of the most popular because of its simplicity is the Gauss-Seidel method. This iterative method can be computed as

$$V_k(i+1) = \frac{1}{Y_{kk}} \left[\frac{P_k - jQ_k}{V_k^*(i)} - \sum_{n=1}^{k-1} Y_{kn} V_n(i+1) - \sum_{n=k+1}^{k-1} Y_{kn} V_n(i) \right], \tag{6.103}$$

$$P_k = V_k(i) \sum_{n=1}^{N} Y_{kn} V_n(i) \cos\left(\delta_k(i) - \delta_n(i) - \theta_{kn}\right), \tag{6.104}$$

$$Q_k = V_k(i) \sum_{n=1}^{N} Y_{kn} V_n(i) \sin\left(\delta_k(i) - \delta_n(i) - \theta_{kn}\right). \tag{6.105}$$

6.4.1 Exercises

This set of exercises requires a connection like the one shown in Diagram 6.3. Obtain the active and reactive power consumed by an RL load. Enter the measured values into the following table:

$L_4 = 1.27\ H$	$V_1\ [V]$	$I_1\ [A]$	$P_1\ [W]$	Q_1 [VAr]	$V_2\ [V]$	$I_2\ [A]$	$\cos\phi_2$
R_1	420	0.57	172.8	374.2	389.3	0.6	0.4
R_2	418.2	0.6	228.1	377.6	386	0.63	0.51
R_3	413.8	0.72	356	372	378.5	0.75	0.7
R_4	408.7	0.87	481	380	371.3	0.89	0.81
$L_5 = 0.9\ H$	$V_1\ [V]$	$I_1\ [A]$	$P_1\ [W]$	Q_1 **[VAr]**	$V_2\ [V]$	$I_2\ [A]$	$\cos\phi_2$
R_1	418.6	0.74	174.7	505.4	378.5	0.77	0.31
R_2	418.1	0.76	226.9	501.7	376.8	0.8	0.41
R_3	412.2	0.85	349	497.5	369.2	0.82	0.59
R_4	408.7	0.97	469	501.2	362.3	0.99	0.71
$L_6 = 0.64\ H$	$V_1\ [V]$	$I_1\ [A]$	$P_1\ [W]$	Q_1 **[VAr]**	$V_2\ [V]$	$I_2\ [A]$	$\cos\phi_2$
R_1	418.2	0.97	177.1	681.3	364.2	1.01	0.24
R_2	417	1.02	225.6	674	361.6	1.02	0.31
R_3	412.1	1.01	338.5	663.2	355.8	1.08	0.47
R_4	407.1	1.13	448.7	659	349.1	1.16	0.59

Now remove the connection to the resistive load and repeat the measurement for $L_4 = 1.27\ [H]$.

$V_1\ [V]$	$I_1\ [A]$	$P_1\ [W]$	Q_1 [VAr]	$V_2\ [V]$	$I_2\ [A]$	cos{ϕ_2}
424.6	0.52	31.82	383.2	394	0.56	0.06

The inductive load also consumes active power due to ohmic resistance and iron losses of the inductor.

Now observe the power consumption due to RC loads. Enter the measured

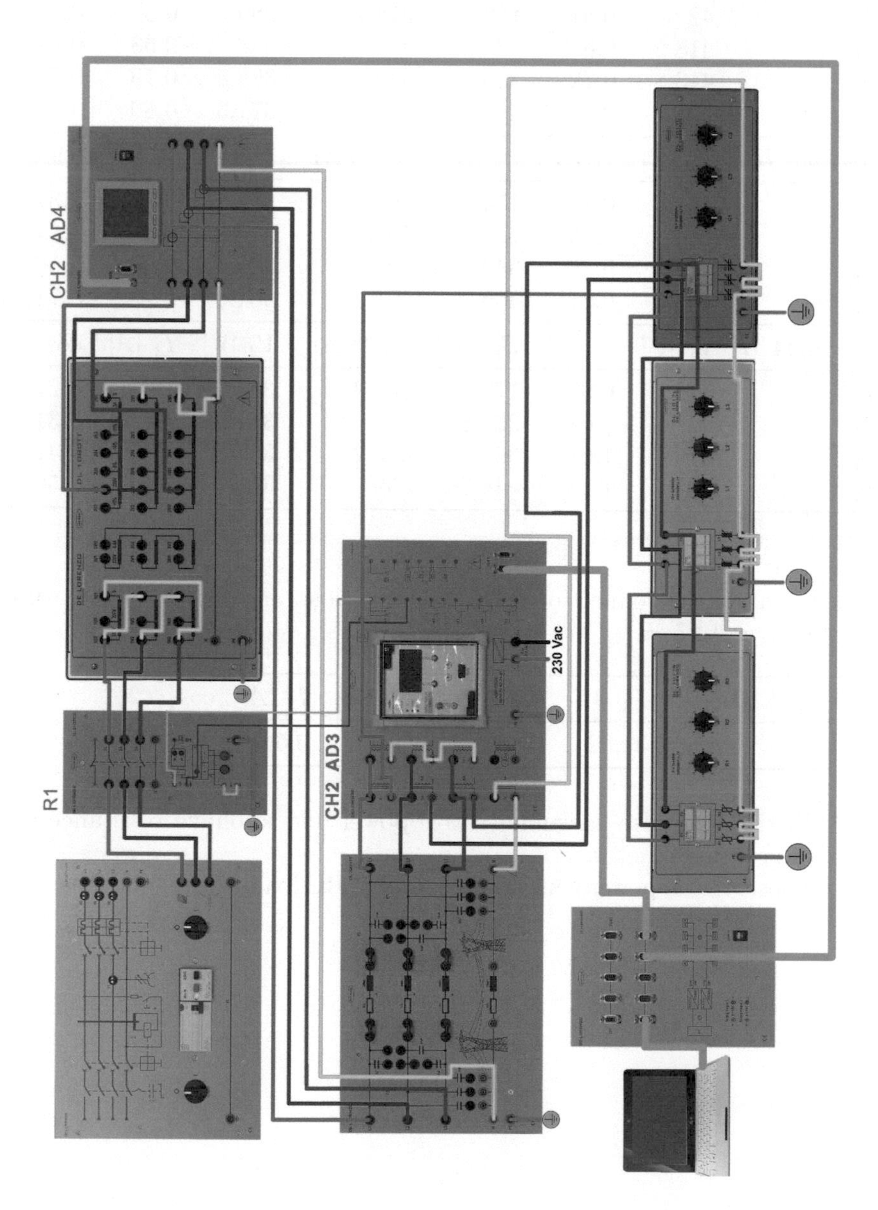

Diagram 6.3
Connections for RC and RL power consumed practice

values into the following table:

$C_1 = 2$ $[\mu F]$	V_1 $[V]$	I_1 $[A]$	P_1 $[W]$	Q_1 **[VAr]**	V_2 $[V]$	I_2 $[A]$	$\cos\phi_2$
R1	423.1	0.29	171.9	128.3	428.2	0.27	0.84
R2	420.2	0.37	237.4	121.2	423.7	0.35	0.91
R3	414.1	0.56	391.5	100.5	414	0.55	0.96
R4	408.8	0.77	540.4	69.9	405	0.76	0.98
$C_2 = 3$ $[\mu F]$	V_1 $[V]$	I_1 $[A]$	P_1 $[W]$	Q_1 **[VAr]**	V_2 $[V]$	I_2 $[A]$	$\cos\phi_2$
R1	423.8	0.36	176.8	195.2	433	0.34	0.7
R2	420.6	0.42	243.7	187.3	428.8	0.4	0.8
R3	414.6	0.6	400.2	162.2	418.5	0.59	0.92
R4	408.8	0.8	551.2	129.7	409.2	0.79	0.95
$C_3 = 5$ $[\mu F]$	V_1 $[V]$	I_1 $[A]$	P_1 $[W]$	Q_1 **[VAr]**	V_2 $[V]$	I_2 $[A]$	$\cos\phi_2$
R1	423.5	0.5	185.1	309.6	441.2	0.46	0.51
R2	420.4	0.54	254.8	299	430.8	0.51	0.64
R3	414.7	0.69	416.4	268.8	427.1	0.67	0.82
R4	407.7	0.87	573.2	228	416.4	0.85	0.89

Remove the connections to the resistive load and repeat the measurements for $C_3 = 5[\mu F]$.

V_1 $[V]$	I_1 $[A]$	P_1 $[W]$	Q_1 [VAr]	V_2 $[V]$	I_2 $[A]$	$\cos\phi_2$
430.5	0.45	3.5	329.3	452.7	0.41	0

Unlike inductors, capacitors demonstrate practically no losses, so almost no active power is consumed.

Recommended equipment and connections:

- Resistive load
- Inductive load
- Capacitive load
- Overhead line model — Long
- Power circuit breaker
- Three-phase power meter
- Three-phase transformer
- Three-phase supply unit
- Feeder manager relay

Naturally, the loads have an inductive and resistive nature. In the industry, it is important to keep the power factor within a certain limit to avoid economic penalties. Capacitor banks are used for this purpose.

Recommended equipment and connections:

- Resistive load
- Inductive load
- Capacitive load
- Overhead line model — Long
- Overhead line model — Medium
- Power circuit breaker
- Three-phase power meter
- Reactive power controller
- Switchable capacitor battery
- Three-phase transformer
- Three-phase supply unit
- Feeder manager relay

Assemble the electric connections shown in Diagram 6.4 and Diagram 6.5. Set all the loads to zero position, power on the three-phase supply unit, and close Relay 3, Relay 1, and Relay 2.

Increment the inductive load in a balanced way, step by step, and observe the $\cos\phi$ variation.

Observe the compensator to know the value of the capacitance connected to the power lines, and calculate the reactive power supplied for each step of the capacitor bank.

Observe the compensator to know the value of the capacitance connected to the power lines, and calculate the reactive power supplied for each step of the capacitor bank.

$$Q = 3 \times 2\pi f C V_{LL}^3.$$

Fill the following table with the measured data:

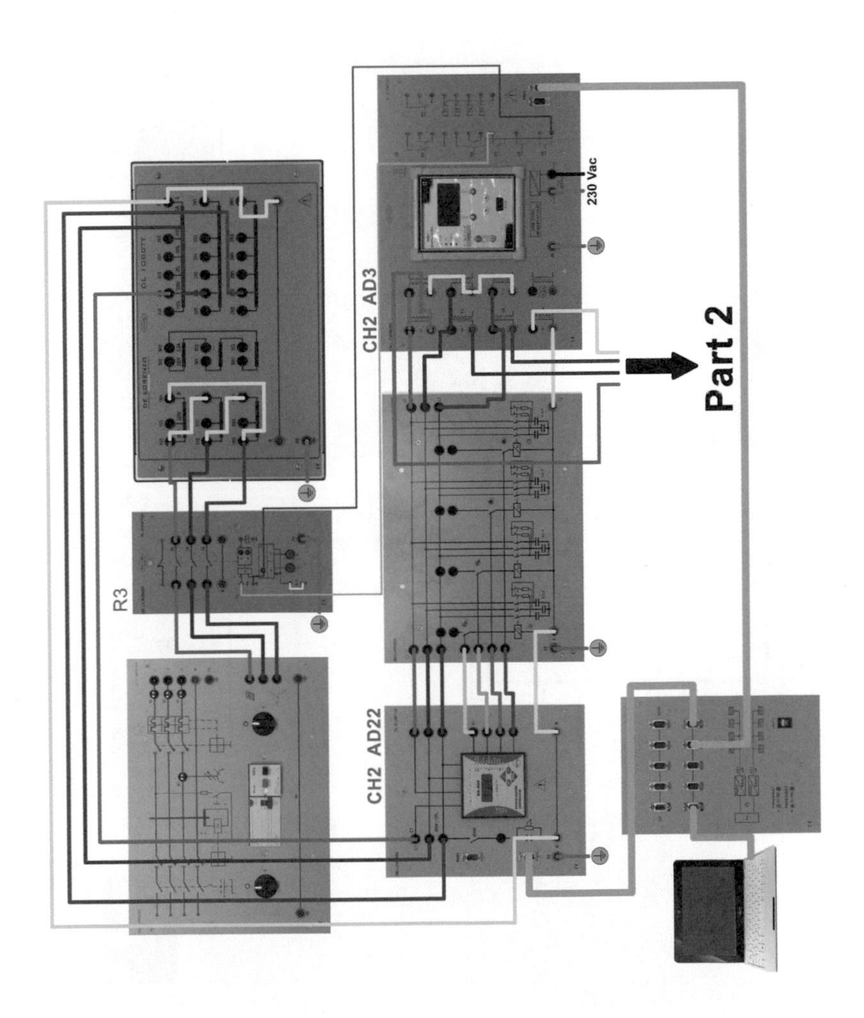

Diagram 6.4
Connections for capacitive compensation 1/2

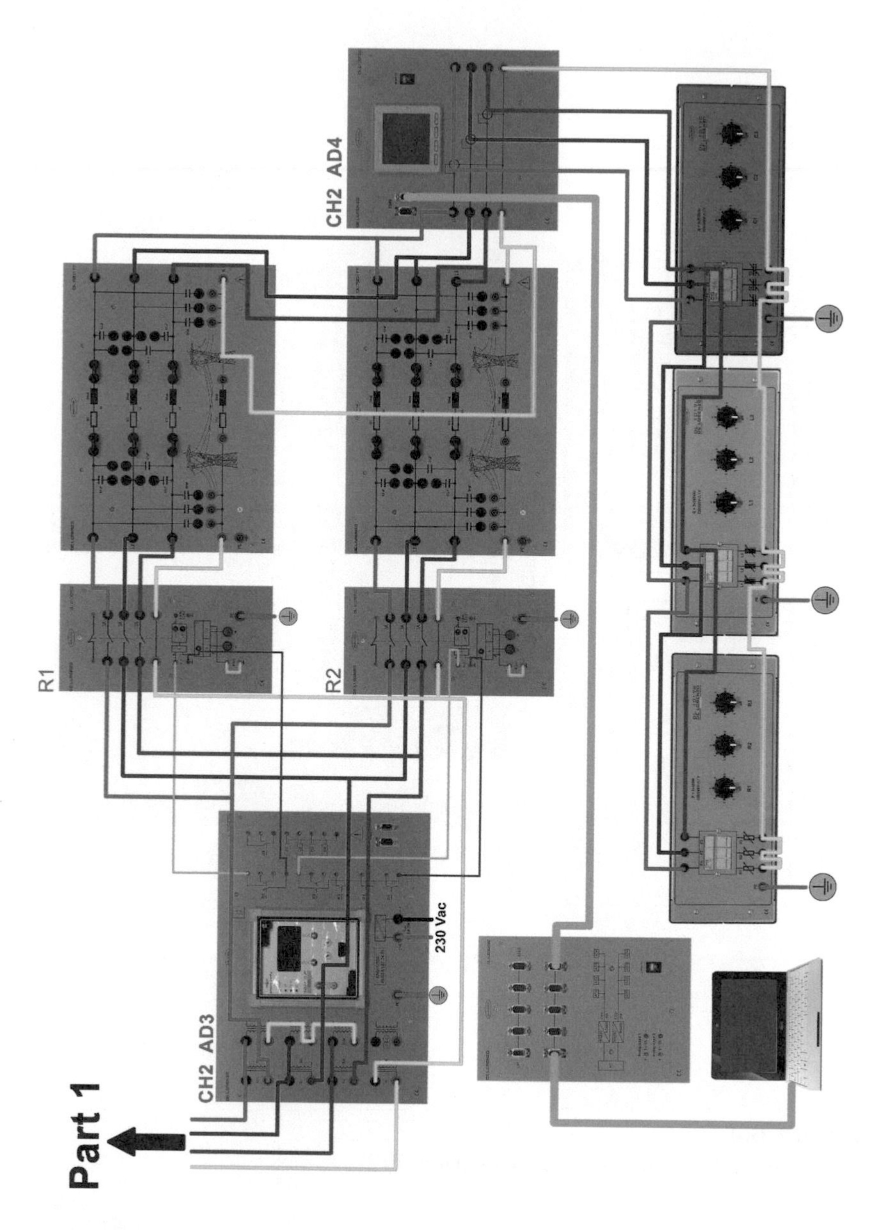

Diagram 6.5
Connections for capacitive compensation 2/2

Resistance	Inductance	Capacitance $[\mu F]$	Line to line voltage $[V]$	Reactive power $[VA]$
	L_2	2	381	328
R_2	L_5	8	363.1	1192.9
	L_6	10	349.87	1384.4

Resistance	Inductance	Capacitance $[\mu F]$	Line to line voltage $[V]$	Reactive power $[VA]$
	L_2	6	336	766
R_7	L_5	10	323.89	1186.4
	L_6	12	316.96	1363.46

6.5 Theoretical problems

Problem 6.1. *Consider a copper stranded conductor formed by 19 strands of* 0.1055 $[in]$ *in diameter. Calculate the DC resistance per unit length in* $[\Omega/km]$ *for this conductor at* $50° \; C$. *Use the hard-drawn copper data of Table 6.1 for the resistivity and the temperature constant* ($\rho_{20°} = 1.78 \; [\Omega m] \times 10^{-8}$, $T = 241.5° \; C$). *Consider 2% increase of electric resistance due to spiraling.*

Solution 6.1. *From Equation (6.2) we obtain the resistivity at* $50° \; C$:

$$\rho_{50°} = 1.78 \times 10^{-8} \; [\Omega/m] \left(\frac{50° + 241.5°}{20° + 241.5°} \right) = 1.98 \times 10^{-8} \; [\Omega/m].$$

The area of the conductor is:

$$A = 19 \times \frac{\pi \left(0.1055 \; [in] \; \frac{1 \; [m]}{39.37 \; [in]} \right)^2}{4} = 1.0716 \times 10^{-4} \; [m^2],$$

and the electric resistance is obtained from (6.1):

$$R_{dc} = \frac{1.98 \times 10^{-8} \; [\Omega/m]}{1.0716 \times 10^{-4} \; [m^2]} = 1.8478 \times 10^{-4} \; \left[\frac{\Omega}{m} \right].$$

Considering 2% of resistance increase and expressing the result in Ω/km:

$$R^*_{dc} = 1.02 \left(1.8478 \times 10^{-4} \; \left[\frac{\Omega}{m} \right] \right) \times \frac{1000 \; [m]}{1 \; [km]} = 0.1884 \; \left[\frac{\Omega}{km} \right].$$

Problem 6.2. *The electric conductor calculated is a 4/0 A.W.G. hard-drawn copper conductor. From datasheets of the conductor we known that the electric resistance at* $50° \; C$ *is as shown below. Calculate the percent increase in resistance at* 60 $[Hz]$ *versus DC.*

Electric Resistance (Ω / *Conductor* / *km*)		
DC	50 $[Hz]$	60 $[Hz]$
0.188	*0.189*	*0.2*

Solution 6.2.

$$\%\ Increase = \frac{0.2}{0.1884} \times 100 = 106.16\%.$$

Problem 6.3. *Consider a complete transposed* 60 $[Hz]$ *bundled transmission line conformed by two ACSR "Drake" conductors with the characteristics below. Calculate the inductance and series reactance per* $[km]$ *for a flat configuration with horizontal spacing between phases of* 10 $[m]$ *(*$ab = 10$ $[m]$*,* $bc = 10$ $[m]$*). Also consider a bundle spacing of* 0.5 $[m]$ *(*$d = 0.5$ $[m]$*).*

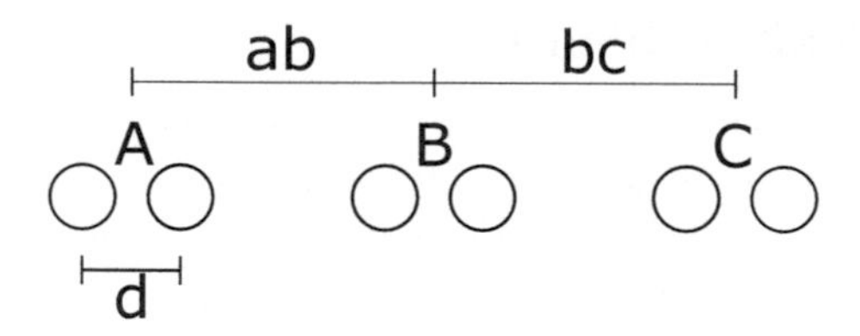

Outside Diameter $[cm]$	***GMR*** $[m]$	***Resistance*** 60 $[Hz]$ 50° *C* ($[\Omega/km]$)	***Current Capacity*** $[A]$
2.8143	*0.01143*	*0.08*	*900*

Solution 6.3. *Let's call the GMD of each single conductor Ds, then the geometric mean distance of each phase bundle is:*

$$GMR = \sqrt[N^2]{\prod_{x=1}^{N}\prod_{n=1}^{N} D_{xn}} = \sqrt[4]{(d \times D_S)^2} = \sqrt{(d \times D_S)} =$$

$$\sqrt{0.5 \times 0.01143} = 0.0756\ [m].$$

From Equation (6.49)

$$L = 2 \times 10^{-7} l \left(\ln \frac{\sqrt[3]{10 \times 10 \times 20}}{0.0756} \right) = 1.024 \times 10^{-6} l\ [H].$$

If we express the result in $[H/km]$*:*

$$L' = 1.024 \times 10^{-3}\ [H/km].$$

The inductive reactance per km is:

$$X'_a = 2\pi f L' = 0.386\ [\Omega/km].$$

The series reactance is formed by the resistance and the inductive reactance:

$$X = 0.08||0.08 + j0.386\ [\Omega/km] = 0.04 + j0.386\ [\Omega/km].$$

Problem 6.4. *Now consider a double circuit line formed by the same conductors with the following distances between conductors. Calculate the inductance and reactance. Calculate the inductance and series reactance per km and obtain the inductive reactance percentage reduction compared with the flat configuration.*

ab $[m]$	***ac'*** $[m]$	***ab'*** $[m]$	***aa'*** $[m]$
10	*10*	*14.14*	*22.36*

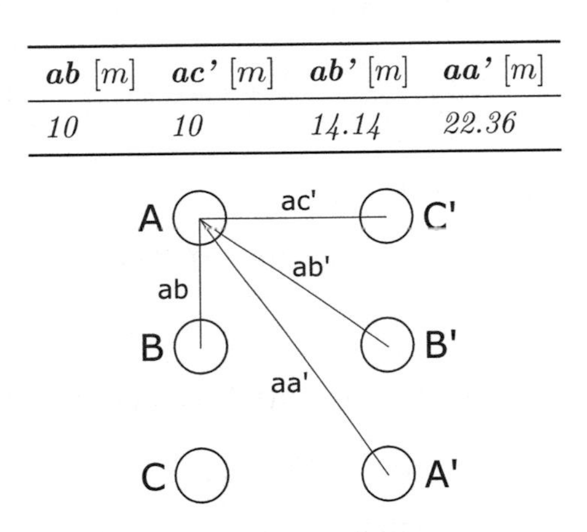

Solution 6.4. *First, let's obtain the GMD from Equations (6.50) to (6.57):*

$$\begin{aligned} GMR_A &= \sqrt{0.01143 \times 22.36} = 0.506\ [m], \\ GMR_B &= \sqrt{0.01143 \times 10} = 0.3381\ [m], \\ GMR_C &= \sqrt{0.01143 \times 22.36} = 0.506\ [m], \\ GMR_{average} &= \sqrt[3]{0.506^2 \times 0.3381} = 0.4424\ [m]. \end{aligned}$$

Finally, the inductance per $[km]$ is:

$$L' = 2 \times 10^{-7}(1000)\left(\ln \frac{12.5971}{0.4424}\right) = 6.698 \times 10^{-4}\ [H/km].$$

The inductive reactance is:

$$X'_a = 2\pi f L' = 0.2525\ [\Omega/km].$$

The percentage of reduction is:

$$\%\ reduction = \left(1 - \frac{0.2525}{0.386}\right) \times 100\% = 34.58\%.$$

Problem 6.5. *For Problem 6.4, obtain the capacitance with reference to neutral in [F/km] and the shunt admittance in [Siemens/km].*

Solution 6.5. *For the flat configuration, the radius of the conductor is:*

$$r = \frac{0.02814\ m}{2} = 0.01407\ [m].$$

Then, the GMR for the bundle is:

$$GMR = \sqrt{d \times r} = \sqrt{0.5 \times 0.01407} = 0.08387\ [m].$$

From Equation (6.72), the capacitance is equal to:

$$C_{an} = \frac{2\pi \times 8.854 \times 10^{-12} \times l}{\ln \frac{\sqrt[3]{10 \times 10 \times 20}}{0.08387}} = 1.1099 \times 10^{-11} \times l\ [F].$$

If the result is expressed in [F/km]:

$$C'_{an} = 1.1099 \times 10^{-8}\ [F/km].$$

The admitance is equal to:

$$Y = j\omega C_{an} = j(2\pi \times 60)(1.1099 \times 10^{-8}) = 4.18422 \times 10^{-6}\ [S/km].$$

For the double circuit line, the $GMR_{average}$ is:

$$\begin{aligned} GMR_A &= \sqrt{0.01407 \times 22.36} = 0.5609\ [m], \\ GMR_B &= \sqrt{0.01407 \times 10} = 0.3751\ [m], \\ GMR_C &= \sqrt{0.01143 \times 22.36} = 0.5609\ [m], \\ GMR_{average} &= \sqrt[3]{0.5609^2 \times 0.3751} = 0.4905\ [m], \end{aligned}$$

From (6.73), the capacitance is:

$$C_{an} = \frac{2\pi \times 8.854 \times 10^{-12} \times l}{\ln \frac{12.5971}{0.4905}} = 1.7139 \times 10^{-11} \times l\ [F].$$

If the result is expressed in [F/km],

$$C'_{an} = 1.7139 \times 10^{-8}\ [F/km].$$

Then, the admitance is equal to:

$$Y = j\omega C_{an} = j(2\pi \times 60)(1.1099 \times 10^{-8}) = 6.4614 \times 10^{-6}\ [S/km].$$

Problem 6.6. *A three-phase system of* 500 *[km] is completely transposed and* *has the following impedance and admittance (solutions of Section 6.2.6 for the double circuit line series impedance and shunt admittance):*

$$z = 0.04 + j0.2525 \left[\frac{\Omega}{km}\right]$$
$$y = j6.4614 \times 10^{-6} \; [S/km].$$

Calculate the exact ABCD model of the transmission line and present the result in matrix form. Note: To obtain the square root of a complex number, it is convenient to employ DeMoivre's identity: $(\cos x + j \sin x)^n = \cos nx + j \sin nx$.

Solution 6.6. *To use Equation (6.91),*

$$\gamma = \sqrt{(j6.4614 \times 10^{-6}) \times (0.04 + j0.2525)} = \sqrt{1.6518 \times 10^{-6}\angle 171^\circ}$$
$$= 1.2852 \times 10^{-3}\angle 85.5^\circ,$$

$$Z_c = \sqrt{\frac{0.04 + j0.2525}{j6.4614 \times 10^{-6}}} = \sqrt{39565.5252\angle -9^\circ} = 198.9108\angle 4.5^\circ.$$

Now, calculating $\cosh(\gamma l)$ *and* $\sinh(\gamma l)$*:*

$$\gamma l = 0.0504 + j0.6406,$$
$$e^{\gamma l} = e^{0.0504} \times e^{j0.6406} = 1.0517 \times (0.8017 + j0.5977) = 0.8431 + j0.6286,$$
$$e^{-\gamma l} = e^{-0.0504} \times e^{-j0.6406} = 0.7623 - j0.5683,$$
$$\cosh \gamma l = \frac{e^{\gamma l} + e^{-\gamma l}}{2} = 0.8027 + j0.03015,$$
$$\sinh \gamma l = \frac{e^{\gamma l} - e^{-\gamma l}}{2} = 0.0404 + j0.59845.$$

Then,

$$\begin{bmatrix} V_S \\ I_S \end{bmatrix} = \begin{bmatrix} 0.8027 + j0.03015 & -1.3284 + j199.3017 \\ (4.3853 + j29.8342) \times 10^{-4} & 0.8027 + j0.03015 \end{bmatrix} \begin{bmatrix} V_R \\ I_R \end{bmatrix}.$$

Problem 6.7. *For Problem 6.6, obtain the nominal* π *circuit and then calculate the exact* π *circuit model. Compare the nominal circuit with the exact model.*

Solution 6.7. *The nominal model is obtained by directly substituting the impedance and shunt admittance of the line.*

$$Z = (0.04 + j0.2525) \times 500 = 20 + j126.51\Omega = 128.08\angle 81.016^\circ \; [\Omega],$$
$$\frac{Y}{2} = (j6.4614 \times 10^{-6}) \times 250 = j1.61535 \times 10^{-3} \; [S] = 1.61535 \times 10^{-3}\angle 90^\circ \; [S],$$
$$Z' = Z\frac{\sinh(\gamma l)}{\gamma l} = 128.08\angle 81.016^\circ \frac{0.0404 + j0.59845}{0.0504 + j0.6406} = 119.56\angle 81.65^\circ.$$

Then,

$$\frac{\gamma l}{2} = 0.0504 + j0.6406 = 0.0252 + j0.3203,$$
$$e^{\gamma l} = e^{0}.0252 \times e^{j0.3203} = 0.9733 + j0.3229,$$
$$e^{-\gamma l} = e^{-0.0252} \times e^{-j0.3203} = 0.9255 - j0.3070.$$

So,

$$\tanh\frac{\gamma l}{2} = \frac{e^{\gamma l} - e^{-\gamma l}}{e^{\gamma l} + e^{-\gamma l}} = 0.02795 + j0.3315,$$
$$\frac{Y'}{2} = \frac{Y}{2}\frac{\tanh(\gamma l/2)}{\gamma l/2} = 1.6726 \times 10^{-3}\angle 89.679°.$$

Let's obtain the percentage of difference between the magnitudes of the exact and approximated models:

$$\% \text{ of difference } Z = \left(\frac{|Z - Z'|}{\frac{Z+Z'}{2}}\right) \times 100\% = 6.881\%,$$

$$\% \text{ of difference } \frac{Y}{2} = \left(\frac{\left|\frac{Y}{2} - \frac{Y'}{2}\right|}{\frac{\frac{Y}{2}+\frac{Y'}{2}}{2}}\right) \times 100\% = 3.48\%.$$

The transmission line modeled is considered to be a long transmission line, and therefore it is not recommended to use the nominal π circuit. Nevertheless, for medium and short length transmission lines, the percentage of difference will be less than 2% and the nominal model can be employed.

6.6 Homework problems

Problem 6.8. *Suppose an annealed copper conductor 4/0 is installed at some building. Calculate the DC resistance if it is known that it will be used at maximum* $60°\ C$*, and that the containing duct is of about* $100\ [m]$.

Solution 6.8. *The 4/0 cable has a diameter of* $0.46\ [in]$*, a* $20°\ C$ *resistivity of* $\rho_{20°\ C} = 1.72 \times 10^{-8}\ [\Omega m]$*, and a temperature constant of* $234.5°\ C$ *(from Table 6.1).*

Then,

$$\rho_{60°\ C} = \rho_{20°\ C}\left(\frac{60° + 234.5°}{20° + 234.5°}\right) = 1.99 \times 10^{-8}\ [\Omega m].$$

In addition,

$$A = \pi r^2 = \pi \left(\frac{0.46 \ [in]}{2} \frac{1 \ [m]}{39.37 \ [in]} \right)^2 = 107.22 \times 10^{-6} \ [m^2],$$
$$R_{dc} = \frac{\rho_{60^\circ \ C}}{A} = \frac{1.99 \times 10^{-8} \ [\Omega m]}{107.22 \times 10^{-6} \ [m^2]} = 185.6 \times 10^{-6} \ \left[\frac{\Omega}{m} \right].$$

So, for 100 $[m]$*, the total resistancec is* 18.56 $[m\Omega]$.

Problem 6.9. *Consider the conductor of Problem 6.8 to be used in a three-phase configuration. All the conductors are individually isolated through a* 0.2 $[in]$ *plastic skin and are touching each other throughout their trajectory. Calculate the inductance of one of these conductors in terms of the distance.*

Solution 6.9. *The conductor's radius is* $0.23 \ [in] + 0.1 \ [in]$. *As all conductors are touching each other, the distance from one of them to any other is* $2(0.23 \ [in] + 0.2 \ [in]) = 0.86 \ [in] = 21.84 \times 10^{-3} \ [m]$. *The equivalent inner radius is then:* $r' = r \exp(-0.25) = 4.55 \times 10^{-3}$ *m. So, it can be directly computed:*

$$\frac{L}{l} = 2 \times 10^{-7} \ln \frac{D}{r'_a} = 314 \ \left[\frac{nH}{m} \right].$$

Problem 6.10. *Consider a conductor to be arranged and installed so it exhibits a DC resistance of* $R_{dc} = 185.6 \times 10^{-6} \ [\Omega/m]$*, and an inductance of* $L = 314 \ [nH/m]$. *Calculate the reactance in terms of the distance of such a conductor for* 50 $[Hz]$ *and* 60 $[Hz]$ *operation. Use the following information to measure resistance:*

DC	50 $[Hz]$	60 $[Hz]$
100%	*100.532%*	*106.383%*

Solution 6.10. *The reactance can be calculated for each frequency as follows:*

$$X_f = R_f + j2\pi f L.$$

Thus,

$$\frac{X_{50}}{m} = 186.59 \times 10^{-6} \ + j98.65 \times 10^{-6} \ [\Omega/m],$$
$$\frac{X_{60}}{m} = 197.45 \times 10^{-6} \ + j118.38 \times 10^{-6} \ [\Omega/m].$$

Problem 6.11. *Consider a transmission line that exhibits the following parameters:*

- *Series impedance* $Z = 0.02 + j0.3 \ [\Omega/km]$

- *Shunt admittance* $Y = j3.21 \times 10^{-6}$ $[S/km]$

If a single resistive load of $R_L = 5$ $[\Omega]$ *is placed at the end of the line (*110 $[km]$*), calculate the voltage at the load's terminals.*

Solution 6.11. *Two mesh equations can be solved, considering the shunt admittance to face the power source:*

$$V_{in} = \frac{i_1 - i_2}{j3.21 \times 10^{-6}\ [S]},$$

$$0 = \frac{i_2 - i_1}{j3.21 \times 10^{-6}\ [S]} + i_2(7.2 + j33\ [\Omega]).$$

The solution leads to:

$$i_2 = 6.31 \times 10^{-3} - j28.92 \times 10^{-3}.$$

So, the voltage at the load will only be of $0.149\ V_{in}$.

Problem 6.12. *For a trasmission line with the following known parameters:*

- *Line resistance of* 13 $[\Omega]$
- *Line inductance of* 290 $[mH]$
- *Earth capacitance* 2 $[\mu F]$ *(both ends)*

Calculate the active power delivered to R_L *at (75, 100, 125, 150, 200)* $[\Omega]$.

Solution 6.12. *The above description leads to a 3-mesh circuit that can be solved by the following equations:*

$$V_{in} = \frac{i_1 - i_2}{j753.98 \times 10^{-6}\ [S]},$$

$$0 = \frac{i_2 - i_1}{j753.98 \times 10^{-6}\ [S]} + i_2(13 + j109.33\ \Omega) + \frac{i_2 - i_3}{j753.98 \times 10^{-6}\ [S]},$$

$$0 = \frac{i_3 - i_2}{j753.98 \times 10^{-6}\ [S]} + i_3 R_L.$$

We can solve for i_3 *only so the resulting equation describes the specific current in relation to active power (fed voltage can be considered unitary).*

$$i_3(R_L) = \frac{1.09 - j0.012}{R_L + 15.44 + j118.99}.$$

As the load is purely resistive, the current's magnitude can be used:

$$\|i_3(R_L)\| = \frac{1.1}{\sqrt{(R_L + 15.44)^2 + 119^2}}.$$

Power can be calculated through

$$\|i_3\|^2 R_L = \frac{1.21 R_L}{(R_L + 15.44)^2 + 119^2}.$$

So, evaluating different values of the load would result in the shown plot in Figure 6.16.

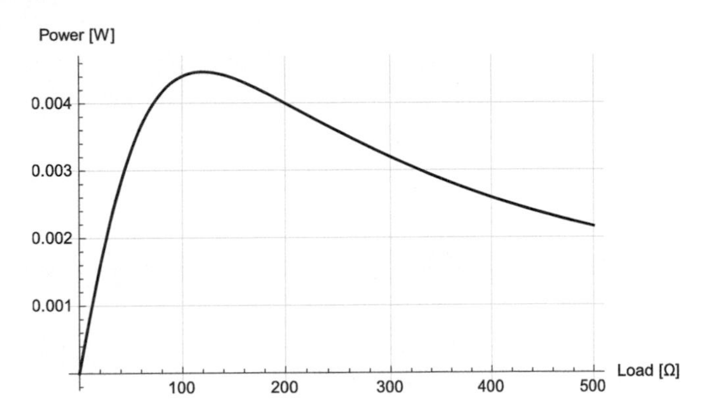

FIGURE 6.16
Power vs load plot

6.7 Simulation

Transmission lines can be seen as a load themselves due to the non-ideal effects they impose on the network see Figure 6.17). A pure resistive load will consume pure active power; however, whenever it is placed after a transmission line, the generators will have demands for reactive power as well. The transmission lines also permit the voltage levels to be different at their terminals, so compensation must occur for the systems to reach desired electrical magnitudes.

Transmission lines increase network complexity and make it more susceptible to failure; they also make an automatic control system to be required. However, electric distribution needs transmission lines despite their shortcomings. Depending on their total length, a generator sees non-ideal loads while the load sees a non-ideal power source. In order to improve network performance, short transmission lines are implemented in micro-grid systems, which generates and consumes energy in a "close" or "local" geographical area.

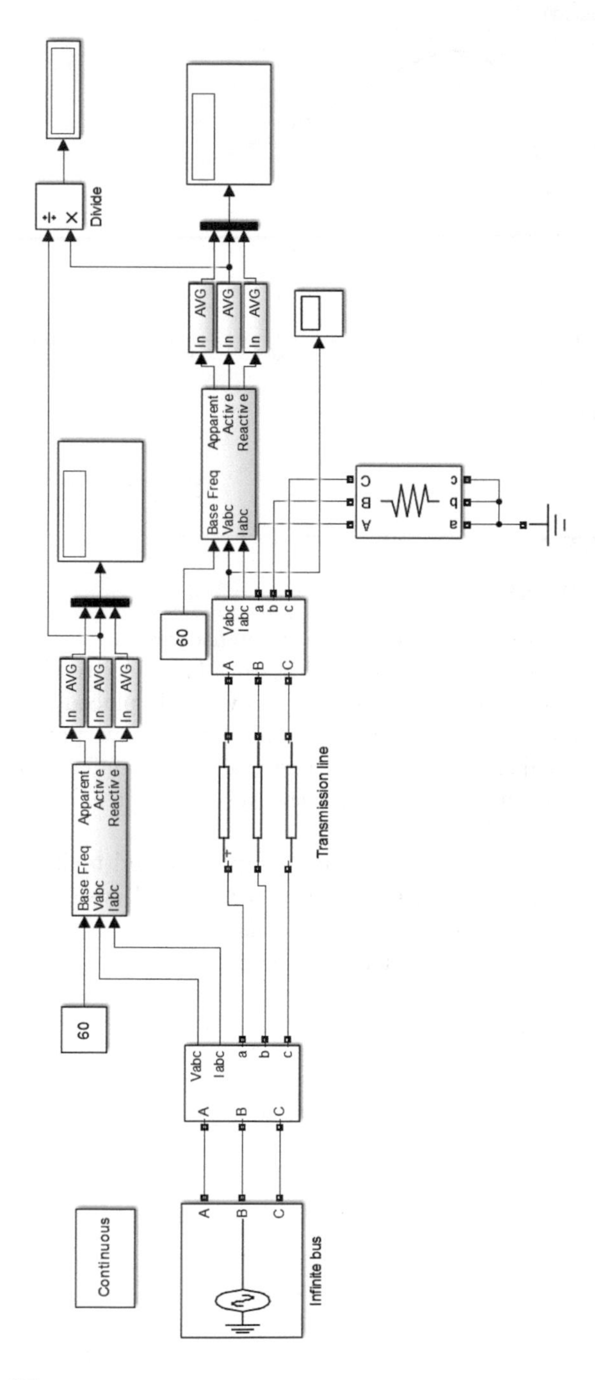

FIGURE 6.17
Transmission lines simulation

7

Power Distribution System, Unsymmetrical Failures, and Power System Protections

Roughly speaking, power systems are composed of generators, and transmission and distribution systems for loads. An electric distribution system aims to provide energy at a minimum voltage variation and as few interruptions as possible if a failure on an element occurs (transmission lines short circuit, transformer's overload, etc.). Distribution systems' networks take electricity from the high voltage transmission lines and condition energy for consumers and are designed to maintain a fixed voltage regardless of the power flow of the load. Moreover, if an emergency condition occurs (transmission line short circuit, transformer's overload, etc.), distribution systems are able to make the adjustments necessary to continue providing energy to the loads.

System planning is a major subject for satisfying the electricity demand of the consumers. The impact of the distribution system is directly related to economical, comfort, and safety issues of consumers. Generation plans are made by electric companies in order to supply reliable well-quality energy and maintain the optimal operation point of generators, which directly affects the cost of production.

The distribution substations are in strategic places so they distribute the power flow at the maximum cost-effectiveness, minimizing transmission losses while considering service reliability. The distribution system design starts at the consumer level by analyzing the demand type, the power factor, and other load characteristics. It continues with distribution transformer, that step down the tension from primary voltage, and then the demands of the primary distribution system are computed and assigned to substations that step down the voltage from the long high-voltage transmission lines. Finally, the last step is the generation: some generators are fixed and always operating at their nominal operation point as coal based power plants, while others are turned on according to the generation plan. Some of the secondary plants are hydroelectrical and diesel-based generators.

The factors that need to be considered for a power system design are many, and some of them are very complex. For example, the designers need to consider transformers and other electric machines' impedances, dispatch of generation, rates of charges for consumers, and an analysis for predicting future levels of load and insulation. There are also more complex issues like time and location of energy demands, duration and frequency of outages, and the always-important economic factors.

In Figure 7.1, the basic structure of a power distribution system is shown. This is the simplest structure where the load is unidirectional from the generators to the loads. The most basic elements of a distribution scheme are also presented: electric generators, transformers, circuit breakers, and disconnector switches. The principal difference between circuit breakers and disconnectors is that the former are used to protect the circuit from overloads and short circuit conditions, while the disconnection switches are employed when a circuit is required to be disconnected from the power supply. This element does not have any protection at high voltages and it requires a circuit breaker before the switch blades are opened.

7.1 Types of distribution systems

There are three basic schemes for power distribution systems:

- Radial networks
- Ring networks
- Mesh networks

Those systems are the most commonly employed, and in the next sections they will be described based on functionality and advantages and disadvantages with respect to other schemes. Distribution systems can also be a combination of schemes, and for large power systems the line diagram will be incredibly complex.

7.1.1 Radial networks

This is the most basic structure where the power flows directly from the source to the load; the path is unique and does not have a return. The power flows to each branch where a load is connected. In Figure 7.2 a basic scheme is shown.

There are many arrangements depending on the requirements of the power system. The advantages of radial topology are: economical cost and little equipment, while the principal disadvantages are the necessity of interrupting the operation of the loads when maintenance is given to equipment like circuit breakers, and the low reliability of this architecture; if a failure occurs at the main power supply this failure will damage all the loads. This topology is installed on both subterranean and aerial circuits, to be explained next.

Aerial radial networks are normally implemented in urban and rural zones (see Figure 7.3). The main power supply comes from the substation and the distribution is made with electricity poles situated near the loads and holding

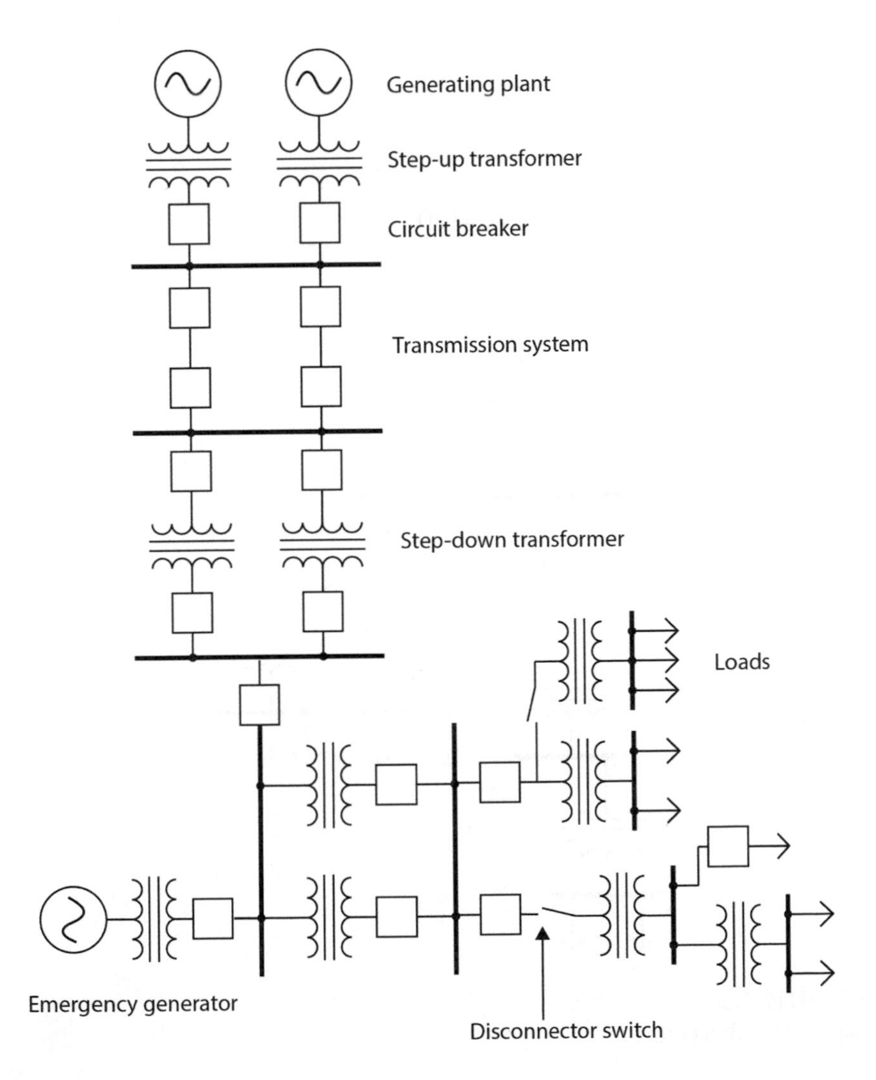

FIGURE 7.1
Basic structure of a power system

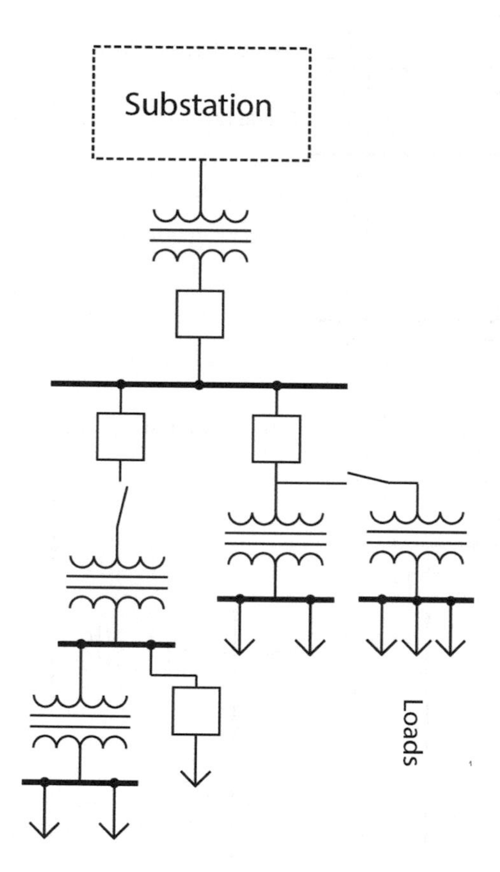

FIGURE 7.2
Basic radial distributions

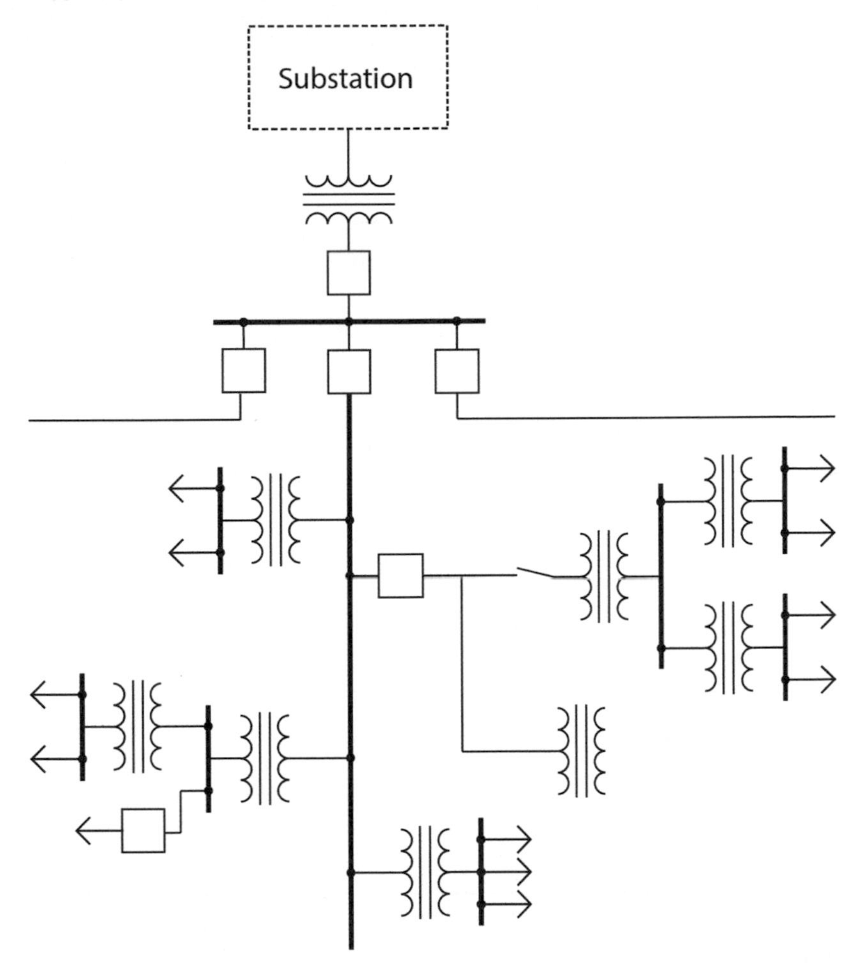

FIGURE 7.3
Aerial radial network

electric transformers. In rural areas the distribution is typically pure radial, while for urban areas with great population density and therefore high load density, interconnecting points are open for use whenever it is necessary to feed the electric lines with a secondary net and continue operating.

Houses do not represent significant loads; therefore, disadvantages of the radial topology can be tempered by the economic advantages of this scheme.

Subterranean radial networks are implemented in urban zones with high or medium population density, and where the environment is not favorable for aerial networks. The subterranean networks have a lower failure rate than aerial networks but the installation and maintenance have higher costs.

When a failure occurs on a subterranean network it is hard to locate, and the repair requires more time and cost than with an aerial network. So, this

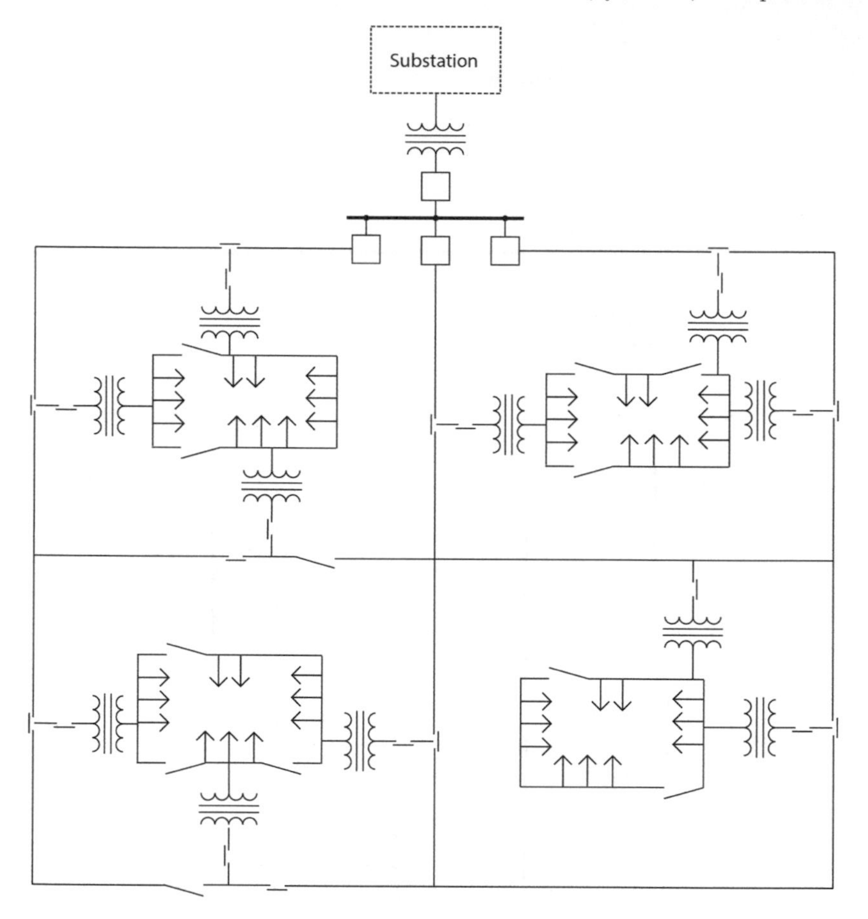

FIGURE 7.4
Subterranean radial networks

kind of network needs switches for interconnecting one branch with the others for maintenance or emergency issues. This flexibility makes subterranean networks more reliable than aerial networks.

The disadvantages of this configuration are the higher cost of installation and maintenance, and overdesign, so that the lines can handle the overload when a contingency occurs.

Nowadays the popularity of subterranean installations has increased for residential suburban areas. Typically, the power supplies consist of ring networks taken from an aerial line. Step-down transformers are connected to the ring and then the loads. In Figure 7.4 a diagram of a subterranean radial network is shown.

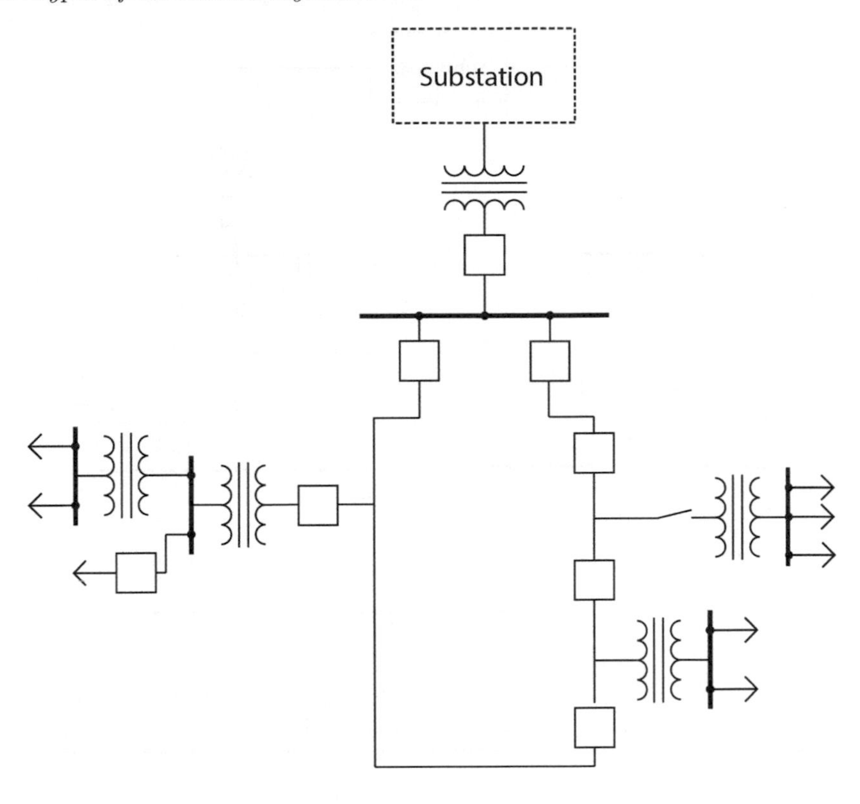

FIGURE 7.5
Ring network

7.1.2 Ring networks

Ring networks surround the complete area where the loads are placed and return to the starting point, which makes it possible to feed the loads from two sides. The complete cycle allows isolating certain zones for giving maintenance, or in case of failure while the rest of the circuit continues to be supplied.

The ring configuration is widely used for industrial networks because it is designed to continue operating in a contingency; furthermore, the voltage regulation is better that on a radial configuration. If a failure occurs on a load, the two adjacent interrupters are opened, but the lines continue to be energized from the two sides, and then the remaining loads continue operating. The principal disadvantage is the major initial cost and the problems in connecting new elements to the circuits, because two new lines need to be prepared.

7.1.3 Mesh networks

The mesh configuration provides greater reliability that the previous ones because two power sources are used for feeding the entire loads. The power

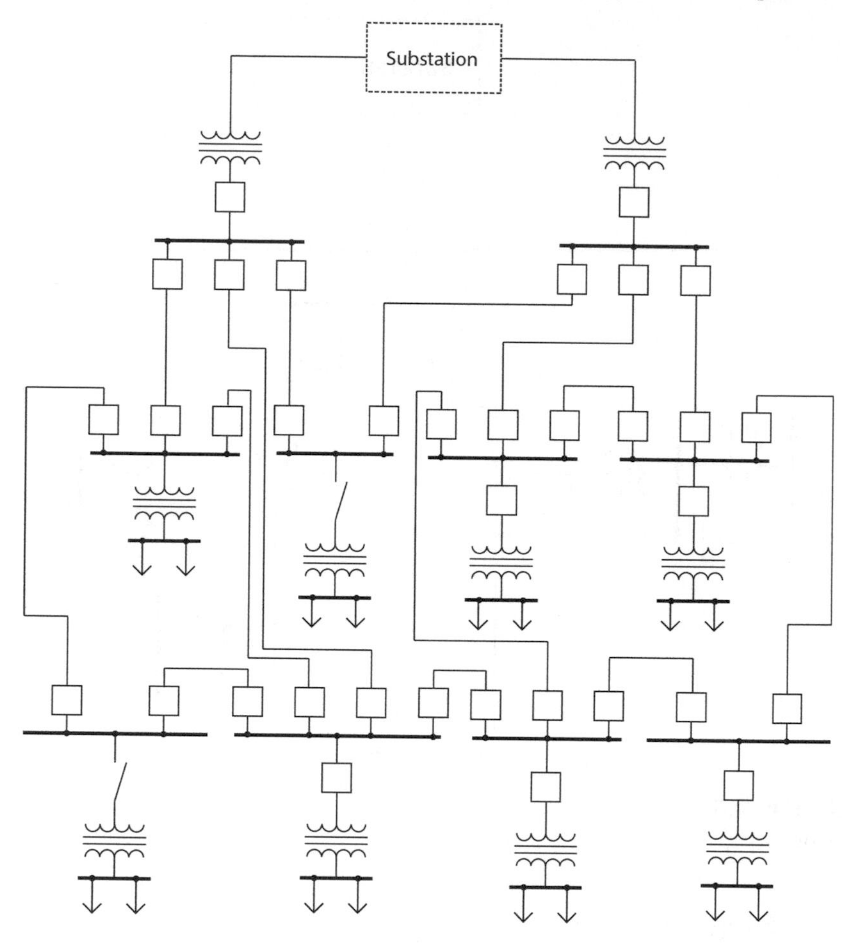

FIGURE 7.6
Mesh network

flow of any of the energy sources can be incremented in order to supply the entire system. This type of configuration is implemented when failures are critical; for example, on a smelter or in a hospital, because all the loads are connected to at least two power sources. The obvious disadvantages are the high cost and overdesign of the power system. Figure 7.6 shows a scheme of a mesh network.

7.1.4 Exercise

Consider the electric installation of a metallurgical plant, where the power supply is critical and the power factor has to be regulated in order to avoid penalty taxes. Build the circuit shown in Diagram 7.1 by connecting as shown

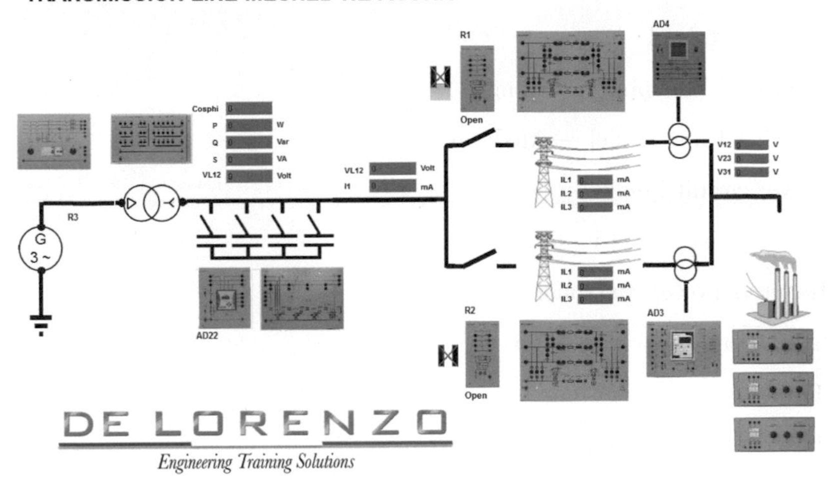

Diagram 7.1
Mesh network

in Diagrams 7.2 and 7.3. Notice that the incoming high voltage has to be reduced. Also observe the connection of the automatic power factor compensator, and take care to have the same order for the phases on both ends of the transmission lines.

Set all loads to zero position, close the relay, and connect the transmission lines in parallel, increment the resistance in a balanced way, and observe the portion of the power that flows for each transmission line:

LOAD	P1 [W]	Q2 [VAr]	P2 [W]	Q2 [VAr]
R1	119	-70	37	71
R2	165	-70	50	70
R3	274	-70	82	68
R4	377	-70	112	66

Now activate the capacitive compensator for power factor correction and increase the inductance load in a balanced way while keeping the resistance fixed and observing the value of correction. Finally, imagine that a power line is undergoing maintenance. Deactivate one power relay and observe the power flow increment in one line.

Recommended equipment and connections:

- Resistive load

- Inductive load
- Capacitive load
- Overhead line model — Long
- Overhead line model — Medium
- Power circuit breaker
- Three-phase power meter
- Reactive power controller
- Switchable capacitor battery
- Three-phase transformer
- Three-phase supply unit
- Feeder manager relay

7.2 Faults on power systems

There are two main ideas behind fault analysis. When designing a power system, all possible scenarios must be foreseen and quantified so proper operation and management is planed prior to daily service. On the other hand, protective systems are based on fault theoretical analysis, so their design and application necessarily follows from it. As three-phase systems are those actually used in most real life application, their analysis is presented next. Common motor faults can be listed (statistically increasing) as follows [36, p. 244]:

- Faulty protection (5%)
- Other causes (19%)
- Rotor or bearing fault (20%)
- Long-time overheating (26%)
- Insulation fault (30%)

Most of the mentioned motor faults (81%) could be avoided if an effective relay was used [36, p. 244]. The main aim of protective relays is to extend the electrical life of machinery and equipment, so understanding the main protection techniques becomes paramount.

Most common electrical failures are due to insulation problems related to short-circuit states among phases and/or ground. They can be summarized as follows [53, pp. 239–240]:

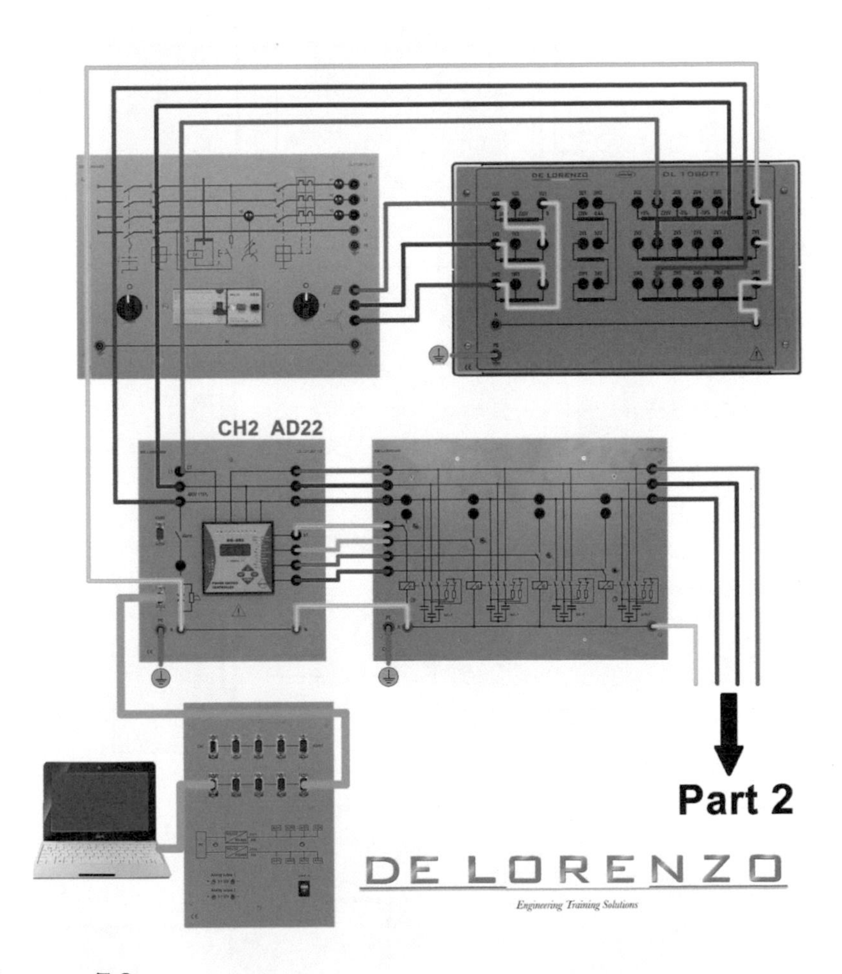

Diagram 7.2
Connections for mesh network 1/2

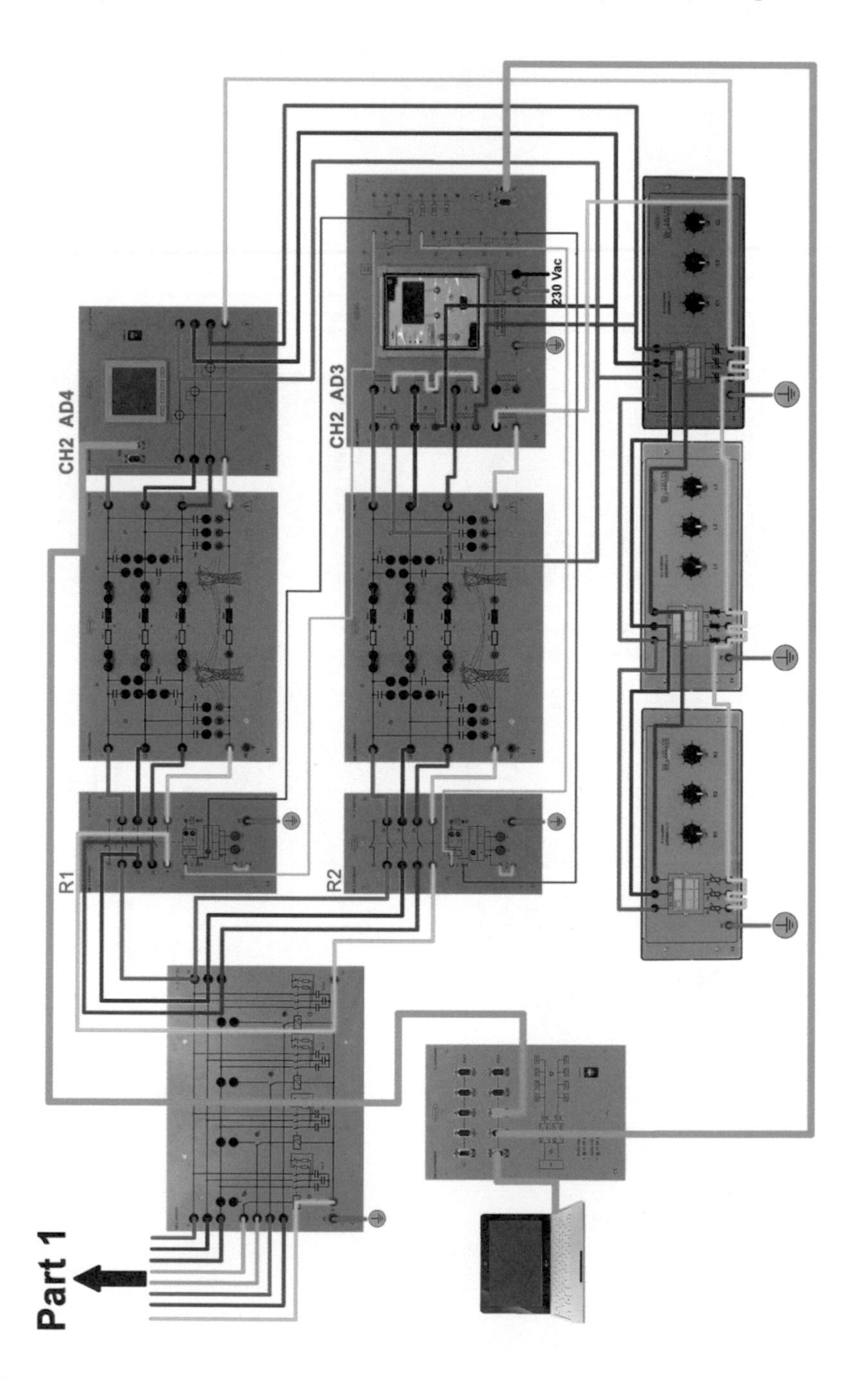

Diagram 7.3
Connections for mesh network 2/2

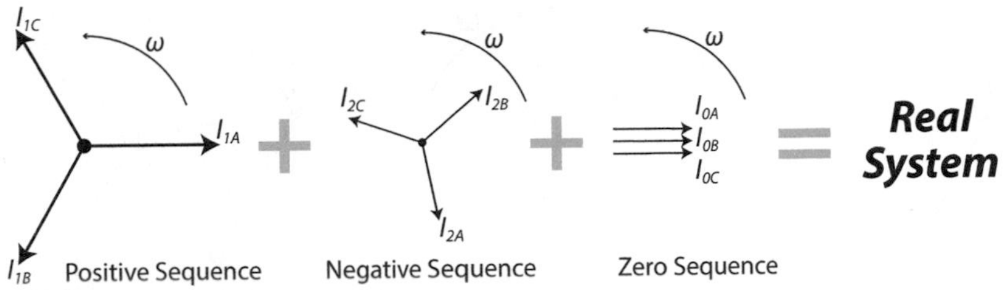

FIGURE 7.7
Symmetrical components explanation

- Phase to ground
- Phase to phase
- Two phases to ground
- Three phases to ground
- Three phases
- Phase to ground with electric arc

7.2.1 Symmetrical components

A commonly used technique to analyze polyphasic unbalanced systems is called symmetrical components. Many protective devices are based on this methodology, so its understanding becomes a paramount tool [49, p. 75]. It was discovered by Charles L. Fortescue while he was investigating unbalanced conditions of induction motors in 1913; however, it was not until 1918 that his technique was published [49, p. 75]. This methods separates a three-phase system into three different components which, when superimposed, form the actual phasor representation of the real system [53, p. 247]. System separation is done as shown in Figure 7.7, where the three components are known as positive sequence, negative sequence, and zero sequence.

- Positive sequence component is a phasor diagram that presents an A-B-C sequence
- Negative sequence component is a phasor diagram that represents an A-C-B sequence
- Zero sequence component is a phasor diagram of three equal-magnitude parallel coaxial phasors

Whether or not the real system is balanced, it will present only a positive

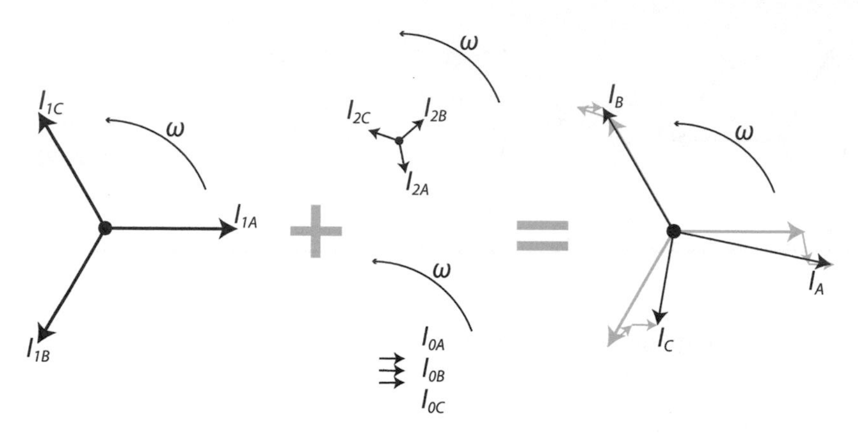

FIGURE 7.8
Equivalence between an unbalanced system and its components

or negative sequence symmetric representation; however, an unbalanced system needs more than one component to be described. Conversion from a real unbalanced system to its components' representation can be understood by the opposite process, as shown in Figure 7.8.

Any phase current or voltage can be derived from its components by using the following equations,

$$\begin{aligned} I_A &= I_{1A} + I_{2A} + I_{0A} = I_1 + I_2 + I_0, \\ I_B &= I_{1B} + I_{2B} + I_{0B} = a^2 I_1 + a I_2 + I_0, \\ I_C &= I_{1B} + I_{2B} + I_{0B} = a I_1 + a^2 I_2 + I_0, \end{aligned} \tag{7.1}$$

where $a = e^{j2\pi/3}$ and $a^2 = a^{j4\pi/3}$ represent phase shifts of 120° and 240°, respectively. Notice that every symmetrical component is equally phase shifted (by 120° multiples) and has equal magnitude I_i; $i = 0, 1, 2$; consequently, construction of real phase characteristics can be made by using a with respect to A phase. The equations shown above are also valid for voltage magnitudes.

In matrix form,

$$T_s = \begin{bmatrix} 1 & 1 & 1 \\ 1 & a^2 & a \\ 1 & a & a^2 \end{bmatrix}, \tag{7.2}$$

where T_s is called the symmetrical component transformation matrix [53, p. 250]. The magnitudes of each component can be computed by inverting T_s as follows:

$$T_s^{-1} = \frac{\mathrm{adj}(T_s)}{|T_s|}, \tag{7.3}$$

where

$$\mathrm{adj}(T_s) = \begin{bmatrix} a^4 - a^2 & a - a^2 & a - a^2 \\ a - a^2 & a^2 - 1 & 1 - a \\ a - a^2 & 1 - a & a^2 - 1 \end{bmatrix}, \tag{7.4}$$

$$|T_s| = a^4 - a^2 - a^2 - a - a - a^2 = a^4 - 3a^2 + 2a.$$

Before performing the calculations to attain the inverse, some properties of a should be taken into account:

$$a^4 = e^{j8\pi/3} = e^{j2\pi/3} = a, \tag{7.5}$$

$$a^3 = e^{j2\pi}. \tag{7.6}$$

Using the properties above, some useful equivalences can be computed,

$$a^4 - a^2 = a - a^2, \tag{7.7}$$

$$a^2(a - a^2) = a^3 - a^4 = 1 - a, \tag{7.8}$$

$$a(a - a^2) = a^2 - a^3 = a^2 - 1. \tag{7.9}$$

Inverse calculation can be restated and easily solved as

$$\begin{bmatrix} I_0 \\ I_1 \\ I_2 \end{bmatrix} = T_s^{-1} \begin{bmatrix} I_A \\ I_B \\ I_C \end{bmatrix} = \frac{1}{3} \begin{bmatrix} 1 & 1 & 1 \\ 1 & a & a^2 \\ a & a^2 & a \end{bmatrix} \begin{bmatrix} I_A \\ I_B \\ I_C \end{bmatrix}, \tag{7.10}$$

$$|T_s| = 3(a - a2).$$

The preceding equation is the basis for determining whether or not the symmetrical components exist, taking as input the direct measurements of phases' electrical magnitudes [49, p. 79].

In three-wire systems, all instantaneous voltages and currents must sum to zero, as there is no neutral connection. Whether or not the neutral connection exists, single-phase currents can flow; otherwise, zero phase-sequence must be substituted as zero in previous equations [53, p. 250]. As single-phase currents flow through the neutral connection, it can be easily calculated as

$$I_0 = \frac{I_A + I_B + I_C}{3} = \frac{I_N}{3}. \tag{7.11}$$

Another analysis permits the inclusion of impedance variations between symmetrical systems,

$$\begin{bmatrix} V_A \\ V_B \\ V_C \end{bmatrix} = \begin{bmatrix} Z_A & 0 & 0 \\ 0 & Z_B & 0 \\ 0 & 0 & Z_c \end{bmatrix} [T_s] \begin{bmatrix} I_0 \\ I_1 \\ I_2 \end{bmatrix}. \tag{7.12}$$

The preceding equation allows the analysis of some faults like phase disconnection (open circuit) [53, p. 250].

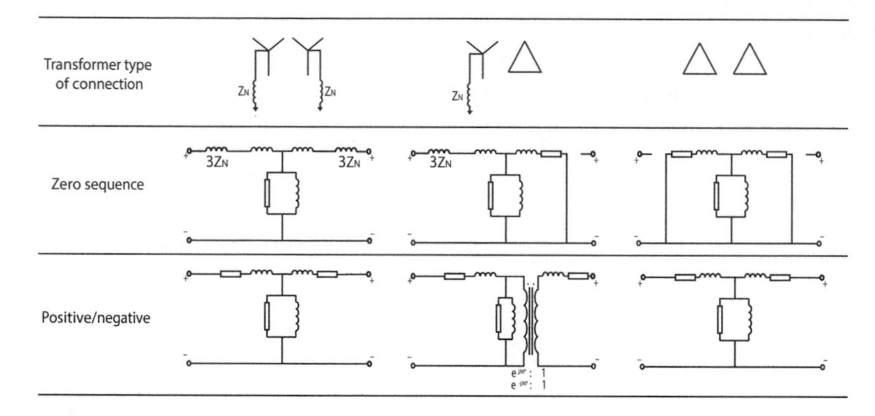

FIGURE 7.9
Sequence networks for Y-Y, Y-Δ, and Δ-Δ transformers

7.2.2 Sequence networks of electric machinery

Based on the Fortescue symmetrical components theory, electric machinery used in power systems can be decoupled of its symmetrical components with the objective of analyzing a complete power system under unbalanced conditions. In this section the decomposition of transformers, transmission lines, a synchronous machine, and an asynchronous machine is presented.

Transformers are the major component of power systems, and their type of connection is essential to determining their behavior in a fault condition. There are essentially three types of connections: Y-Y, Y-Δ, and Δ-Δ. For Y-Y and Δ-Δ transformers, the phase angle between voltage exchanges is 0°. In Figure 7.9, voltage phasor diagrams are represented as a per-unit system. As the reader can see, the zero positive and negative sequence are the same (equivalent circuit of a transformer), which is always valid for non-rotating machines.

For a Y-Δ transformer for the positive sequence, the high voltage side leads the corresponding low voltage side by 30°, while for the negative sequence the high voltage lags the low voltage, as shown in Figure 7.10. In power system notation it is common to label the high voltage side as H and the low voltage as X. The displacement is obtained from phasor analysis when the Δ side is converted into an equivalent Y for a simpler model. Figure 7.10 shows how if the actual Δ is converted to an equivalent Y, the resulting voltage will lag the high voltage side by 30°. This is not a special problem because it can be ignored when exhaust analysis is not required; for example: for calculating the value of circuit breakers and fuses the approximation is good enough. Indeed, we will not learn anything valuable by shifting the quantities in phase by ±30° [15].

For a Δ the zero sequence currents cannot enter or leave the windings, so the equivalent circuit is represented as an open circuit. Usually, the impedance

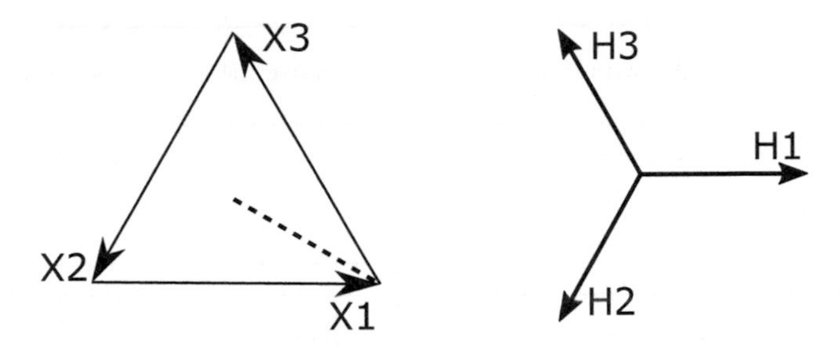

FIGURE 7.10
Voltage phasor diagram for a Δ-Y step-up transformer with HV side leading by 30°

due to the magnetizing currents and reactance are ignored because they are much larger than leakage reactance and resistance of windings.

In Figure 7.11 the sequence networks for transformers are presented.

Synchronous generators are usually grounded through an impedance Z_n. The impedances of electric machines are not equal for all the sequence networks as for transformers; an important problem emerges, since rotating machines are complex to describe mathematically and many assumptions are made for delivering impedances' values. To give the reader a sense, the phenomena considered for obtaining the equivalent circuits are the degree of saturation of the core, the linearity of the magnetic circuit, and the rotating speed, and others must be considered.

For the positive-sequence network, the magnetomotive force produced by the positive sequence current rotates at the synchronous speed, and a high quantity of flux crosses the rotor, such that the saturation provokes a high value positive-sequence impedance. For the negative-sequence network, the magnetomotive force rotates opposite to the synchronous speed and then its speed with respect to the rotor is twice the synchronous speed. In this case, currents are induced for preventing the magnetic flux from penetrating the rotor, and the negative sequence impedance is smaller than the positive sequence impedance. The zero sequence impedance is generated because of leakage flux, harmonic flux, and end turns, and its magnitude is smaller than positive and negative sequence impedances. Exhaustive analysis can be found in many classic books on the subject [46, 64, 34].

Also, it is important to know that the impedances needed for analyzing failure conditions in rotating machines are the subtransient ones, which are smaller than synchronous and transient impedances. The synchronous impedances are used for normal, steady-state conditions analysis while the transient impedances are used for stability studies. For transmission lines, it is obvious that the negative and positive sequence networks are equal because those are passive, bilateral devices. As explained previously, for complete transposed lines, the average impedance of the phases is equal. For

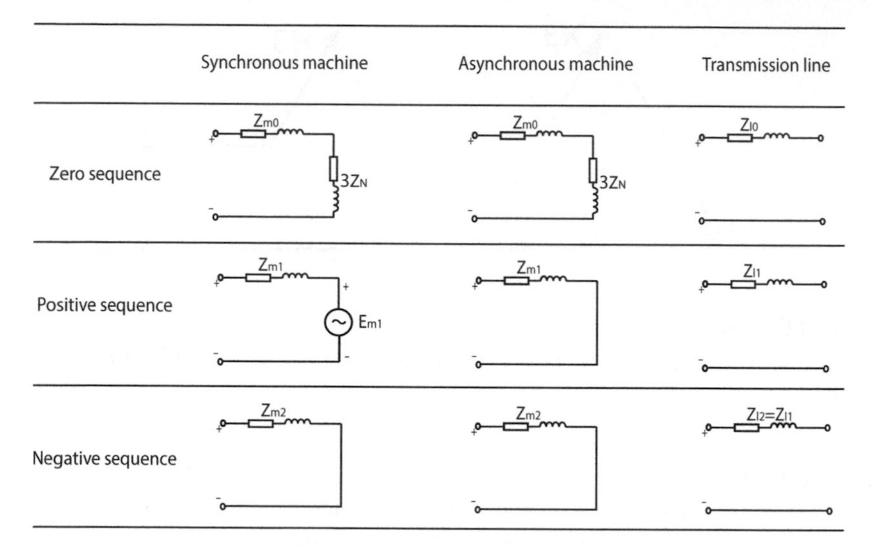

FIGURE 7.11
Sequence networks for rotating machinery and transmission lines

the zero sequence network the currents in each phase are identical in magnitude and phase. Those currents have to return by both the ground and the ground wires grounded at towers. The magnetic field patterns caused by zero-sequence currents is very different from those caused by positive or negative sequences where the currents have a phase difference of 120°. The zero sequence impedance of transmission lines usually ranges from 2 to 3.4 times the positive sequence impedance [30].

Figure 7.11 shows simplified models for network sequences, which do not consider saturation effects, magnetic losses, machine saliency, etc. Nevertheless, these networks are accurate enough for power system studies [38].

Once we have transformed a power system into three-sequence equivalent circuits, we can obtain the Thévenin equivalent on the failure point of the power system. As we have studied, the zero and the negative sequence do not have power sources, and the voltage sources of generators are only present in the positive sequence. So the Thévenin equivalents of negative and positive are only an impedance, while for the positive there is a voltage source in series with an impedance, as shown in Figure 7.12.

7.2.3 Coupled circuits for unsymmetrical faults

A very common application of symmetrical components is the calculation of fault currents, which, because of the nature of the faults, generate a non-balanced system.

For doing this, the interconnected sequence network formed from the

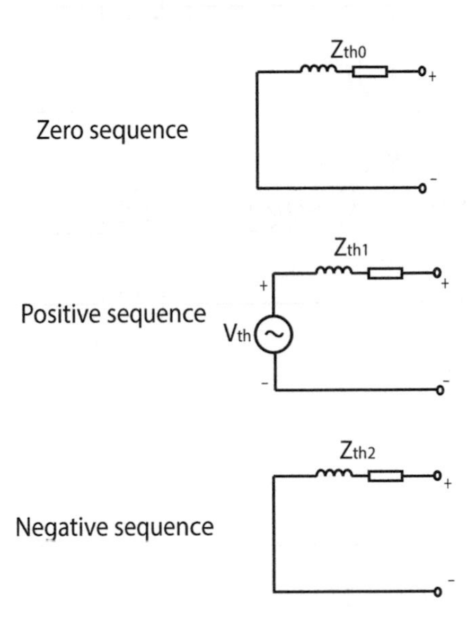

FIGURE 7.12
Thevénin equivalents of sequence networks

Thévenin equivalents of the sequence networks are used. The currents I_{a0}, I_{a1}, and I_{a2} are calculated and then transformed to phase currents by Equation (7.1):

$$\begin{bmatrix} I_a'' \\ I_b'' \\ I_c'' \end{bmatrix} = \begin{bmatrix} 1 & 1 & 1 \\ 1 & a^2 & a \\ 1 & a & a2 \end{bmatrix} \begin{bmatrix} I_{a0} \\ I_{a1} \\ I_{a2} \end{bmatrix} v(t) = V_m \cos(\omega t + \theta).$$

In this section the most common symmetrical faults of coupled circuits are presented.

7.2.3.1 Single line-to-ground fault

Phase-to-earth fault implies one of the phases is short-circuited to ground, and can be represented as in Figure 7.13, where the other phases are open-circuited. The approach shown is like the one in [53, pp. 253–254].

From the equations derived above,

$$\begin{bmatrix} I_0 \\ I_1 \\ I_2 \end{bmatrix} = \frac{1}{3} \begin{bmatrix} 1 & 1 & 1 \\ 1 & a^2 & a \\ 1 & a & a2 \end{bmatrix} \begin{bmatrix} I_A \\ 0 \\ 0 \end{bmatrix} = \frac{1}{3} \begin{bmatrix} I_A \\ I_A \\ I_A \end{bmatrix}. \tag{7.13}$$

Also,

$$V_A = E - I_1 Z_1 - I_2 Z_2 - I_0 Z_0 = E - I_1(Z_1 + Z_2 + Z_0) = 0, \tag{7.14}$$

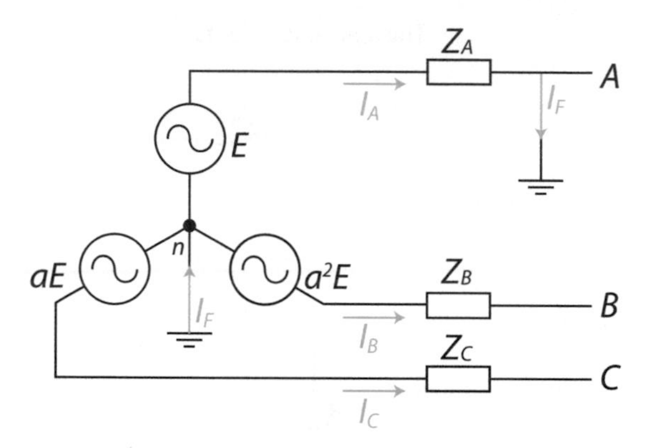

FIGURE 7.13
Phase-to-ground fault

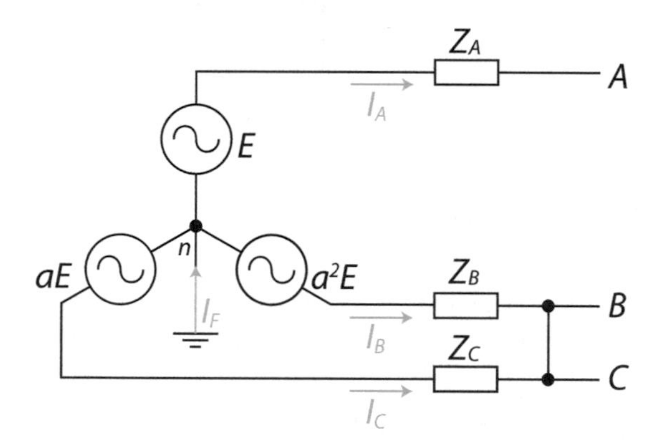

FIGURE 7.14
Phase-to-phase fault

leading to

$$I_1 = \frac{E}{Z_1 + Z_2 + Z_3}. \tag{7.15}$$

The fault current is that flowing through phase A, which in this case is

$$I_f = I_A = 3I_1. \tag{7.16}$$

7.2.3.2 Line-to-line fault

Another common fault is the phase-to-phase fault, which consists of two of the three phases short-circuited (Figure 7.14). Again, the approach made is similar to the one in [53, p. 256].

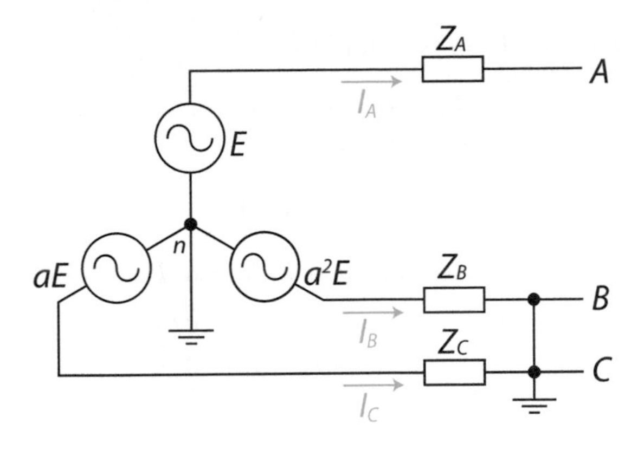

FIGURE 7.15
Phase-to-phase to ground fault

It can be quickly seen that

$$I_A = 0, \tag{7.17}$$
$$I_B = -I_C, \tag{7.18}$$
$$V_B = V_C \neq 0. \tag{7.19}$$

So,

$$I_0 = 0, \tag{7.20}$$
$$I_1 = \frac{1}{3} I_B (a - a^2), \tag{7.21}$$
$$I_2 = \frac{1}{3} I_B (a^2 - a), \tag{7.22}$$
$$I_1 = -I_2. \tag{7.23}$$

Matching phases voltages:

$$a^2 E - a^2 I_1 Z_1 - a I_2 Z_2 = aE - a I_1 Z_1 - a^2 I_2 Z_2, \tag{7.24}$$
$$E(a^2 - a) = I_1(-aZ_1 + a^2 Z_2 + a^2 Z_1 - aZ_2)$$
$$= I_1(Z_1(a^2) + Z_2(a^2 - a)). \tag{7.25}$$

Leading to:

$$I_1 = \frac{E}{Z_1 + Z_2}. \tag{7.26}$$

7.2.3.3 Double line to ground

If both phases are not only short-circuited between them but with ground, the analysis becomes different (Figure 7.15) [53, pp. 257–258].

Statement of the problem is as follows:

$$I_A = I_1 + I_2 + I_0 = 0, \tag{7.27}$$
$$V_B = V_C = 0. \tag{7.28}$$

So, a linear equation system can be built by:

$$a^2 Z_1 I_1 + a Z_2 I_2 + Z_0 I_0 = a^2 E, \tag{7.29}$$
$$a Z_1 I_1 + a^2 Z_2 I_2 + Z_0 I_0 = a E c, \tag{7.30}$$
$$I_1 + I_2 + I_3 = 0. \tag{7.31}$$

Solving for each current:

$$I_1 = \frac{E}{Z_1 + \left[\frac{Z_2 Z_0}{Z_2 + Z_0}\right]}, \tag{7.32}$$
$$I_2 = -\frac{I_1 Z_0}{Z_2 + Z_0}, \tag{7.33}$$
$$I_0 = -\frac{I_1 Z_2}{Z_2 + Z_0}. \tag{7.34}$$

7.2.3.4 Summary of coupled circuits

Based on the analysis made in the three previous sub-sections, it is convenient to obtain the Thévenin equivalent of each of the sequence networks and re-couple them for calculating the current of the symmetrical component, and then calculate the actual failure current.

In Figure 7.16 the schemes of the most common faults and the interconnected network are presented for the most common types of faults. In addition to the single and double line-to-ground and line-to-line failure, the three-phase or symmetrical fault is represented. Obviously, as this is not an unsymmetrical failure, the zero, negative and positive sequence networks are completely decoupled.

7.2.4 Exercises

7.2.4.1 Line-to-earth fault protection

In this exercise, the feeder manager relay will be programmed in order to break the circuit when a fault condition appears: in this case a solid fault-to-earth fault in one phase of the transmission line.

Connect Diagram 7.4 proposed for the fault-to-earth and start the practice shown in Figure 7.17.

Close contactor 1 and notice the measured voltages of the capacitors' currents. Close contactor 2 and observe that the voltage in one phase becomes zero with respect to earth, as will occur in a phase-to-earth fault condition.

The next step is to configure the feeder manager relay for opening the

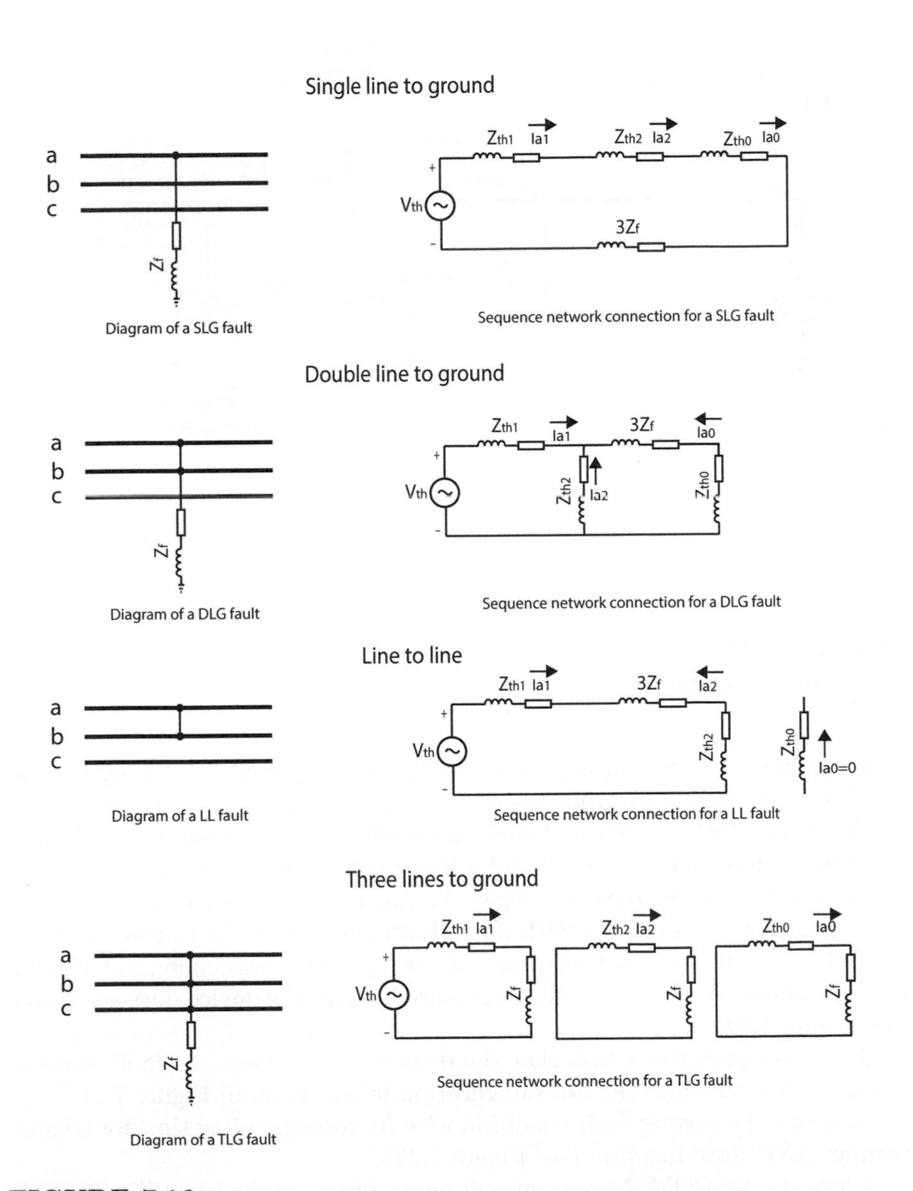

FIGURE 7.16
Diagram of the electric fault and interconnected sequence network for most commons faults

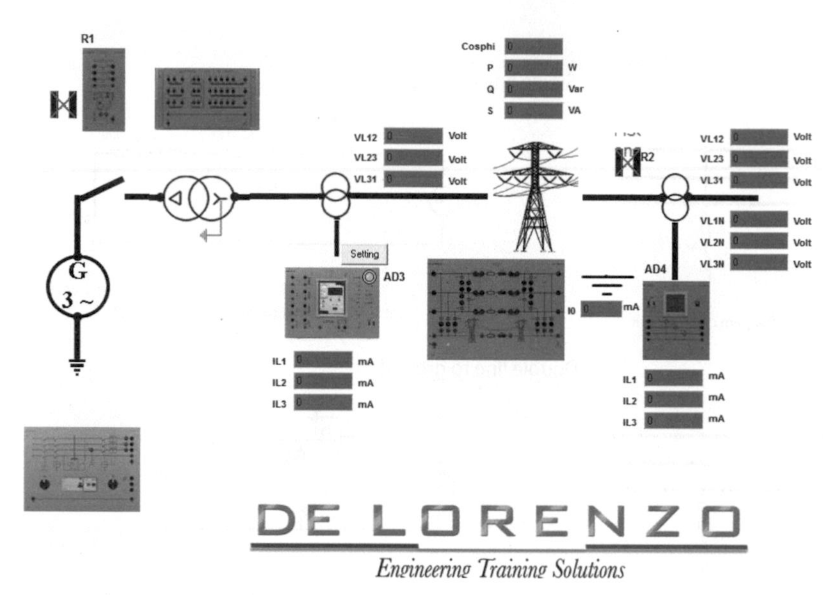

FIGURE 7.17
Connection diagram

power supply when a fault occurs. Close the smart grid SCADA program for avoiding communication problems with the MSCom configuration program.

Open the MSCom 2.exe and click on search − > on the serial port. In the emerging window select COM 3 and a Baud rate of 9600 (see Figure 7.18).

Once the device is connected, open the function settings (see Figure 7.19).

Configure the 2Is current with the following info; this allows you to set up the failure condition as a percentage of the I_n value (also configurable) and the time of continuous failure state measured before the device does an action (see Figure 7.20).

Then configure the action that the device will do when the fault state is triggered. You can find the do configuration in the menu in Figure 7.21.

Now link the current fault condition with its corresponding time for trigger with an AND logic function (see Figure 7.22).

Close the MSCOM 2 program and again open the smart grid's SCADA program, and connect relays 1 and 2. Observe how the device starts counting and then open the circuit (see Figure 7.23).

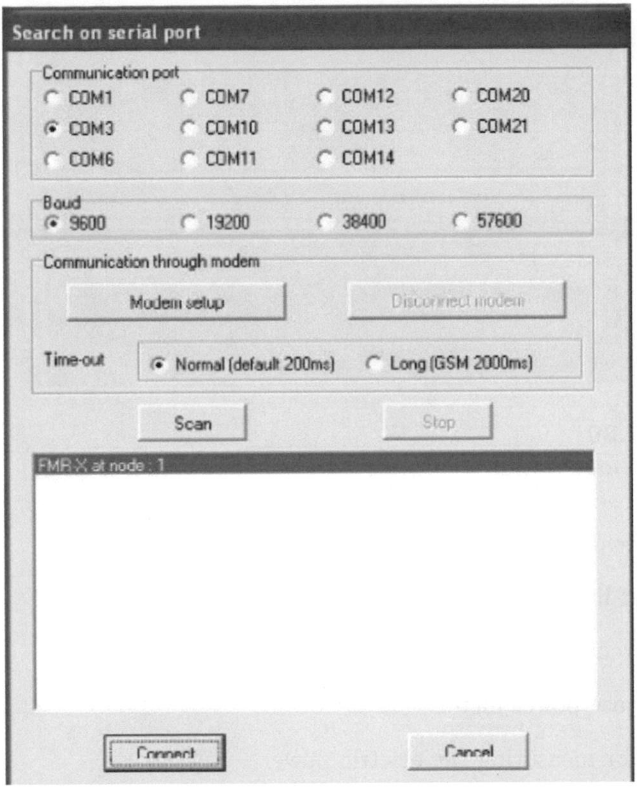

Search on serial port
Communication port
COM1 COM7 COM12 COM20
COM3 COM10 COM13 COM21
COM6 COM11 COM14
Baud
9600 19200 38400 57600
Communication through modem
Modem setup
Disconnect modem
Time-out
Normal (default 200ms)
Long (GSM 2000ms)
Scan
Stop
FMR-X at node : 1
Connect
Cancel

FIGURE 7.18
Baud rate configuration

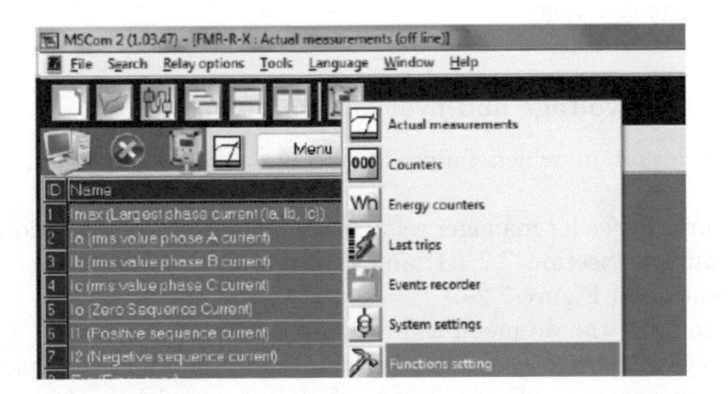

FIGURE 7.19
Function settings

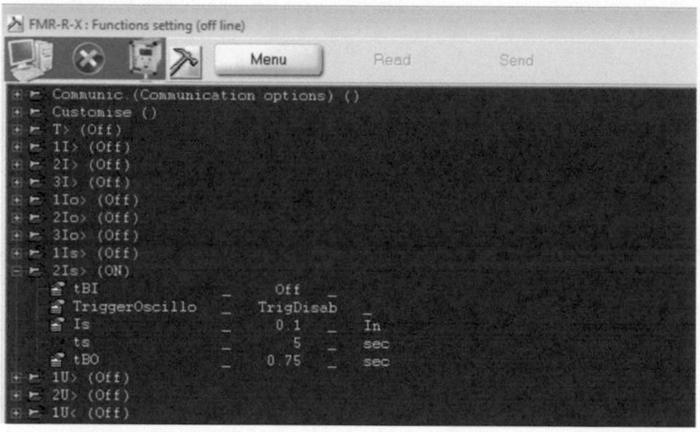

FIGURE 7.20
2Is current configuration

Recommended equipment and connections:

- Overhead line model — Long
- Power circuit breaker
- Three-phase power meter
- Module for measuring the electric power
- Three-phase transformer
- Three-phase supply unit
- Feeder manager relay

7.2.4.2 Undervoltage and overvoltage protection

Connect a circuit in which failure conditions can be triggered, as in Diagram 7.5.

Configure the feeder manager relay for an undervoltage condition following the procedure in Section 7.2.4.1, and configuring the function settings with the information in Figure 7.24.

Also configure the do menu as in Figure 7.25.

Now feed the power system and generate an undervoltage condition. For example, you can take off one bridge of the line model as shown in Figure 7.26.

Finally, observe how the device breaks the circuit when the time has elapsed. For the overvoltage condition, configure the function settings and the do menus as in Figure 7.27 and Figure 7.28.

In order to generate the overvoltage, you can carefully disconnect one phase

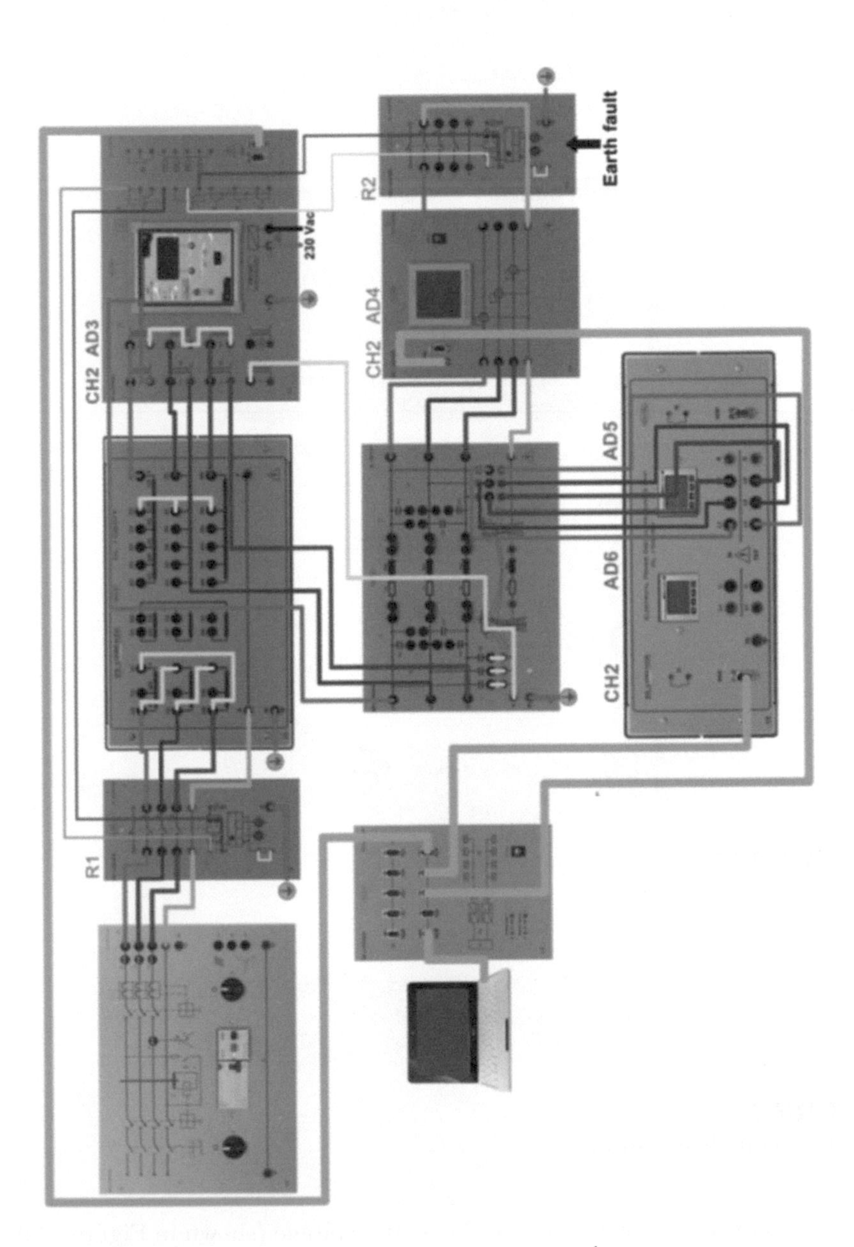

Diagram 7.4
Connections for line-to-ground fault

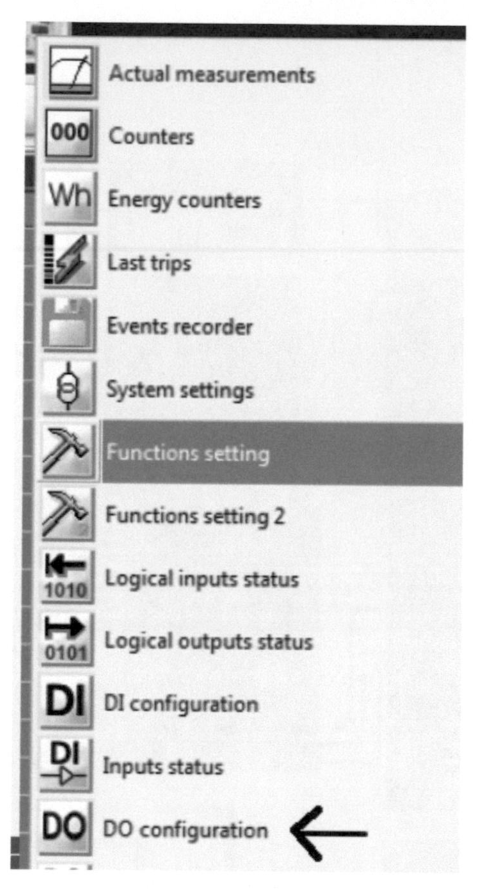

FIGURE 7.21
Configuring the action

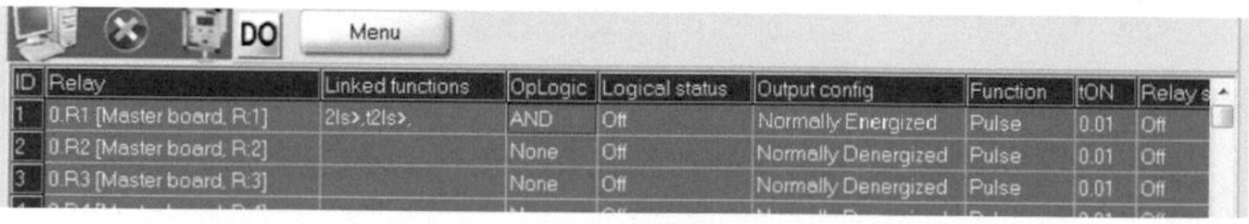

DO Menu

ID	Relay	Linked functions	OpLogic	Logical status	Output config	Function	tON	Relay s
1	0.R1 [Master board, R:1]	2Is>,t2Is>,	AND	Off	Normally Energized	Pulse	0.01	Off
2	0.R2 [Master board, R:2]		None	Off	Normally Denergized	Pulse	0.01	Off
3	0.R3 [Master board, R:3]		None	Off	Normally Denergized	Pulse	0.01	Off

FIGURE 7.22
Linking the fault condition

of the transformer and connect it to the +10% voltage (shown in Figure 7.29). Observe how the failure condition is triggered.

Recommended equipment and connections:

- Resistive load
- Inductive load

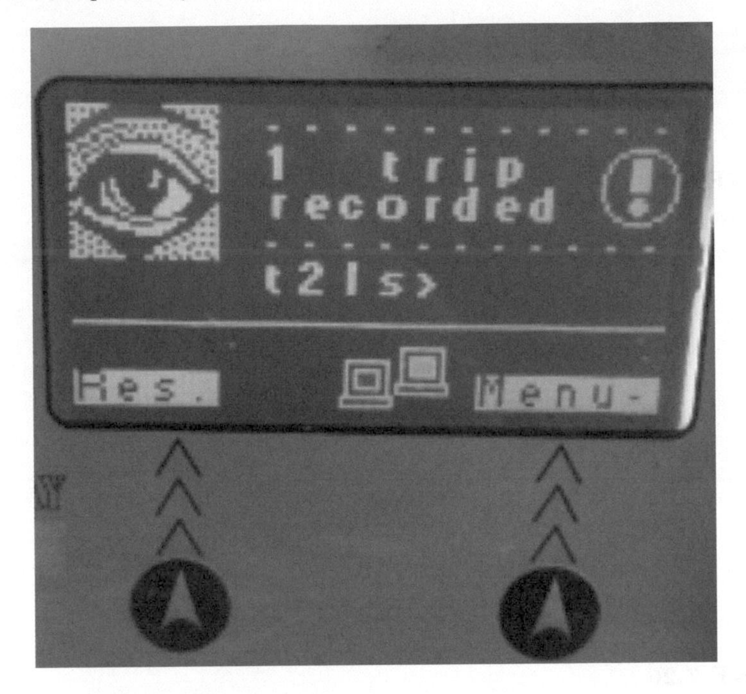

FIGURE 7.23
Device tripped

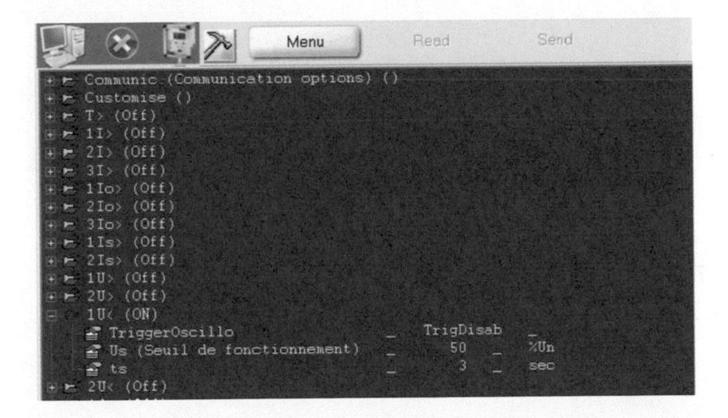

FIGURE 7.24
Configuring function settings

- Capacitive load
- Overhead line model — Long
- Overhead line model — Medium
- Power circuit breaker

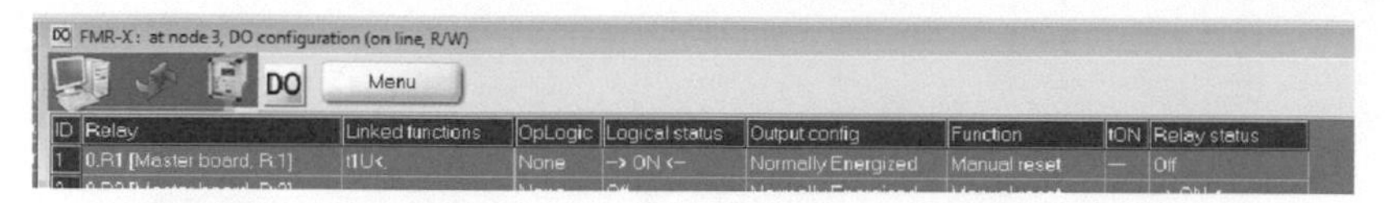

FIGURE 7.25
Configuration of do menu

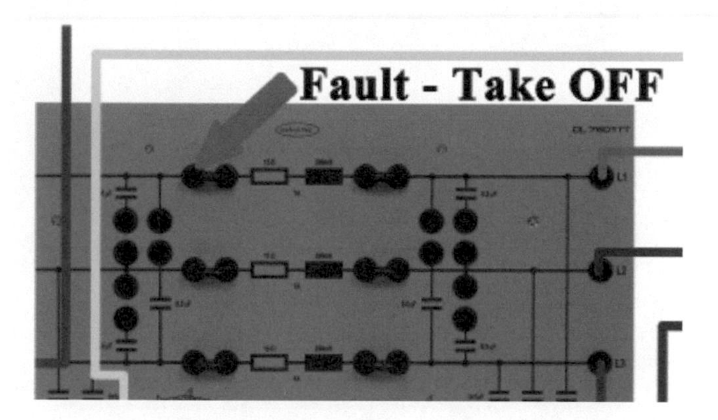

FIGURE 7.26
Example

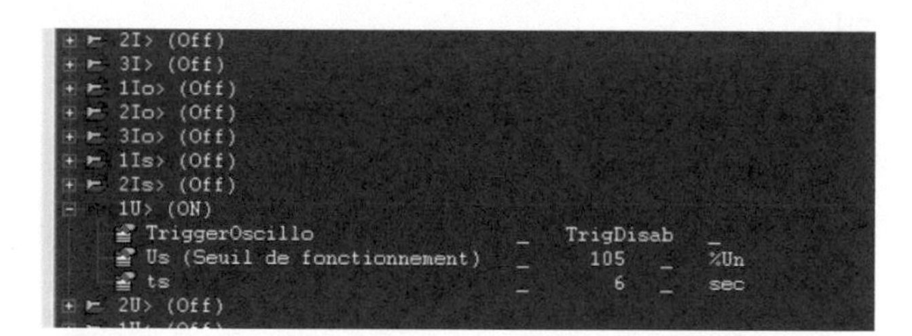

FIGURE 7.27
Configuring function settings

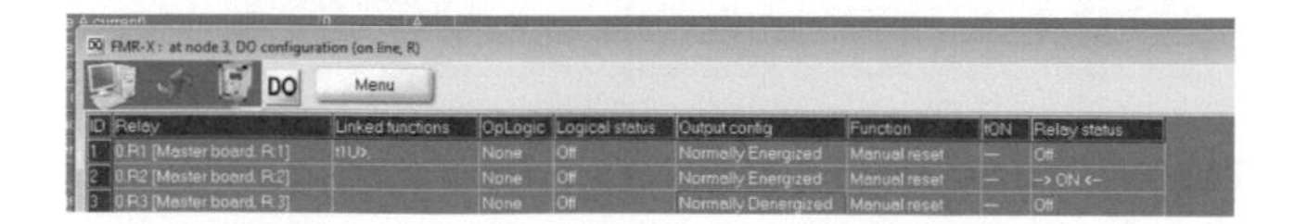

FIGURE 7.28
DO configuration

- Three-phase power meter
- Three-phase transformer

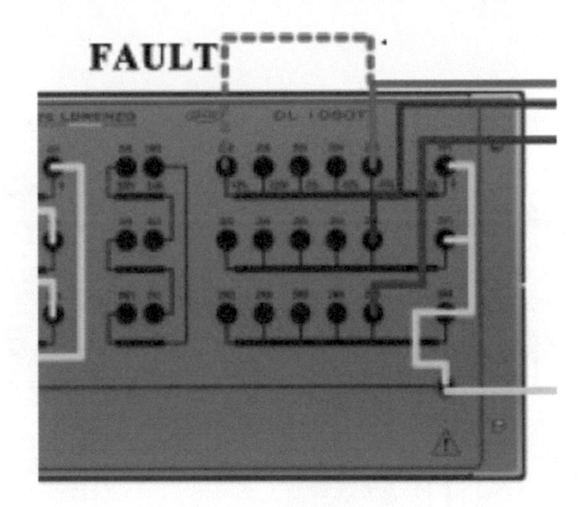

FIGURE 7.29
Generating overvoltage

- Three-phase supply unit
- Feeder manager relay

7.3 Electric protections

Every step in the power system requires meeting public demands in sufficient quantity, and being reliable and safe, so the huge investment first made can be effectively exploited by supplying a continuous service [36, p. 1]. In order to maintain these requirements without major problems, there are two possible paths to take:

- Implement a system using only non-failing components, which require nil maintenance
- Foresee possible failures, design the whole system with isolable modules, and install detection equipment to manage it

The first presented option is obviously infeasible (if not impossible) and highly expensive. The second option pretends to use a simpler system that

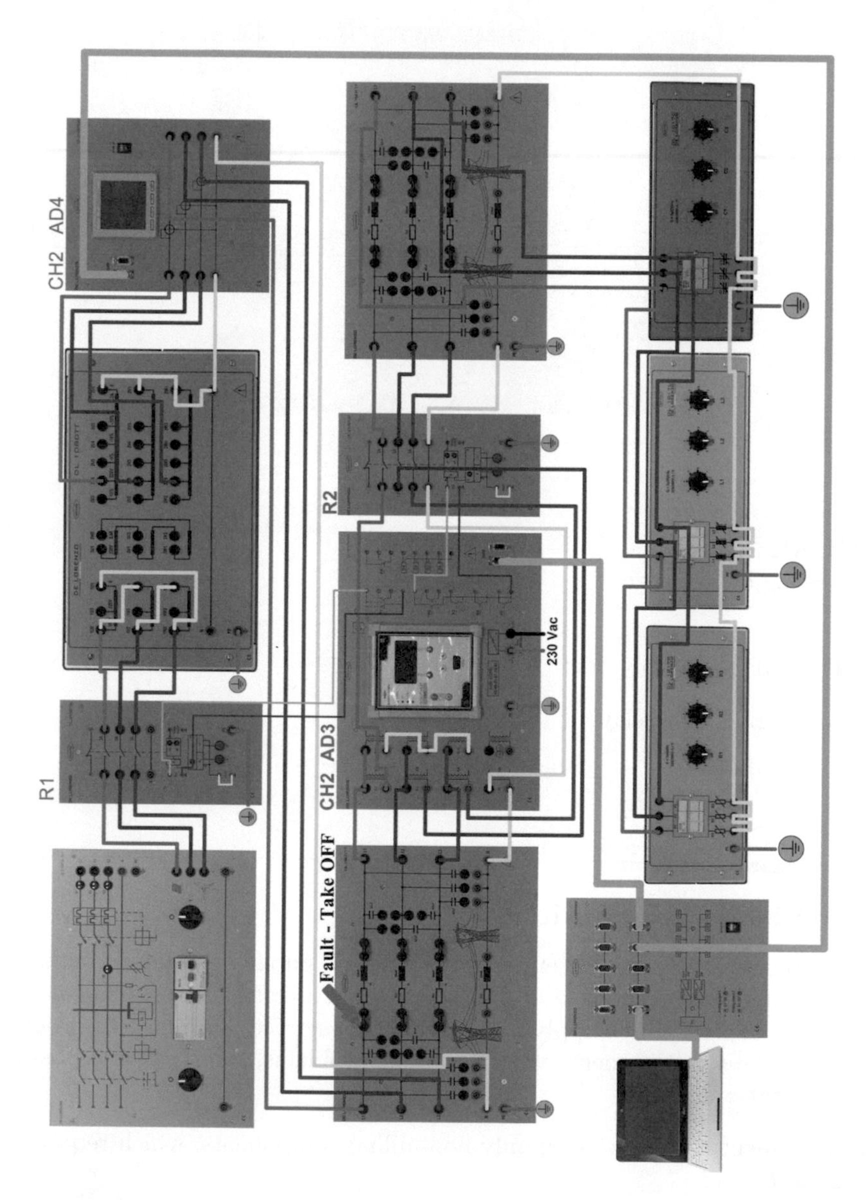

Diagram 7.5
Connections for undervoltage and overvoltage faults

cannot guarantee whole-time operation; however, the use of protective equipment (also referred to as protective relay) enables many possibilities regarding security issues:

- Maintenance can be effectively programmed and given
- Big failures can be avoided by isolating the faulty module
- Small failures can be avoided by parallel devices or control modules
- Service can be provided continuously
- The system can be fully monitored
- Resulting system is safe to a high extent
- Quick restoration time after failure

A protective relay is a device that instructs the system to disconnect a faulty module, letting other parts still operate without risk of being affected by the former failure. As the protective equipment allows the power system to be operative with minimum interruptions, and enhancing restoration time, its use is evidently very vital [36, p. 1]. In other words, its "fundamental objective is to provide isolation of a problem area in the power system quickly, so that the shock to the rest of the system is minimized and as much as possible is left to continue service" [49, p. 18].

The perspective of protective devices is not to anticipate faults or troubles; they act only at the presence of abnormal operation whenever the operation limits have been surpassed. Although protective equipment implies the "prevention" of fatal failures, it cannot actually prevent (anticipate) the unwanted state from happening. It can only minimize and limit further damage in equipment, facilities, or employees. [49, p. 18].

Protective equipment must have the following characteristics [49, p. 18], [36, p. 2]:

Reliability: A certain degree of confidence in the correct behavior of the protection.

Dependable: It must depend on the trip conditions

Secure: It must not trip if there is no faulty scenario

Selectivity: Detection and isolation of the problematic module.

Speed: Minimum time between the occurrence of the fault and the protection action.

Sensitivity: To react even towards the smallest fault.

Simplicity: Minimum equipment and equipment requirements to achieve protection goals.

Stability: Guarantee continuous operation of adjacent system modules.

Economics: Maximum protection at minimal cost

The characteristics listed above imply a challenge for designers at all levels, from the electronics and instrumentation layers to the integration at automation works. One of the most recurrent obstacles to system protection is that it is not directly refundable as it is not required for normal system operation and failures are infrequent [49, p. 23]. However, occurrence and magnitude of failures are unpredictable and could force a new big investment if the system is damaged; in this way, protection is needed to minimize the failure effects, hoping it will never be used.

Some systems different from relays are also used to manage power flow as per a modern trend and pursuit of stable service and energy management. Flexible AC transmission systems (FACTS) are able to provide a real-time control of power flows [53, p. 467] by incorporating condensers in series/shunts to avoid voltage drops in transmission lines as well as to correct power factor in an automatic manner. A FACTS can be defined as "a power electronic based system and other static equipment that provide control of one or more AC transmission system parameters to enhance controllability and increase power transfer capability" [10].

A particular type of FACTS is the STATCOM (Static Synchronous Compensator), also known as a static synchronous condenser, as it is based on a DC capacitor that provides power through a voltage source converter (VSC). Whenever the STATCOM voltage is under the bus voltage, it sinks reactive power; otherwise, it provides reactive power. Controlling the VSC based on measured voltages allows the STATCOM to regulate power flow.

While FACTS are not precisely protective, they are actually preventive, especially because of their real-time compensating capabilities. Similar systems are considered to be a new trend capable of changing the conventional philosophy in power systems operation and development [77, p. 466].

7.3.1 Types of faults and relays

Faults can be classified depending on three parameters: immediacy of current flow, consequences' permanence, and the polyphasic conditions [36, pp. 7–9].

Active: Implies current flowing from one phase to other phase or to ground.

- **Solid:** It results from an immediate insulation breakdown. It is destructive and is usually accompanied by explosions and flashing: e.g., a rockfall crushes a cable.
- **Incipient:** It starts mildly. If the fault develops further with time it can become solid: corona effect.

Passive: They are not actually faults but conditions that will lead to them.

Commonly related to taking the system to operating points it was not designed to attain. Some examples are:

- Overload
- Overvoltage

Transient: They do not imply permanent damage to the system or its insulation. The system can be used again after a transient fault.

Permanent: System insulation is damaged and its operation is compromised. Repairs must be made before the system can be re-energized.

Symmetrical: In three-phase systems a symmetrical fault is balanced, so it can be considered to be at steady state.

Asymmetrical: If the fault shows offsets and decays, an unbalance condition is assumed, so the system is called to present an asymmetrical fault.

The most common way to classify relays is by their function [49, pp. 25–26].

Protective: They become active if the system faces an intolerable condition.

Regulating: They attempt to control the electric levels whose variation is due to load changes. They are not used for fault conditions.

Reclosing, Synchronism Check: They are used to reconnect the system after an outage.

Monitoring: They do not directly detect the faults but their related variables. Lamps and alarms could be considered monitoring relays.

Auxiliary: They can take the effect of a given relay to apply over other modules of the system. They provide other relays with the ability to reach multiple objectives or higher voltages/currents.

Protective relays can also be classified depending on the technology they use as the main means of energy flow interruption. In a board sense, they are electromechanical or static, where static implies the omission of moving mechanical parts [36, p. 108]. Both analogic and digital electronic relays are static and have displaced the traditional ones for many reasons. They are more cost-convenient, accurate, flexible, sensible, fast, and reliable; moreover, recent improvements in silicon integration and power management ICs (integrated circuits) have minimized sizes and enabled designers to integrate various protection systems within a single board. Usage of traditional relays together with static ones is a common practice that attempts to incorporate a cost-benefit congruent protection system.

Along with analogic electronics improvements, microprocessors and microcontrollers are quickly incorporating more functions and capabilities, allowing designers to determine the relay's behavior through pre-programmed instructions and calculations (algorithms) [36, p. 108]. Digital relaying implies data acquisition and conversion (analog to digital), numerical processing, and power electronics coupling. Depending on the application, the time needed to acquire the signals and process the instructions will become paramount when using a microprocessor-based relay; if the desired characteristic to capture is fast, like current peaks, the time restriction became important for the relay's correct operation. However, a typical sampling time of 1 $[ms]$ is enough for taking 10 samples of every half cycle of a 50 $[Hz]$ wave, which is convenient in many protection applications. Whenever the reaction time is desired to be short, solid-state electronics protections are preferred.

Digital protections are commonly attached to output mechanical relays that take $10 - 15$ $[ms]$ to respond [36, p. 113]. This implies that the fault will be attended tardily from its detection; nonetheless, it is totally comparable to traditional protections with standard performance, also including several better characteristics like self-supervision, communication protocols, and configurability (magnitudes/times).

The term intelligent electronic device IED is commonly used for digital microcontroller-based power systems that achieve protection and regulation of electric magnitudes. A single IED can integrate around a dozen functions, including configurability, communication protocols, and automation capabilities.

7.3.2 Relay operation

One principal protection paradigm is differential protection; the electrical magnitudes that go in and out from a black-box system are compared in order to find differences that reveal malfunctioning inside the system. Although equal magnitudes are not to be sensed as the system has losses, a fault can be detected if this difference exceeds a certain threshold (Figure 7.30). The system can effectively detect faults as current changes drastically whenever a fault condition is met, so the associated relays can be triggered [49, p. 181].

The preceding principle can be applied to individual modules or sets of modules, making the fault detection capable of localizing the fault sources. The relay shown in Figure 7.30 represents a coil whose activation depends on current flow; its contact can be placed in line with the module input so a failure detection makes the contact open.

There are some basic electromechanical relays that have been used since the early twentieth century [49, p. 186], and that represent the protective relay area foundations. While their use has been relegated to modern solid state and digital units, knowing their operating principles makes it easier to understand how the problem was solved traditionally and why they are being replaced. Beyond the classical still-in-use clapper and plunger relays, the overcurrent relay (based on induction disc) and the cylinder relay (based on induction cup) were once the main solution to the protective relay problem.

The induction disc shown in Figure 7.31 receives an AC voltage or current

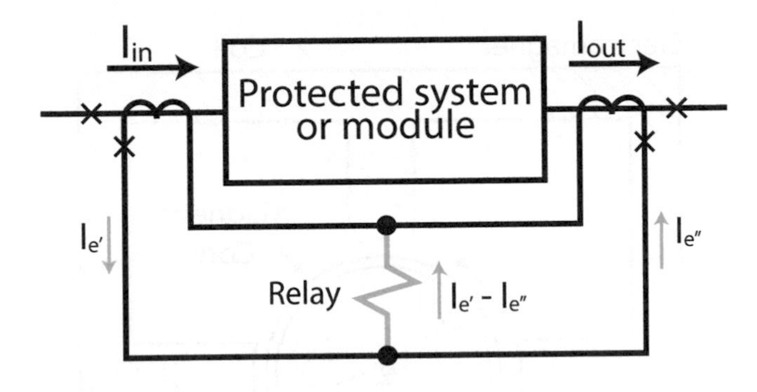

FIGURE 7.30
Differential principle

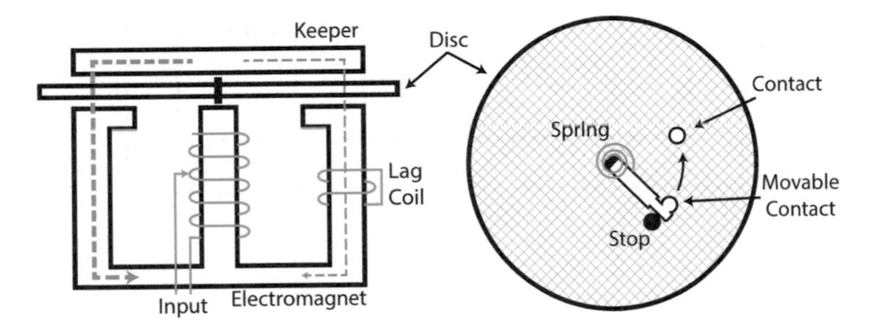

FIGURE 7.31
Induction disc relay

on its input coil, which generates magnetic flux along the electromagnet and the keeper. The lag short-circuited coil present in one of the electromagnet legs lags the magnetic flux linked to it, making the movable disc rotate. The disc is attached to a movable contact on its top, which moves radially with disc rotation. If the fault condition takes long enough to allow both contacts to join, the relay is triggered and further equipment can be enabled or disabled. The disc includes a radial spring on its center that opposes contact movement and returns to the stop condition if the fault is not present anymore [49, pp. 186–189].

These relays were commonly known as time-overcurrent relays. Their operation can be understood quite straightforwardly for overcurrent conditions, where a high current would move the disc quickly while a low current would take some time before the contact reaches the active state. The stop position can also be used as a normally closed contact, so two actions can be performed by the same relay at once.

There are cases in which it is important to know the direction of the current or the relation towards a certain reference. Whenever the measured electrical magnitude's direction is desired as a means of protection, a cylinder

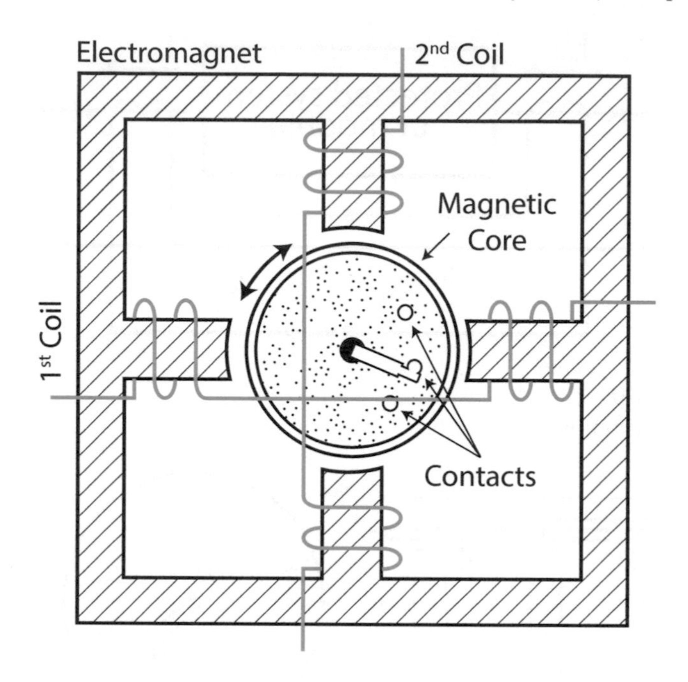

FIGURE 7.32
Cylinder relay

relay based on an induction cup (Figure 7.32) can be used. Imagine that for a given system, the current is known to lag by 30° phase A voltage; whether or not voltage is connected to the first coil and current to the second coil of cylindrical relay, the magnetic core at the center will orientate axially to the magnetic flux generated, moving a contact to a certain operation position. Whenever the current changes in magnitude or phase, the moving contact will also move itself towards a contact, announcing the fault condition [49, pp. 191–192].

Cylinder relays were extensively used for directional sensing, which includes among its applications phase and polarity fault detection. Its use together with time-overcurrent relays covered many fault scenarios and was the prime option in protective relaying during the first half of the twentieth century.

Besides the two most common and useful relays mentioned above, there is another that presents a way to compare magnitudes and calculate tolerable impedances. Distance relays (Figure 7.33) compare two magnitudes between its contacts through an electromagnetic bar that is influenced by both of them, and can tilt depending on the magnitude of each one over a fixed pivot. If voltage and current are connected to the relay and it is adjusted so it reaches equilibrium at a certain threshold condition, a change in impedance can be detected as $Z = V/I$ [49, pp. 192–193].

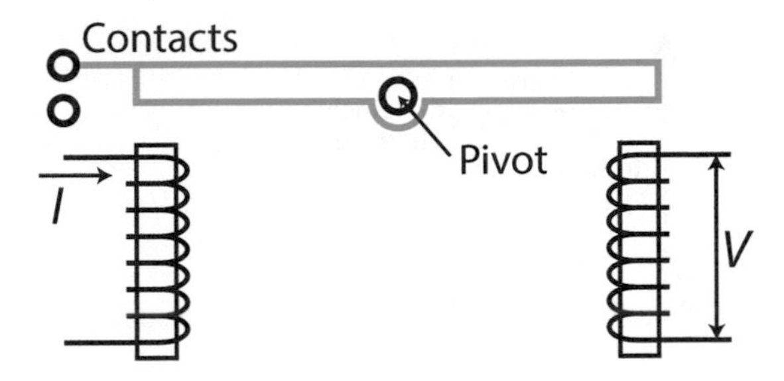

FIGURE 7.33
Distance relay

As mentioned before, traditional relays that allow measurement and comparison are no longer used and have been replaced by static systems; however, electromechanical devices in continuity applications are commonly attached to sensing/monitoring systems to intervene in electric line connection.

7.3.3 Overcurrent protection

Most transmission line protection equipment includes coverage of different faulty scenarios, making it possible to address other electrical issues differently from distance protection. Besides the trip polygon's being capable of detecting many electrical problems on the overhead lines, differentiation over fault origin allows an improved performance decision-making, thus providing the system with alternative paths through localized disconnection, a history log of specific events, and a better understanding of the causes and consequences of the fault, among other things.

In this particular case, overcurrent protection will be our first step into the operation of the distance protection equipment in terms of configuration, as well as connection and integration with the other smart grid modules.

This type of relay is the simplest and cheapest type of protection. However, when the system changes, they need to be readjusted, or in the worst scenario, replaced. The typical applications are phase and ground fault protection on distribution circuits, and electric utility in industrial and household applications where the cost of distance relay is not justified. Overcurrent relays have been slowly replaced by distance relaying, but they continue to be extensively used; therefore it is worthwhile to study those systems.

The most important reasons for suiting them to overcurrent relaying on distribution systems are its low price and simplicity, the fact that very often relaying does not need to be directional, and that reactor or capacitor tripping may be used.

As the name implies, overcurrent relays are designed to operate when more than a specified amount of current flows through the protection, and there are two types of overcurrent relays: time delay and instantaneous. Instantaneous

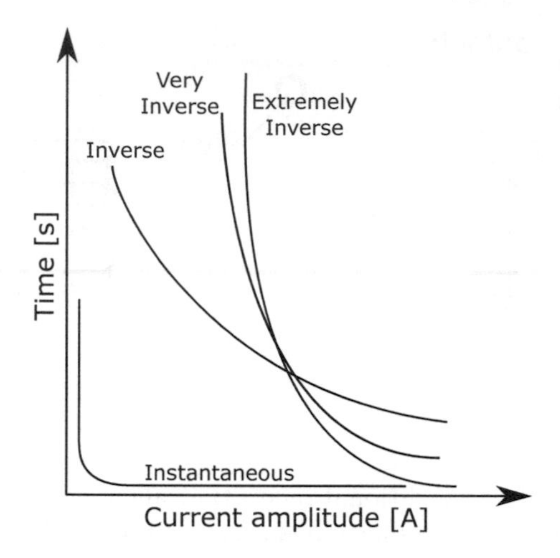

FIGURE 7.34
Distance relay

needs to be as quick as possible for commercial protection. The operation time may be from 0.016 [s] to 0.1 [s] and the characteristics are illustrated in Figure 7.34.

Time-delay protection has the characteristic of a trigger inversely to the current flowing in the relay. This typical characteristic is shown in Figure 7.34. The most common time-delay relay's characteristics are: inverse, very inverse, and extremely inverse, which describe the rate as the time delay decreases with the increase of the current.

Overcurrent relays are used primarily on distribution systems where the low cost of the components is important. They can be installed along the distribution systems, placing the fastest relays near to loads and time-delay relays near the distribution substation.

Instantaneous relaying is used mainly in situations when the short circuit current is substantially greater than any other possible situation. The adjustment of sensitivity is based on the calculation of the short circuit current of the far end of the protected line, and configured for a 25% higher current. Usually, the instantaneous relay configured as previously described will cover 80% of the line section.

An important disadvantage of overcurrent relays is that during the low periods of power consumption, diesel and gas-based turbines are usually shut down, while at high-load periods the system may split into several parts. Whatever the case, the short-circuit current will vary with the amount of generators.

Pickup and time-delay settings need to be configured on time-delay relays; those characteristics refer to:

Pickup settings: The pickup value is the minimum value of the operating

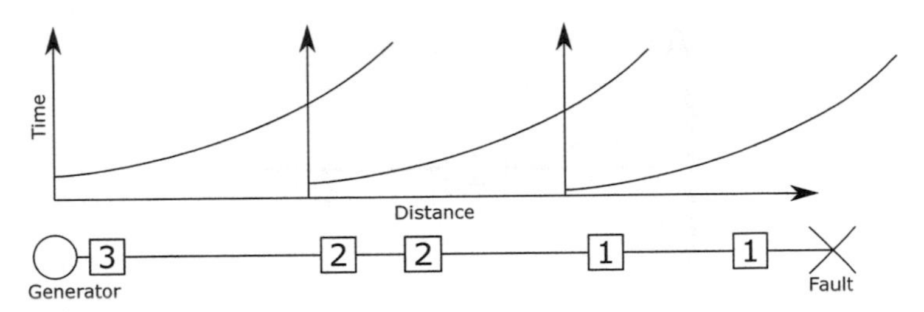

FIGURE 7.35
Time-delay in trigger characteristic of overcurrent relays

current or other input quantity that is considered hazardous to operation of the line. First we need to determine which values of minimum current trigger the relay, and the maximum current that shall trigger the circuit breaker immediately. Those values must be calculated for the protection of all short circuit types in the line section protected. Whenever possible, the settings should give additional protection to the next lines or adjoining equipment, but it is important to remember that the primary function is to protect the line itself and not compromise backup features.

Time-delay settings: The purpose of time delay is to coordinate relays with each other, so the closest relay to the fault will open first, and if the problem is not fixed, the next circuit breaker should be triggered. As there are more line sections in series, the tripping time increases at the source end. This is done to avoid catastrophic blackouts in the power system. Figure 7.35 shows the time response of the relays according to the distance at which the fault is generated. As can be seen, relay 1 is configured to trigger immediately, the second one has a small time delay, and the relay closest to the generator has a bigger time delay. This characteristic of triggering was traditionally based on a rotating disc that closes a contact. The larger the area that the rotating disc traveled, the larger the triggering time. Modern relays use a digital timer to reach this action.

Example 7.1. *Determine the CT ratio, pickup, and time-delay settings for the transformer at circuit breaker 1 assuming non-pilot relaying. Assume that nominal current is 60 [A], minimum fault current is 300 [A], and maximum fault current is 600 [A]. Select the CT ratio in order to give 1 [A] at the secondary current for the maximum load (see Figure 7.36).*

The current ratio shall be 60:1, since we want to reduce the trigger signal from 60 [A] to 1 [A]. Some authors recommend configuring the relay settings as twice the maximum load and one third of the minimum fault divided by 60. The resulting configuration is: $[1.66, 20]$.

The operation current is 60 [A] and the minimum current of triggering is set to be $1.66 \times 60 = 100$, *giving* 1.66 *pu to prevent false operation. Assuming no coordination, the relay shall trigger at its fastest configuration.*

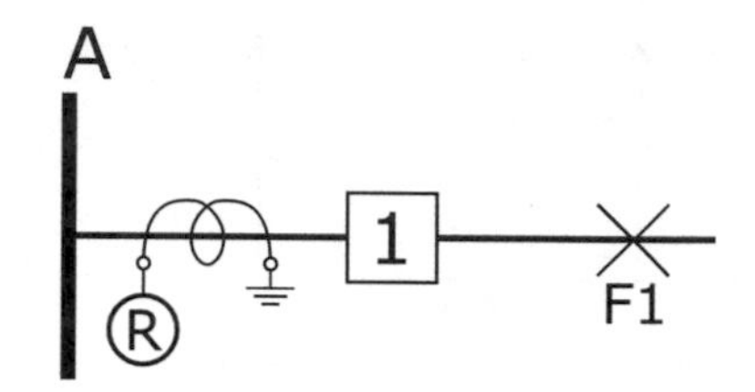

FIGURE 7.36
Example diagram

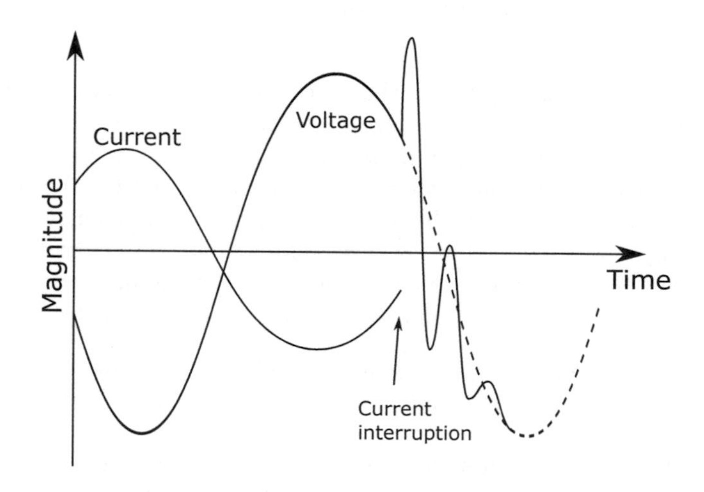

FIGURE 7.37
Overvoltage caused by current interruption in a transmission line

7.3.4 Overvoltage protection

Whenever the grid system presents an abnormal or unexpected rise in voltage, this can be detected by parallel connected protection equipment so that the grid can be disconnected in order to prevent damage to connected loads and generators.

In this lab, a slight overvoltage condition will be generated to test protection against this particular condition, considering a different relay-logic approach, and the "reset level" parameter commonly found in protection equipment.

Internal overvoltage occurs when overvoltage is generated by the system itself; for example, in switching operations, fault conditions, resonance phenomena, and sudden changes in system parameters.

Switching operation: When transformers or other electric machines are switched out, the magnetizing current may remain before the instantaneous value of the current reaches zero, because it cannot be dissipated immediately. Instead, the energy will be dissipated in a damped oscillation due to the reactive nature of the transformer together with the capacitance of the connections. The magnetic energy at the instant of chopping the current is $(Im^2 \times L/2)$, where Im is the magnetizing current and L is the electric machine inductance. In a similar way, the electric energy is equal to $V_0^2 \times C/2$, where V_0 is the peak value of the voltage and C is the capacitance of the windings and connections to the electric machine. At the moment of chopping, the voltage induced magnitude is: $V_0 = Im\sqrt{(L/C)}$ with a natural frequency of $f_n = 1/(2\pi\sqrt{LC})$. Figure 7.37 illustrates the induced voltage at the moment of current chopping.

Resonance: This phenomenon occurs in an electric circuit when the admittance and reactance of a series RLC circuit have the same magnitude and therefore cancel each other because of the 180° phase displacement. A situation like that may arise in a partially open circuit (faulty circuit breaker or broken conductor) when one phase remains connected while the other two are opened. As shown in Figure 7.38, if the phase c remains connected while the other two are opened, a resulting circuit and the inductive reactance $(\omega L + \omega L/2)$ are equal to the capacitive reactance $(1/2\omega C)$. It is important to notice that a condition for a possible resonance is not to clamp the neutral point of the load, despite the fact that the neutral point of the generator is clamped.

Sudden loss of load: The most common reason for overvoltage in power systems is the sudden disconnection of some heavy loads. Since the current of the synchronous generator decreases, the voltage at the terminals will increase because the internal voltage generated is dependent on the speed of the machine and the excitation current of the field. Considering that there is no regulation device for the field current, the internal voltage of the synchronous generator will remain constant; therefore, the voltage at the end terminals will increase. Although this problem is easy to solve, if no action is taken, overvoltage may damage the transformers, and other equipment connected to the generator. Figure 7.39 shows a synchronous generator as a simplified static model to illustrate the issue described.

Most electric machinery operates close to the knee point of ferromagnetic material. Therefore, a small overvoltage is reflected in a large increase in excitation current that may cause overheating and permanent damage to equipment.

Table 7.1 summarizes the typical permissible overvoltage at no load for generators and transformers, with data taken from [49]:

It is important to avoid long overvoltage conditions. Traditional overvoltage relays were designed based on inverse time tripping characteristics, as shown in Figure 7.40.

Example 7.2. *Review questions*

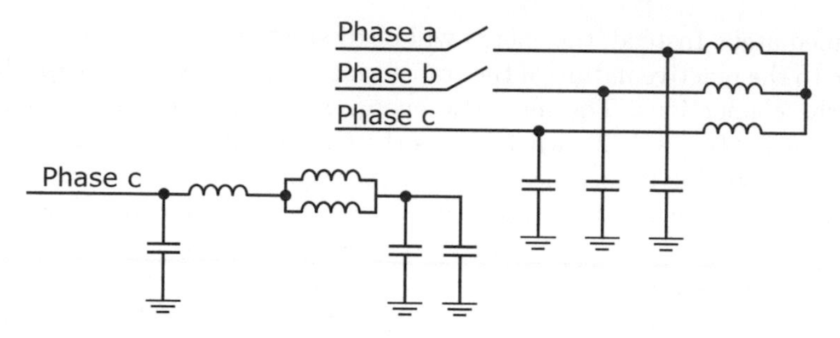

FIGURE 7.38
Faulted circuit breaker and resulting remaining circuit

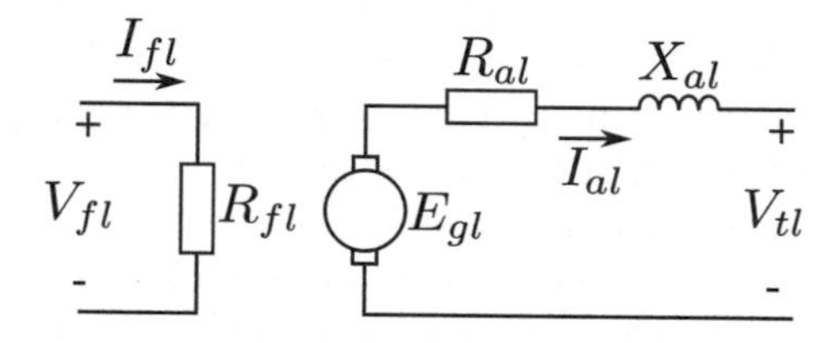

FIGURE 7.39
Synchronous generator equivalent circuit

TABLE 7.1
Electric machinery maximum operation time at overvoltage conditions

Generators		Transformers	
Percentage of overvoltage	Permissible time of operation	Percentage of overvoltage	Permissible time of operation
105	Continuous	110	Continuous
110	30 minutes	115	30 minutes
115	5 minutes	120	5 minutes
125	2 minutes	130	3 minutes

1. *What does the "Time Delay" parameter imply? It makes the system wait for the amount of time specified to trip after a fault is detected.*

2. *What is the use of a trip delay on a protection system? It permits the system to continue working if the fault detected was momentary or a bad measurement. If the system remains in a faulty condition for a given amount of time, then there will be no doubt of a faulty scenario. It is like a digital anti-bump device.*

3. *What does the "ResetRatio" parameter achieve at step 9? It works like a hysteresis latch by permitting the system to release the trip only if the measured magnitude is below a predefined percent from the trip condition.*

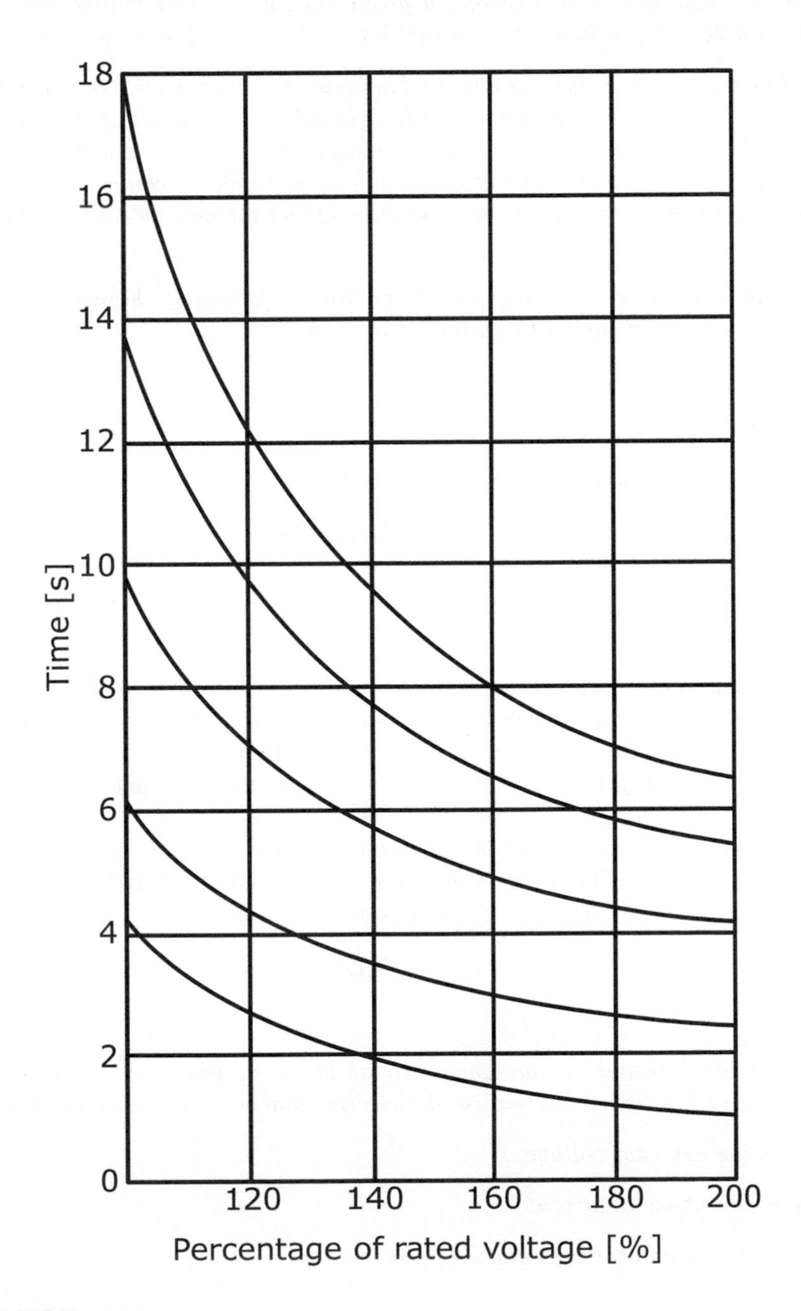

FIGURE 7.40
Tripping characteristics of overvoltage relay

So, the system will trip beyond a given voltage but will release the fault state only if the voltage is measured below the "ResetRatio" percent.

4. *What are your conclusions about relay logic in this exercise? How would it be changed if the relay used was R3? Applied relay logic allows the system to hold the contactor open even if a malfunction occurs, and disconnects the grid whenever the protection module is not able to continue working. R3 has an opposite normal state, so it would let the grid be connected even in fault scenarios.*

Example 7.3. *A synchronous generator with the following characteristics is being used as power supply to an industrial plant:*

- *480* $[V]$ *(*Δ *connected)*
- *60* $[Hz]$
- *Synchronous impedance:* $0.015 + j0.1$ $[\Omega]$
- *Full load current: 1200* $[A]$ *at 0.8 inductive PF*
- *Friction losses: 40* $[kW]$
- *Core losses: 30* $[kW]$

If the generator is supplying current at full load and then the load is suddenly disconnected, how much voltage will be at the generator terminals in volts and in percentage of the nominal voltage? According to the recommendations in the text, for how much time will the generator be able to continue operating?

If the load is disconnected, the terminals voltage of the generator will be the internal voltage of the electric machine, since the current will stop flowing and no voltage will be dropt at the impedance of the generator:

$$E_A = V_\phi + I_A(Z_A),$$

where:

- E_A *is the internal voltage generated at phase a, which will also be the voltage at the terminals because the electric machine is connected in* Δ
- V_ϕ *is the external voltage*
- I_A *is the phase a current*
- Z_A *is the generator impedance of phase a*

$$I_A = \frac{1200}{\sqrt{3}}\angle - \arccos 0.8A = 692.82\angle - 36.87°,$$

$$E_A = 480 + (692.82\angle - 36.87°) \times (0.015 + j0.1) = 532.161\angle 5.304°.$$

The percentage of overvoltage is:

$$\%_{overvoltage} = \frac{532.16V}{480V} = 1.1086.$$

The overvoltage will be 10%, so the generator could continue rotation during 30 minutes without permanent damage.

7.3.5 Undervoltage protection

It is possible that the generating system suddenly disconnects one or more lines or simply reduces the voltage magnitude in one of them due to unexpected failures, such as a short-circuit to ground or between lines.

In this lab, an undervoltage condition will be generated to test protection against phases' disconnection or abnormal voltage reduction.

Undervoltage protection is important in motors, since low voltage operation results in high current demand or failure to start, running at low speed, and even pullout. The torque equation of the motor is directly proportional to the voltage applied to the terminals; the reduction in torque by low voltage is described by the following equation:

$$\tau_{avg} = \left(\frac{V}{V_R}\right)\tau_R - \tau_L,$$

where:

- τ_{avg} is the average torque produced
- V is the applied voltage
- V_R is the rated voltage of the electric machine
- τ_R is the rated torque
- τ_L is the load torque

It is clear that if the applied voltage is small enough, the load torque may be higher than the average torque, resulting in full braking of the electric machine, high current demand (locked rotor), and possible damage to the motor windings. When undervoltage is presented for several seconds, motors should be disconnected, making undervoltage relays ideal for tripping motors. For 50 to 70% of rated voltage, instantaneous tripping is recommended, since according to the previous equation, the output torque for a 50% applied voltage will be just 25% of the rated torque.

The main reason for using undervoltage protection in industrial applications is to ensure tripping of circuit breakers on a loss of supply. If the motors are not disconnected, the system may overload in the moment of reconnection because of the starting of several induction machines at the same time. High starting demand of induction motors may trip overcurrent protection of feeder transformers and overheat power conductors.

Example 7.4. *Review questions:*

1. *What happens whenever any phase is disconnected from the measuring unit of the protection (VT4)? As soon as one phase is loose, the other two voltage measuring transformers report wrong voltages.*

2. *What difference does it make to configure the system as "2 out of 3" or "1 out of 3"? As voltage measurements are unbalanced and incorrect, the system trips another phase along with the one actually disconnected, so both configurations do the same.*

3. *If the generator is subjected to load, what happens with its output voltage? It is reduced, so more excitation voltage is needed to maintain the desired output level.*

Example 7.5. *A 6-pole induction motor connected in Y with a nominal tension of 440 [V] and 60 [Hz] has the equivalent circuit shown in Figure 7.41 (all values are in [Ω]).*

The motor is used for pumping oil and has a constant load of 50 [Nm]. At nominal voltage, the torque speed characteristic is shown in Figure 7.42.

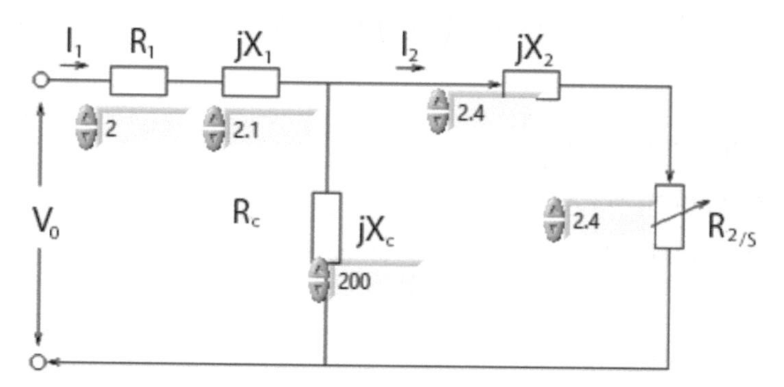

FIGURE 7.41
Equivalent circuit

What will be the minimum voltage that can be applied to the motor before pullout?

The pullout torque is 110 [Nm] with the conditions imposed, applying the torque reduction formula of the text:

$$\tau_{avg} = \left(\frac{V}{V_R}\right)\tau_R - \tau_L.$$

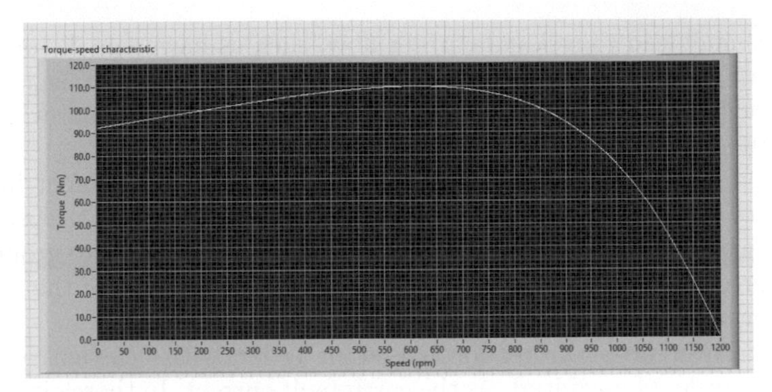

FIGURE 7.42
Torque speed characteristic

For a pullout condition:

$$0 = \left(\frac{V}{V_R}\right) 110 - 50,$$

$$\frac{V}{V_R} = \sqrt{\frac{50}{110}} = 0.674.$$

So, with an applied voltage of $V = 0.674 \times 440 = 296.65$*, the pullout torque of the motor will be 50* $[Nm]$*, run-in verifying the torque speed characteristic shown in Figure 7.43.*

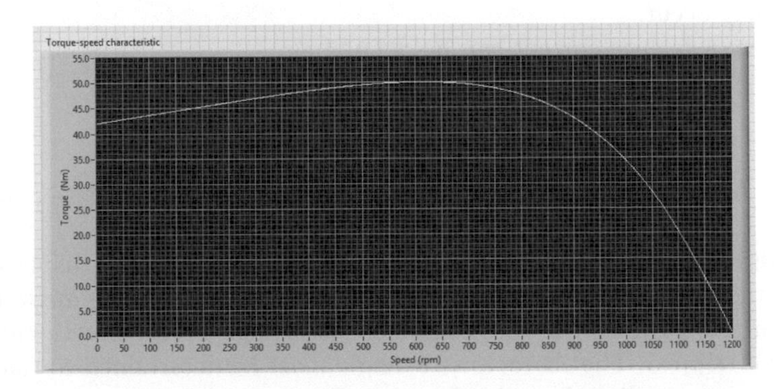

FIGURE 7.43
Torque speed characteristic

7.3.6 Load unbalance protection

Whenever there is phase unbalance from the generation system or the triphasic load, a parasitic flow of current appears at the neutral connection. Other conditions may lead to residual current flow, such as harmonic distortion; its effects must be detected and restricted as representing a power leakage derived from unwanted variations on energy consumption and generation, which may damage the system.

This protection is used in equipment or a networks component in which direction of flow could change as a result of a fault in the power system; for example, when a short circuit occurs between phases or phases to ground, the natural direction of the current may change since the location of the fault has zero voltage.

Residual overcurrent protection is used for feeders with a capacitive current of the same order of magnitude as the earth fault current. In those circuits, the phase-to-earth capacitance is high enough for a zero-sequence current to be detected by a non-directional protection when a short circuit occurs. Therefore, directional protection and overcurrent protection are complementary.

In order to determine the direction of a fault, a reference (or polarizing quantity) needs to be provided to the relay. The polarizing variable can be a zero-sequence voltage or zero-sequence current. In this case we are studying the current polarization.

Current polarization: The current in the neutral line of a power circuit can be used as a polarization value. This type of polarization has the important advantage of only requiring one current transformer, while the voltage polarization requires 3 VT. Another advantage is the possibility of measuring high currents with a much more economical device. The main disadvantages are: the saturation of CT when a short circuit occurs, producing a false value in the false residual current. When there are mutually coupled lines, current reversal may occur.

The zero-sequence current can be defined as:

$$I_0 = \frac{1}{3}(I_a + I_b + I_c),$$

where I_a, I_b, and I_c are the currents of phase a, phase b, and phase c, respectively.

The direction of the fault can be determined with the phase displacement between the current and the zero sequence current. When the angle of the phase's currents exceeds 90° of the characteristic angle of the directional relay, the relay is tripped. The characteristic angle must be determined so that only changes in current direction trip the relay. Figure 7.44 shows a characteristic curve of a directional overcurrent protection device.

Example 7.6. *Review questions:*

1. *Why does a new current appear when the loads are unbalanced? The new current is the one that flows through the neutral line due to load unbalancing.*

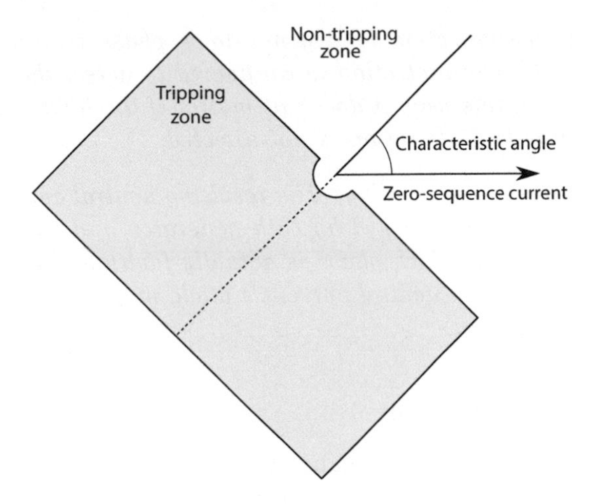

FIGURE 7.44
Tripping characteristic of directional relay

2. *How can you know if the unbalancing has occurred on the generator/load side? The phase (angle) of the current is opposed to that of its corresponding phase whether or not the fault is found on the load side; on the other hand, the neutral current can have the same direction (angle) as the corresponding phase's current if the fault is found on the generator's side.*

3. *Why does the disconnection of a generator's phase trigger the residual overcurrent fault? The protection is configured to detect the magnitude of neutral current; in this way, it does not matter if the fault is found at any side. Briefly, this fault detection is not directed.*

4. *What happens at s 8 and 9? Why? The resulting neutral current is directed to a composite angle conformed by both generator and load unbalancing. This angle depends on which phase is actually faulty and on which side it is. Knowledge about the neutral current's angle may lead to a fully operative directed protection.*

5. *Why does a new current appear when the loads are unbalanced? The new current is the one that flows through the neutral line due to load unbalancing.*

6. *How can you know if the unbalancing has occurred on the generator or the load side? The phase (angle) of the current is opposed to that of its corresponding phase whether or not the fault is found on the load side; on the other hand, the neutral current can have the same direction (angle) as the corresponding phase's current if the fault is found on the generator's side.*

7. *Why does the disconnection of a generator's phase trigger the residual overcurrent fault? The protection is configured to detect the magnitude of neutral current; in this way, it does not matter if the fault is found at any side. Briefly, this fault detection is not directed.*

8. *What happens at s 8 and 9? Why? The resulting neutral current is directed to a composite angle conformed by both generator and load unbalancing. This angle depends on which phase is actually faulty and on which side it is. Knowledge about the neutral current's angle may lead to a fully operative directed protection.*

7.3.7 Distance protection

Detection of a given fault is not enough; the approximate location of the fault is also expected to be known in order to make decisions that let the actual fault be isolated while its surrounding grid elements remain unaltered.

Additionally, the direction of the fault with respect to the protection location is also important: Imagine that a new generator is added in parallel to a previously installed one. In that case, current conditions will change, making a certain protection consider erroneous data; hence, a protection aware of the fault's direction and distance, and with isolation capabilities, is needed.

This type of protection is reliable when the distance between line terminals is relatively far. The principle of operation is to measure the impedance between the relay location and the fault location. Since the impedance of transmission lines is almost constant when a failure occurs, this can be detected as a change in the impedance of the faulted line. For distance impedance the R-X (real and imaginary part of impedance) diagram is the basic tool. The distance protection is not as accurate as comparing the current entering a circuit and leaving the circuit, since both voltage and current change dramatically during a fault. However, distance protection is economically viable for long transmission lines since it does not require current measurements at every load point and communication between power measurement devices. Those are the principal reasons for the extended application of distance relaying protection.

The typical settings of the distance relay depend on the programed zone of protection; see Figure 7.34.

Zone 1: This is also called the instantaneous zone because the relay opens as fast as it can when a measured impedance is within the protection zone of the R-X diagram. For modern digital distance relays, this zone is configured for a distance of about 85% of the total length until the next relay. This is made in order to avoid overreaching the protected line due to possible errors in measuring of current and voltage.

Zone 2: The remaining 15% of the line must be covered by the second zone, as seen in Figure 7.45. It is recommended that the second zone be configured to at least 120% of the protected line. In the traditional relays, zone 2

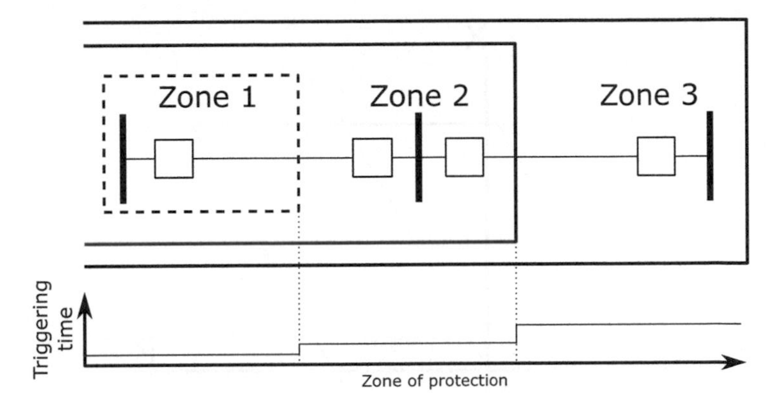

FIGURE 7.45
Zones of protection and time to trigger the circuit breaker

was activated by a timer signal trigger, but in the modern devices this is implemented in software. Some authors recommend setting the zone to up to 50% of the length of the adjacent transmission line, to ensure that zone 2 does not extend beyond zone 1 of the next relay.

Zone 3: Also known as remote back-up protection, usually set at least 1.2 times the impedance that will be measured for a fault at the remote end of the second line section. When there are interconnected power systems, the impedance measured will be much greater than the actual fault impedance; therefore, more than three protection zones will be unpractical.

In polygonal relays it is possible to work with zero and negative sequence voltages and currents, making fault resistance results less sensitive to the positive sequence load. There is no advantage to extending the R2 zone to the left as in MHO relays; although for convenience, many times it is defined as equal to R1, the impedance of the line is set in the center of the relay's X-R characteristic, and the quadrilateral is aligned with the line angle. The resistance lines are parallel to the line angle because the fault resistance in a quadrilateral relay is computed independently of the line resistance.

Quadrilateral elements can detect phase and ground faults by applying mathematical computation. They can be configured in all directions to provide the desired sensitivity and can be combined with conventional MHO relays to maintain the advantages of the well-known MHO relay and provide additional protection to the arc resistance zone. Differences in the source and line impedances may cause the reactance element to overreach or underreach. It is common to reduce the reactive reach for covering high fault resistance and use the MHO characteristics for giving a nice protection along the line and avoiding overreaching and underreaching (see Figure 7.46).

Example 7.7. *Review questions:*

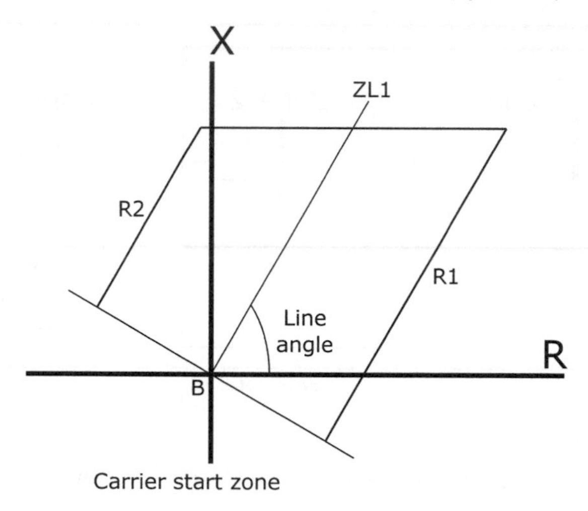

FIGURE 7.46
Quadrilateral relay R-X characteristic

1. *What is the main difference between overcurrent protection and distance protection? Overcurrent protection measures the current directly and reacts depending on current thresholds, while distance protection calculates the complex load based on voltage and current. Calculating the directed resistance and reactance can account for different faults as they reach the inner area of the polygon, even at different particular points. Coverage is bonded in this way.*

2. *The process of step 4 is not possible for real overhead lines, so how can this be done on real lines? Lines information is available so its resistance and reactance can be calculated at a certain distance; thus, you can configure the zone polygon. In addition, equivalent arc resistance or humidity fault conditions can be integrated to the calculus.*

3. *Measured impedances are not the same as known resistance and reactance of lines; why? The current and voltage ratios of instrumentation transformers change the numerical value of calculated impedance. Indeed, the calculation loops of module DL 2108T22 are designed to measure line-to-line magnitudes.*

4. *Measured distance to fault is not consistent with expected results; why? We are using an inductor to emulate phase-to-ground faults. A short-circuit condition assumes much lower resistances/reactances between faulty lines, so calculations will never be accurate. Moreover, an inductor adds an unexpected reactance to the line measurements, hence distance is not properly calculated.*

Example 7.8. *Consider a power system represented by Figure 7.47.*

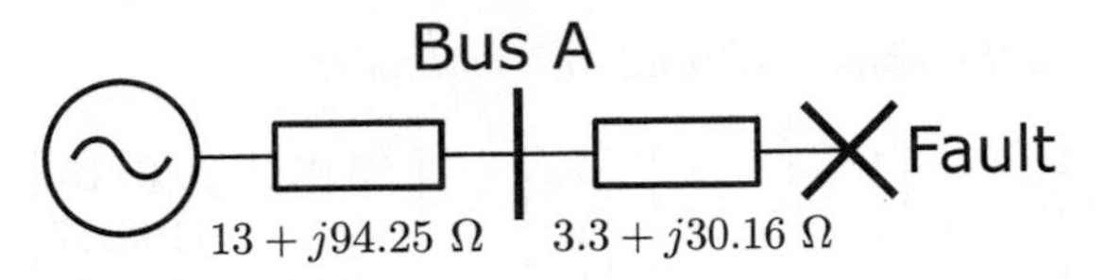

FIGURE 7.47
Electric diagram

At bus A there is a distance relay. The nominal voltage of the system is 380 [V] and the nominal apparent power is 1.1 [kVA]. The transmission line impedances are $13 + j109$ $[\Omega]$ *from the power source to the relay at bus A and* $3.3 + j30.16$ $[\Omega]$ *from bus A to the fault location. Consider that positive and negative sequence impedances are equal, and zero sequence impedances are* $30 + j190$ $[\Omega]$ *and* $10 + j60$ $[\Omega]$ *for the first and second transmission lines, respectively.*

For a three-phase fault, if the system is balanced, only the positive sequence current exists. So, calculating the phase "a" current considering a solid three-phase to ground fault:

$$I_a = I_1 = \frac{\frac{380}{\sqrt{3}}}{(13 + j94.25)(3.3 + j30.16)} = 0.227 - j1.734 \quad [A].$$

The voltage of phase "a" referred to neutral at the relay location is equal to:

$$V_a = \frac{380}{\sqrt{3}} - (0.227 - j1.734) \times (13 + j94.25) = 53.038 + j1.129 \quad [V].$$

So, the impedance seen by the relay at point A is equal to:

$$Z_f = \frac{V_a}{I_a} = 3.3 + j30.16.$$

Considering that a solid line-to-line fault occurs between phase "a" and phase "b," the positive and negative sequence currents are equal to:

$$I_1 = -I_2 = \frac{380/\sqrt{3}}{2 \times (13 + j94.25)(3.3 + j30.16)} = 0.1135 - j0.867 \quad [A].$$

Applying Fortescue's transformation to obtain the phase currents:

$$\begin{bmatrix} I_a \\ I_b \\ I_c \end{bmatrix} = \begin{bmatrix} 1 & 1 & 1 \\ 1 & a^2 & a \\ 1 & a & a^2 \end{bmatrix} \begin{bmatrix} I_0 \\ I_1 \\ I_2 \end{bmatrix} = \begin{bmatrix} 0 \\ -1.5017 - j0.1966 \\ 1.5017 + j0.1966 \end{bmatrix}.$$

Calculating the fault sequence voltages:

$$V_1 = \frac{380}{\sqrt{3}} - (0.1135 - j0.867) \times (13 + j94.25) = 136.20 + j0.57362 \quad [V],$$
$$V_2 = (-0.1135 + j0.867) \times (13 + j94.25) = -83.1902 + j0.57362 \quad [V],$$

and obtaining the phase "a" and "b" voltages:

$$\begin{bmatrix} V_a \\ V_b \\ V_c \end{bmatrix} = \begin{bmatrix} 1 & 1 & 1 \\ 1 & a^2 & a \\ 1 & a & a^2 \end{bmatrix} \begin{bmatrix} V_0 \\ V_1 \\ V_2 \end{bmatrix} = \begin{bmatrix} 53.01 + j0.0115 \\ -26.5 - j190.57 \\ -26.5 + j124.39 \end{bmatrix}$$

$$Z_f = \frac{V_b - V_c}{I_b - I_c} = \frac{-j380}{2 \times (-1.5017 - 0.1966j)} = 16.285 + j124.39.$$

7.4 Power line carrier

The power transmission system provides an extremely large interconnected network, so the potential of employing this hardware for communication issues has been a very attractive research topic for electric and telecommunications engineers around the world. The power line carrier (PLC) was introduced in 1919 as a communication technique based on high-frequency signals (between 20 [kHz] and 300 [kHz]) flowing through the power transmission system. Yet the explosion of high-bandwidth technology and optic fiber made PLC a dying technology for many years. Nowadays, this communication system is mainly employed for protective relaying, voice and signal transmission, and SCADA systems.

If only communication is needed for protective relaying, PLC is the obvious choice. Maintenance is one of the principal disadvantages for PLC applications because in some economic architectures the transmission line must stop operating before maintenance service starts.

Amplitude modulation (AM), frequency modulation (FM), single side band (SSB), and frequency shift keying (FSK) are the most common techniques implemented in PLC to encode information. Typically, the transmission takes place via coupling equipment to the transmission lines where carrier signals of 10 [ms] blocks of carrier and 10 [ms] blocks of no carrier are transmitted. The receiver needs to decouple the communication signal from the combined signal. For doing this, a coupling capacitor and tuner separate the desired frequency-range for obtaining the communication signal, and a demodulator extracts the information. For large transmission lines, line traps are needed to prevent the carrier from travelling down undesired paths.

The development of the complex frequency modulation process and the advances in highly integrated inexpensive electronics has enabled us to reach data transmission rates of about 1 [Mbps], which is about 12 times faster than the ISDN capacity. The correct implementation of this technique will make it possible to transmit phone and Internet signals over transmission lines and provide customers with a reliable communication service. The potential of this technology goes far beyond this; other interesting applications are long-distance monitoring of power consumption, control of intelligent household appliances, and monitoring of air-conditioning systems, alarm systems, and many others.

The advancing privatization and the international energy market competition, added to the liberalism of the telecommunication market, will enhance the commercialization of power lines for communication issues, and traditional network operators will find a very strong competitor.

7.5 Theoretical problems

The single line diagram of a power system is shown below. The neutral line of the generator is solidly grounded, while the motor is grounded through a reactance of $X_n = 0.02835$ pu; the transformers are both solidly grounded. The prefault voltage at bus 2 is 1.00 pu. The motor is an induction machine. The following table sums up the principal characteristics of the electric machinery:

Element	Power and voltage base	X_1	X_2	X_0
Generator 1	1000 $[MVA]$, 20 $[kV]$	0.1 pu	0.15 pu	0.05 pu
Motor 1	800 $[MVA]$, 18 $[kV]$	0.162 pu	0.1785 pu	0.0648 pu
Transformer 1	1000 $[MVA]$, 20 $[kV]\Delta/Y$ 400 $[kV]$	0.1 pu	0.1 pu	0.1 pu
Transformer 2	800 $[MVA]$, 400 $[kV]Y/\Delta$ 18 $[kV]$	0.081 pu	0.081 pu	0.081 pu
Transmission Line		20 $[\Omega]$	20 $[\Omega]$	60 $[\Omega]$

The line diagram of the power system is shown in Figure 7.48.

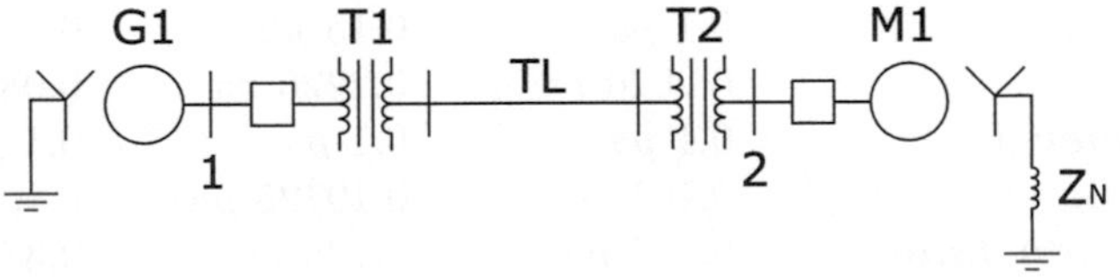

FIGURE 7.48
Line diagram

Problem 7.1. *Obtain the per-unit equivalent circuit for the zero, positive, and negative sequence networks taking the generator's base.*

Solution 7.1. *First, let's obtain the impedances at a common base with the following equation:*

$$X_{new} = X_{old}\left(\frac{V_{old}}{V_{new}}\right)^2 \times \left(\frac{S_{new}}{S_{old}}\right),$$

where "new" refers to the desired base for the voltage and power, and "old" is the original base.

$$X_{m1} = 0.162\left(\frac{18\ [kV]}{18\ [kV]}\right)^2 \times \left(\frac{1000\ [MVA]}{800\ [MVA]}\right) = 0.2025,$$
$$X_n = 0.02835\left(\frac{18\ [kV]}{18\ [kV]}\right)^2 \times \left(\frac{1000\ [MVA]}{800\ [MVA]}\right) = 0.03544,$$
$$X_{T1} = 0.1\left(\frac{20\ [kV]}{20\ [kV]}\right)^2 \times \left(\frac{1000\ [MVA]}{1000\ [MVA]}\right) = 0.1,$$
$$X_{T2} = 0.081\left(\frac{18\ [kV]}{18\ [kV]}\right)^2 \times \left(\frac{1000\ [MVA]}{800\ [MVA]}\right) = 0.10125.$$

For the transmission line, the impedance base is at the high voltage side:

$$Z_{base} = \frac{V_{base}^2}{S_{base}} = \frac{400\ [kV]^2}{1000\ [MVA]} = 160\ [\Omega],$$
$$X_{TL} = \frac{20}{160} = 0.125.$$

Calculating the resting impedances and updating the table's parameters to a 1000 $[MVA]$, 20 $[kV]$ base:

Base: 1000 $[MVA], 20\ [kV]$	***Impedance*** X_1	***Impedance*** X_2	***Impedance*** X_0
Generator 1	0.1 pu	0.15 pu	0.05 pu
Motor 1	0.2020 pu	0.2226 pu	0.081 pu
Transformer 1	0.1 pu	0.1 pu	0.1 pu
Transformer 2	0.10125 pu	0.10125 pu	0.10125 pu
Transmission Line	0.125 pu	0.125 pu	0.375 pu

Problem 7.2. *Obtain the Thévenin equivalent for each of the sequences seen from bus 2.*

Solution 7.2. *The sequence networks can be obtained from Figure 7.9 and Figure 7.11. The Thévenin equivalent for the zero sequence is $Z_{th0} = j0.081 + j0.03544$.*

For the positive sequence: $Z_{th1} = j0.2020(j0.1+j0.1+j0.125+j0.10125)$, *while the Thévenin voltage is always the prefault voltage at the failure point. Considering the no-load condition, the voltage will be equal to the generator voltage* $= 1$ *pu.*

The negative sequence impedance is calculated as follows: $Z_{th2} = j0.2226(j0.15+j0.1+j0.125+j0.10125)$. *The resulting systems are shown in Figure 7.49.*

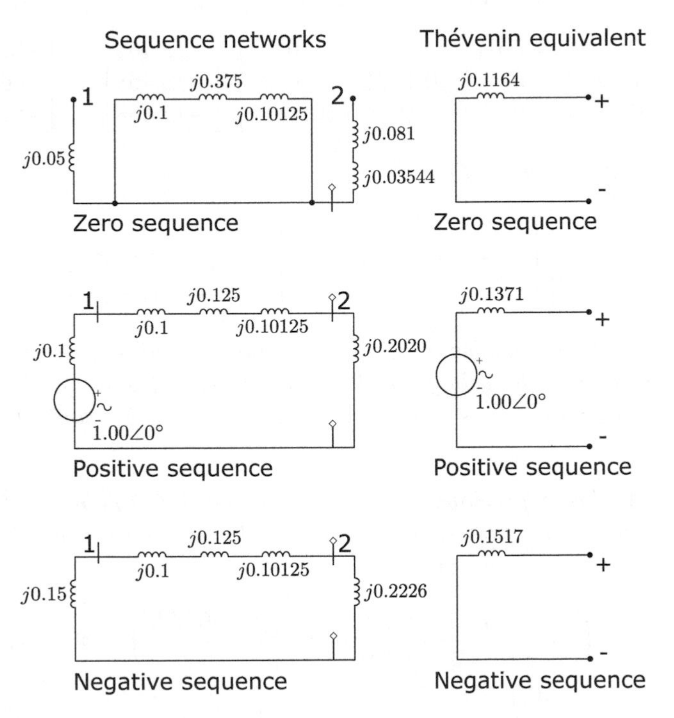

FIGURE 7.49
Negative sequence impedance

Problem 7.3. *If a single line-to-ground fault from phase "a" occurs at bus 2 of the system described in Problem 7.2, calculate the fault current in per unit and in amperes, and the voltages at bus 2.*

Solution 7.3. *From the previous exercise we know the Thévenin equivalent for each sequence network viewed from bus 2. From Figure 7.16 we know that:*

$$I_{a0} = I_{a1} = I_{a2} = \frac{1}{(0.1516 + 0.1371 + 0.1164)j} = -j2.468.$$

From Equation 7.1, the subransient currents are:

$$\begin{bmatrix} I_a \\ I_b \\ I_c \end{bmatrix} = \begin{bmatrix} 1 & 1 & 1 \\ 1 & a^2 & a \\ 1 & a & a^2 \end{bmatrix} \begin{bmatrix} -j2.468 \\ -j2.468 \\ -j2.468 \end{bmatrix} = \begin{bmatrix} -j7.406 \\ 0 \\ 0 \end{bmatrix}.$$

The current base at failure point is:

$$I_{base} = \frac{S}{\sqrt{3}\,[V]} = \frac{1000\ [MVA]}{\sqrt{3} \times 18\ [kV]} = 32075.01\ [A],$$
$$I_a = j7.406 \times 32075.01\ [A] = 237.547\ [kA].$$

The voltages in the sequence domain are:

$$\begin{bmatrix} V_0 \\ V_1 \\ V_2 \end{bmatrix} = \begin{bmatrix} 0 \\ 1 \\ 0 \end{bmatrix} - \begin{bmatrix} j0.1164 & 0 & 0 \\ 0 & j0.1371 & 0 \\ 0 & 0 & j0.1517 \end{bmatrix} \begin{bmatrix} -j2.468 \\ -j2.468 \\ -j2.468 \end{bmatrix} = \begin{bmatrix} -0.2873 \\ 0.6616 \\ -0.3744 \end{bmatrix}.$$

Transforming to phase domain:

$$\begin{bmatrix} V_a \\ V_b \\ V_c \end{bmatrix} = \begin{bmatrix} 1 & 1 & 1 \\ 1 & a^2 & a \\ 1 & a & a^2 \end{bmatrix} \begin{bmatrix} -0.2873 \\ 0.6616 \\ -0.3744 \end{bmatrix} = \begin{bmatrix} 0 \\ 1\angle{-115.65^\circ} \\ 1\angle 115.65^\circ \end{bmatrix} pu.$$

Problem 7.4. *Now consider a solid line-to-line fault at half the transmission line length. Calculate the voltages of the fault at buses 1 and 2 and the current of the fault in pu. Note: It is convenient to create a new bus halfway along the transmission line.*

Solution 7.4. *From previous problems we know the sequence networks for the power system. The first step is to obtain the Thévenin equivalent viewed from half the transmission line:*

$$Z_{th0} = \left(j0.1 + \frac{j0.375}{2}\right) \Bigg\| \left(j0.10125 + \frac{j0.375}{2}\right) = j0.1441,$$
$$Z_{th1} = \left(j0.2 + \frac{j0.125}{2}\right) \Bigg\| \left(j0.2020 + j0.10125 + \frac{j0.125}{2}\right) = j0.1528,$$
$$Z_{th2} = \left(j0.25 + \frac{j0.125}{2}\right) \Bigg\| \left(j0.2226 + j0.10125 + \frac{j0.125}{2}\right) = j0.1728.$$

From Figure 7.16, we know that

$$I_{a0} = 0,$$
$$I_{a1} = -I_{a2} = \frac{1}{(0.1528 + 0.1728)j} = -j3.0712.$$

The subransient currents in the phase domain are:

$$\begin{bmatrix} I_a \\ I_b \\ I_c \end{bmatrix} = \begin{bmatrix} 1 & 1 & 1 \\ 1 & a^2 & a \\ 1 & a & a^2 \end{bmatrix} \begin{bmatrix} 0 \\ -j3.0712 \\ j3.0712 \end{bmatrix} = \begin{bmatrix} 0 \\ -5.32 \\ 5.32 \end{bmatrix}.$$

The current base at the failure point is:

$$I_{base} = \frac{S}{\sqrt{3}\,[V]} = \frac{1000\ [MVA]}{\sqrt{3}} \times 400\ [kV] = 1443.37\ [A],$$
$$I_b = -5.32 \times 1443.37\ [A] = -7678.76\ [A] = -I_c.$$

The voltages in the sequence domain are:

$$\begin{bmatrix} V_0 \\ V_1 \\ V_2 \end{bmatrix} = \begin{bmatrix} 0 \\ 1 \\ 0 \end{bmatrix} - \begin{bmatrix} j0.1441 & 0 & 0 \\ 0 & j0.1528 & 0 \\ 0 & 0 & j0.1728 \end{bmatrix} \begin{bmatrix} 0 \\ -j3.0712 \\ -j3.0712 \end{bmatrix} = \begin{bmatrix} 0 \\ 0.5307 \\ 0.5307 \end{bmatrix}.$$

Transforming to phase domain:

$$\begin{bmatrix} V_a \\ V_b \\ V_c \end{bmatrix} = \begin{bmatrix} 1 & 1 & 1 \\ 1 & a^2 & a \\ 1 & a & a^2 \end{bmatrix} \begin{bmatrix} 0 \\ 0.5307 \\ 0.5307 \end{bmatrix} = \begin{bmatrix} 1.06 \\ 0.53\angle 180^\circ \\ 0.53\angle 180^\circ \end{bmatrix} pu.$$

7.6 Homework problems

Problem 7.5. *Consider an unbalanced system whose phase currents are known to be:*

- $IA = 3.42\angle 2^\circ\ [A]$
- $IB = 3.3\angle 121.23^\circ\ [A]$
- $IC = 3.6\angle 242^\circ\ [A]$

Calculate the symmetrical components.

Solution 7.5. *By directly applying the inverse Fortesquieu transform:*

$$\begin{bmatrix} I_0 \\ I_1 \\ I_2 \end{bmatrix} = \frac{1}{3} \begin{bmatrix} 1 & 1 & 1 \\ 1 & a & a^2 \\ 1 & a^2 & a \end{bmatrix} \begin{bmatrix} I_A \\ I_B \\ I_C \end{bmatrix},$$

the result can be computed as:

$$\begin{bmatrix} I_0 \\ I_1 \\ I_2 \end{bmatrix} = \begin{bmatrix} 79 \times 10^{-3}\angle -86^\circ \\ 97 \times 10^{-3}\angle -74^\circ \\ 3.44\angle 1.75^\circ \end{bmatrix}.$$

Problem 7.6. *The symmetrical current components were acquired for some loads and are provided below. If those currents were acquired for* 28 $[\Omega]$ *loads and the substation voltage is known to be a stiff* 120 $[V_{rms}]$ *source, estimate the distribution reactances responsible for such unbalancing.*

$$\begin{bmatrix} I_0 \\ I_1 \\ I_2 \end{bmatrix} = \begin{bmatrix} 79 \times 10^{-3} \angle -86^\circ \\ 97 \times 10^{-3} \angle -74^\circ \\ 3.44 \angle 1.75^\circ \end{bmatrix}.$$

Solution 7.6. *This can be computed directly from the following equation:*

$$\begin{bmatrix} V_A \\ V_B \\ V_C \end{bmatrix} = \begin{bmatrix} Z_A & 0 & 0 \\ 0 & Z_B & 0 \\ 0 & 0 & Z_C \end{bmatrix} T_s \begin{bmatrix} I_0 \\ I_1 \\ I_2 \end{bmatrix} = \begin{bmatrix} Z_A + 28 & 0 & 0 \\ 0 & Z_B + 28 & 0 \\ 0 & 0 & Z_C + 28 \end{bmatrix} \begin{bmatrix} 3.6 \angle 242^\circ \\ 3.3 \angle 121.23^\circ \\ 3.6 \angle 242^\circ \end{bmatrix}.$$

It is necessary to solve as follows:

$$\begin{bmatrix} 120 \angle 0^\circ \\ 120 \angle 120^\circ \\ 120 \angle 240^\circ \end{bmatrix} = \begin{bmatrix} Z_A + 28 & 0 & 0 \\ 0 & Z_B + 28 & 0 \\ 0 & 0 & Z_C + 28 \end{bmatrix} \begin{bmatrix} 3.6 \angle 242^\circ \\ 3.3 \angle 121.23^\circ \\ 3.6 \angle 242^\circ \end{bmatrix}.$$

The above system leads to:

$$Z_A \rightarrow 7.07 - j1.22[\Omega]$$
$$Z_B \rightarrow 8.36 - j0.78[\Omega]$$
$$Z_C \rightarrow 5.31 - j1.16[\Omega].$$

Problem 7.7. *The load unbalancing of a building has led to an abnormal voltage distribution among phases. Those voltages are measured to be:*

- $V_A = 124 \angle 0^\circ$
- $V_B = 128 \angle 123^\circ$
- $V_C = 126 \angle 241^\circ$

If the neutral terminal is known to have a reactance of $0.03 + j0.12$ $[\Omega]$, *calculate the power lost through it.*

Solution 7.7.

$$\begin{bmatrix} V_0 \\ V_1 \\ V_2 \end{bmatrix} = \frac{1}{3} \begin{bmatrix} 1 & 1 & 1 \\ 1 & a & a^2 \\ 1 & a^2 & a \end{bmatrix} \begin{bmatrix} V_A \\ V_B \\ V_C \end{bmatrix}.$$

So, $V_0 = -2.27 - j0.95$ $[V]$, *and the power is*

$$P = \frac{V_0^2}{X_n^*} = 48.84 \angle -58.5^\circ \ [VA].$$

Problem 7.8. *Consider a system to be furnished to trip a current protection whenever the neutral line reaches a current of* 5 $[A_{rms}]$. *Starting from a three-phase balanced reactance of* $10 + j3$ $[\Omega]$, *find how* Z_B *should decrease (both parts proportionally) so the protection trips.*

Solution 7.8. *It is necessary to calculate the current of the neutral line from the symmetrical components analysis equations:*

$$\begin{bmatrix} I_0 \\ I_1 \\ I_2 \end{bmatrix} = \frac{1}{3}\begin{bmatrix} 1 & 1 & 1 \\ 1 & a & a^2 \\ 1 & a^2 & a \end{bmatrix}\begin{bmatrix} I_A \\ I_B \\ I_C \end{bmatrix}.$$

$$I_0 = \frac{I_A + I_B + I_C}{3} = \frac{\left(\frac{V_A}{10+j3\ [\Omega]}\right) + \left(\frac{V_B}{(10+j3\ [\Omega])p}\right) + \left(\frac{V_C}{10+j3\ [\Omega]}\right)}{3} =$$
$$0.93\left(1 - \frac{1}{p}\right) + j3.95\left(\frac{1}{p} - 1\right).$$

RMS current can be obtained as the magnitude of the above calculated current, so:

$$\|I_0\| = 5 = \sqrt{\left(0.93\left(\frac{p-1}{p}\right)\right)^2 + \left(3.95\left(\frac{1-p}{p}\right)\right)^2},$$
$$25 = 16.46(1 - 2p^{-1} + 1p^{-2}).$$

The B phase load must be of $p = 0.448$ its original value to trigger the protection.

Problem 7.9. *Suppose that some symmetrical analysis triggers due to a fault. After checking the system's status, all symmetric components are the same. What happened to the system before triggering?*

Solution 7.9. *The only possibility is a phase-to-earth fault, as:*

$$\begin{bmatrix} I_0 \\ I_1 \\ I_2 \end{bmatrix} = \frac{1}{3}T_S^{-1}\begin{bmatrix} I_A \\ 0 \\ 0 \end{bmatrix} = \frac{1}{3}\begin{bmatrix} I_A \\ I_A \\ I_A \end{bmatrix}.$$

7.7 Simulation

Electric networks, besides being quite complex, are also subject to failures. There are some common types of faults, mostly related to transmission lines, which involve the short-circuit of a combination of conductors. Once a short-circuit appears, the system must disconnect the faulty section of the grid in order to prevent further damage to loads and generators.

It is important to notice that current protections and power analyzers are not fully capable of identifying the fault's origin, and sometimes, a distance protection is needed. After all, the full model of the grid must be taken into account in order to effectively detect the type of fault, its location, and the correction that can be performed in order to keep the power flowing to users (see Figure 7.50).

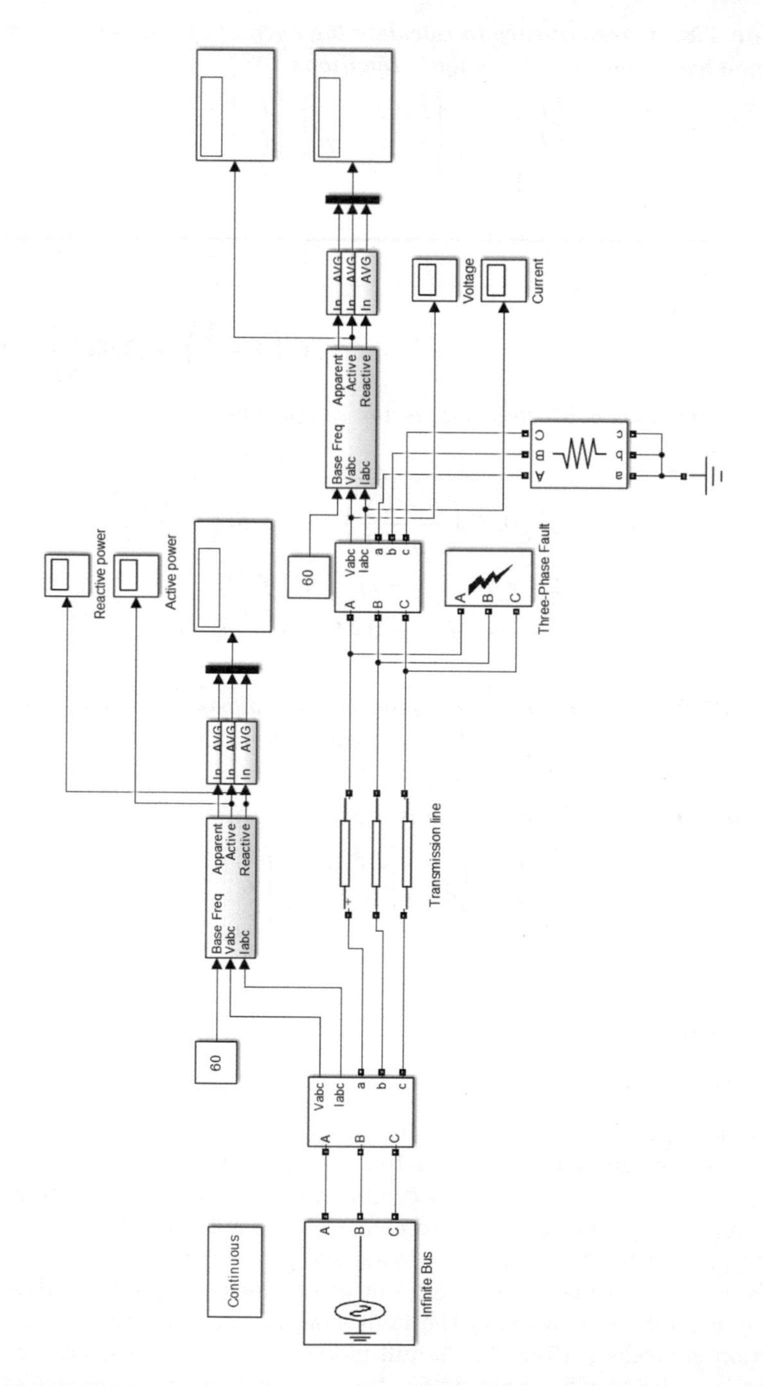

FIGURE 7.50
Simulation of transmission faults

8

Renewable Energy

The provision of energy or related energy services involves a huge variety of environmental impacts that are increasingly less tolerated by society. This is why the energy problem in conjunction with the underlying environmental problem continues to be a major topic in energy engineering, as well as in the energy and environmental policies around the world. From the current viewpoint, this attitude is not expected to change within the near future; the worldwide controversy about the potential risks of the anthropogenic greenhouse effect is only one example. On the other side, in view of the increasing knowledge and recognition of the effects associated with energy utilization in the broadest sense of the term, increasing complexity has to be expected.

According to Max Planck, energy is defined as the ability of a system to cause external action. Many forms of energy are distinguished: mechanical energy, thermal, electric and chemical energy, nuclear energy, and solar energy. In practical energy appliances the ability to perform work becomes visible by force, heat, and light. The ability to perform work from chemical energy, as well as nuclear and solar energy, is only given if these forms of energy are transformed into mechanical and thermal energy.

The term energy carrier, thus a carrier of the above-defined energy, is a substance that could be used to produce useful energy, either directly or by one or several conversion processes. According to the degree of conversion, energy carriers are classified as primary or secondary energy carriers and as final energy carriers, explained below:

- Primary energy carriers are substances that have not yet undergone any technical conversion, whereby the term primary energy refers to the energy content of the primary energy carriers and the primary energy flows. From this energy, a secondary energy carrier can either be produced directly or by one or several conversion steps.

- Secondary energy carriers are energy carriers that are produced from primary energy carriers, either directly or by several steps, whereby the secondary energy refers to the energy content of the secondary energy carrier and the corresponding energy flow. This processing of primary energy is subject to conversion and distribution losses. Secondary energy carriers and secondary energies are available to be converted into other secondary or final energy carriers or energies by the consumers.

- Final energy carriers and final energy, respectively, are energy streams di-

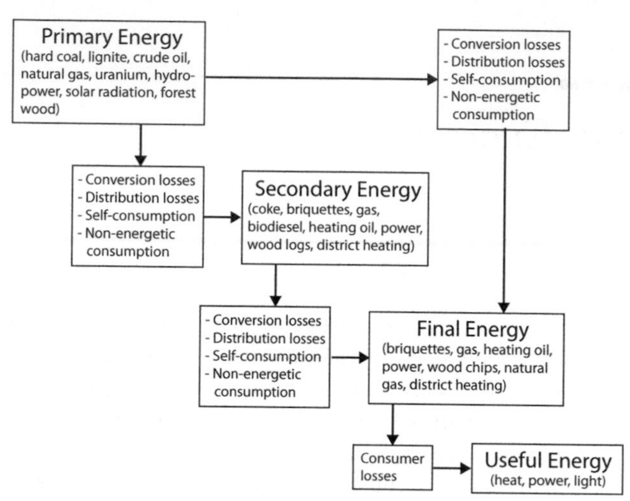

FIGURE 8.1
Energy conversion chain

rectly consumed by the final user. They result from secondary and possibly from primary energy carriers, or energies, minus conversion and distribution losses, self-consumption of the conversion system, and non-energetic consumption. They are available for the conversion into useful energy.

- Useful energy refers to the energy available to the consumer after the last conversion step to satisfy the respective requirements or energy demands. It is produced from the final energy carrier, reduced by losses of this last conversion.

The entire energy quantity available to humans is referred to as energy basis. It is composed of the energy of the energy resources and the energy sources. In Figure 8.1 the scheme of the energy carriers is presented.

8.1 Energy sources and resources

In terms of energy resources, generally fossil and recent resources are distinguished. Fossil energy resources are stocks of energy that have formed during ancient geologic ages by biologic and geologic processes. They are further subdivided into fossil biogenous energy resources and fossil mineral energy resources. The former include, among others, hard coal, natural gas, and crude oil deposits, whereas the latter comprise, for instance, the energy contents of uranium deposits and resources to be used for nuclear fusion processes. Recent resources are energy resources that are currently generated, for instance, by biological processes.

Energy sources, by contrast, provide energy streams over a long period of time; they are thus regarded as almost inexhaustible in terms of human time. But these energy flows are released by natural and technically uncontrollable processes from exhaustible fossil energy resources. Even if such processes take place within very long time periods, and are unlimited in human time, they are nevertheless exhaustible. The available energies can be further subdivided into fossil biogenous, fossil mineral, and renewable energy carriers.

8.2 Types of energy

There are many possible energy sources for the prime mover. The most common sources are water, coal, natural gas, oil, and nuclear energy. Water power is perhaps the best source, since it is nonpolluting and there is no cost for the fuel. Hydroelectric power is a renewable resource. Almost all desirable sites for dams have already been developed, so there is not much scope for future growth in water power.

Nuclear power is also an excellent source of nonpolluting energy. Nuclear power plants are extremely expensive to build, and they require elaborate safety systems and expensive training. Once built, though, the cost of the nuclear fuel is low and nuclear power produces no hydrocarbon emissions to pollute the atmosphere. Unfortunately, the future of nuclear power is clouded by concerns over nuclear reactor safety. Few nuclear plants are being built, and that situation will not change for the foreseeable future.

Coal is the most common source of energy for electrical power generation. Coal is a relatively cheap fuel, but it is also one of the most polluting fuel sources. Natural gas is a much better and cleaner energy source than coal. It is relatively cheap and it burns cleanly with little pollution. The principal disadvantage of natural gas is that it is relatively hard to transport over long distances. Oil is a bit more polluting than natural gas, but it is easier to transport; however, it is also much more expensive. Coal, oil, and gas all have the additional disadvantage that they are nonrenewable energy sources.

Other sources of electric power include wind turbines and solar energy. These are both renewable resources, but they are not yet economical compared to other sources of electrical energy. Fuel cells and nuclear fusion are also possibilities on the distant horizon.

Unfortunately, there is no perfect source of energy for electricity generation. All possible sources have their advantages and disadvantages, and all will be used in varying degrees for the foreseeable future.

8.3 Types of renewable energy

Provision of final or useful energy using renewable energies is based on energy flows originated by the movement and gravitation of planets, heat stored and released by the earth, and in particular, energy radiated by the sun. There is thus a great variety of renewable energies in terms of energy density, variations of the available forms of energy and the related secondary or final energy carriers, and final energy to be provided. Each technical option for utilizing the described renewable energy flows must be adapted to the corresponding characteristics of the available renewable energy. There is a broad range of current and also future technical processes and methods to exploit the renewable energy options most successfully.

The three sources of renewable energy give rise to a multitude of very different energy flows and carriers due to various energy conversion processes occurring in nature. In this respect, for instance, wind energy and hydropower, as well as ocean current energy and solid or liquid biofuels, all represent, more or less, conversions of solar energy.

The energy flows available on earth that directly or indirectly result from these renewable energy sources vary, for instance, in terms of energy density or with regard to spatial and time variations. In general, the renewable energies are:

- Solar radiation
- Wind energy
- Hydropower
- Bioenergy
- Photosynthetically fixed energy
- Geothermal energy
- Gravitation and motion of planets, tides and waves

Solar energy is referred to as renewable or sustainable energy because it will be available as long as the sun continues to shine. Estimates for the remaining life of the main stage of the sun are another 4 to 5 billion years. The energy from the sun, electromagnetic radiation, is referred to as insolation. Wind energy is derived from the uneven heating of the surface of the earth due to more heat input at the equator, with the accompanying transfer of water and thermal energy by evaporation and precipitation. In this sense, rivers and dams for hydro energy are stored solar energy. The third major aspect of solar energy is the conversion of solar energy into biomass by photosynthesis.

Animal products such as oil from fat and biogas from manure are derived from solar energy. Another renewable energy is geothermal energy due to heat from the earth from decay of radioactive particles, and residual heat from gravitation during formation of the earth. Volcanoes are fiery examples of geothermal energy reaching the surface from the interior, which is hotter than the surface. Tidal energy is primarily due to the gravitational interaction of the earth and the moon.

Overall, 14% of the world's energy comes from bioenergy, primarily wood and charcoal, but also crop residue and even animal dung for cooking and some heating. This contributes to deforestation and the loss of topsoil in developing countries. Production of ethanol from biomass is now a contributor to liquid fuels for transportation. In contrast, fossil fuels are stored solar energy from past geological ages. Even though the quantities of oil, natural gas, and coal are so large they are infinite and for the long term of hundreds of years, they are not sustainable.

Due to the great variety of possibilities for using renewable energy sources with the aim of fulfilling the demand for end or useful energy, it is very difficult to present the different possibilities in a similar manner.

The possibilities and boundaries for converting renewable energies into end or useful energy largely depend on the respective physical and technical conditions. Whenever possible, also among others the theoretically maximum achievable efficiencies and technical availabilities among others will also be indicated. These technical key figures are defined below:

- Efficiency, which is defined as the ratio of useful power output to the power input, depends on the respective operating conditions of the conversion plant, as well as a series of other factors, which vary over time.
- Utilization ratio is defined as the ratio of the total output of useful energy to the total energy input within a certain period of time. The observed time periods may include part load periods and breaks as well as start-up and shutdown times. Utilization ratios are thus usually smaller compared to the efficiency of conversion plants indicated to the design point at full load.
- Technical availability describes that portion of the time period under observation, within which a plant has actually been available for its intended purpose and thus considers time periods during which the plant has been unavailable due to malfunctions.

8.4 Advantages and disadvantages

The advantages of renewable energy are that they are sustainable (nondepletable), ubiquitous (found everywhere across the world), and essentially nonpolluting. Note that wind turbines and photovoltaic panels do not need water for the generation of electricity, in contrast to steam plants fired by fossil fuels and nuclear power.

The disadvantages of renewable energy are variability and low density, which in general results in higher initial cost. For different forms of renewable energy, other disadvantages or perceived problems are visual pollution, odor from biomass, avian and bat mortality with wind turbines, and brine from geothermal energy. Wherever a large renewable facility is to be located, there will be perceived and real problems for the local people. For conventional power plans using fossil fuels, for nuclear energy, and even for renewable energy, there is the problem of "not in my backyard."

8.5 Storage

Energy on demand means stored energy, and the most common means of storage are water in dams, batteries, and biomass. Fossil fuels are stored solar energy from past geological ages. Storage is a billion dollar idea. Economical storage would mean no new electric power plants would have to be constructed for many years, as energy could be stored from existing power plants during periods of low demand.

Energy cannot be created or destroyed, only transformed from one form to another, so in reality there are only two forms for storage: as kinetic energy and potential energy.

Storage as kinetic energy could be as flywheels, and thermal storage could be as heat. For example, a passive solar home would use concrete or rock and maybe water for 2–4 days of thermal storage, and a ground source heat pump would use the earth as seasonal thermal storage. Compressed air is king of a mixture; it is mechanical, but there is a thermal change. Super flywheels with high revolutions per minute and composite materials for strength have been designed and are used in uninterrupted power supplies and prototypes on buses.

Potential energy is due to generalized interactions, of which we consider gravitational and electromagnetic for storage systems. The gravitational potential energy is primarily water: dams, tidal basins, and pumped storage. The electromagnetic interactions include chemical, phase change, magnetic, electric, and mechanical interactions. Chemical storage is by batteries, photosynthesis, production of methane and hydrogen, fertilizer, and other types. The storage of gas requires high pressure, converting it to liquid or as a chemical compound, or, for example, storing hydrogen in metal hydrides. Of course, lots of solar energy is stored in chemical compounds as food and fiber.

Thermal storage could be thermal mass, ice, molten salt, cryogenic, earth, solar pound, and phase change. Molten salt for thermal storage is being used in combined heat and power systems. The main components for consideration of different storage systems are: energy density, efficiency and rate of charge/discharge, duration or lifetime (number of cycles), and economics.

Different storage technologies can be utilized for power or energy. Energy density and then size and weight for some applications are important factors. Liquid fuels have large energy density, while hydrogen gas has low energy density. In general, storage efficiencies range from 50% to 80%, and the lifetimes vary widely, from 100 years for dams, to 5 to 10 years for lead acid batteries in a photovoltaic system, to less than an hour for nonrechargeable batteries.

The maximum rate and best rate of charging and discharging the storage is related to type and use of storage. Whether energy storage is included in an application and the type of storage are driven by economics and specific power and energy. An electric car that takes 6 hours to charge its batteries could not be used on a cross-country trip, even if charging stations were available. Thus, it can be seen how versatile liquid fuels are for transportation. Also, the storage requirements for high power over a short time is different from energy storage for a few days. For utilities, the only large storage systems are pumped storage and compressed air energy systems; however, battery systems for power shaving, conditioning and reducing the variability from renewable energy sources have been installed. Both pumped hydro and compressed systems have long life and a large number of cycles. For remote village power systems and stand-alone systems, batteries are the most common means of storage.

The electricity produced by a generator cannot be stored; energy minus energy losses is the demand, so the generation supplies that amount of demand, which varies by time of day and season. In addition, a utility system must meet peak demands and have spinning reserve for unforeseen conditions. If demand exceeds capacity, then users are taken off the grid. In some parts of the world, there are rolling blackouts, or electricity is only available for certain time periods. Finally, extreme events may force shutdown of the total grid. A good source for information on storage for electricity is provided by the Electricity Storage Association.

8.5.1 Pumped hydro

Pumped storage has two reservoirs, and the same motor/generator can be used to pump water to the upper reservoir during periods of low demand and then generate electricity during periods of high demand, just as any other hydroelectric plant with a reservoir. The pumped hydro levels the load for other generators on the grid. The need for a pump-priming head places the motor/generator below the water level of the lower reservoir. Pumped storage systems generally have a high head to reduce the size of the reservoirs. Pumped storage can respond to full power within minutes, and if operated in the spinning mode, which uses less than 1% of their rated power, they can be changed to a pump or generator mode within 10s. Nuclear power plants can only change load slowly, and other base load plants can be operated at maximum efficiency through the use of pumped storage to absorb their output at night during low demand. There are motor/generator, friction, and evaporation losses, so the overall efficiency of a storage cycle is around 75%.

The advantages are: improved energy regulation and operation of the supply grid, ancillary services such as standby and reserve, black-station star, frequency control, and flexible reactive loading. The disadvantages are the high capital cost due to fairly large reservoirs, and, as for other dams, the land area needed, and the fact that collapse could happen.

8.5.2 Compressed air

Compressed air energy storage (CAES) is a peaking power plant that consumes 40% less natural gas than a conventional gas turbine, which uses about two-thirds of the input fuel compressed air. In a CAES plant, air is compressed during off-peak periods and then is utilized during peak periods. The compressed air can be stored in underground mines or salt caverns, which take 1.5 to 2 years to create by dissolving the salt. For an ideal gas, the amount of energy for an isothermal process from a pressure difference is

$$E = (mR)(T)ln\left(\frac{P_A}{P_R}\right) = (P)(V)ln\left(\frac{P_A}{P_R}\right)\ [J], \tag{8.1}$$

where P is the absolute pressure, V is the volume, m is the amount of gas, R is the ideal gas constant, and T is the absolute temperature in degrees kelvin. The approximation is $(100)(P_A/P_B)\ \left[\frac{kJ}{m^3}\right]$ for gas at around atmospheric pressure.

8.5.3 Flywheels

Flywheels store energy due to rotational kinetic energy, which is proportional to the mass and the square of the rotational speed,

$$E = (0.5)(I)(\omega^2)\ [J], \tag{8.2}$$

where I is the moment of inertia ($kg\ m_2$), and ω is the angular velocity ($radians/s$). For a mass M and radius R, the moment of inertia for a ring is $I = MR^2$, and for a homogeneous disk it is

$$I = (0.5)(M)(R^2). \tag{8.3}$$

Increasing the revolutions per minute increases the energy density, so high-speed flywheels have revolutions per minute in tens of thousands. Low-speed flywheels are made from steel, and high-speed flywheels are made from carbon fiber or fiberglass. High-speed flywheels are housed in a low vacuum and use magnetic bearings to reduce or eliminate frictional losses. Advances in power electronics, magnetic bearings, and materials have resulted in direct current

(DC) flywheels. Note that if there is material failure, the container has to retain that energy inside. Cycle efficiency is around 80%. Flywheels have been used in trains, cars, and buses; however, they were primarily experimental or prototypes. In the past, there were flywheels on tractors to smooth out the rotation of the crankshaft of two-cylinder engines. A hybrid vehicle could use flywheels for acceleration in conjunction with a smaller internal combustion engine, much like hybrid vehicles with batteries. In a car, the flywheel is a gyroscope, which will change the handling.

8.5.4 Batteries

Batteries are common all over the world, as lead-acid batteries are used for vehicles. Batteries are used for low-power and low-energy applications for lights, radio, TV, and electronic devices. Photovoltaic (PV) with rechargeable batteries has now replaced batteries for very low-power applications. Batteries convert chemical energy into electrical energy using electrodes immersed in a medium (liquids, gels, and even solids) that supports the transport of ions or electrolyte reactions at the two electrodes. Individual cells are placed in series for higher voltage and in parallel for higher current. Since there is an internal resistance and because of other factors, there are losses in charging and discharging a battery.

The power is the product of the voltage and current; however, batteries are generally specified by volts and storage capacity (battery charge CB), which is related to stored energy. CB is the amount of charge that a battery can deliver to a load. It is not an exact number because it depends on the age of the battery (number of cycles), temperature, state of charge, and rate of discharge. If a lead-acid battery is discharged to essentially zero a few times, it drastically reduced its lifetime. As a first approximation, the energy is

$$E = (V)(C_B)\ [J], \tag{8.4}$$

where V is the voltage and C_B is the battery capacity $[Ah]$. Decreased temperature results in less battery capacity, and for a lead-acid battery, storage capacity decreases around 1% for every $1°C$ drop in temperature. Explosions due to short circuits and generation of hydrogen plus disposal of used batteries and toxic chemicals are problems.

8.5.5 Rokkasho-Futamata Wind Farm

This energy storage farm has its location in Rokkasho, Aomori, Japan, and it is the largest and first combined wind generation plant, which produces about 51 [MW]. Also, its energy storage goes to about 34 [MW], so that the sum of both quantities gives 85 [MW] of total energy storage. This facility is one of the world's largest sodium sulfur battery assemblies. This project was integrally developed with energy storage capacity of 34 [MW], using NGK Insulators' sodium-sulfur (NaS) batteries for load leveling, enabling the storage of low-cost off-peak wind power for sale and/or distribution during peak demand times. The sodium sulfur battery sets are charged primarily at night when demand for power is lower. The stored electricity is then supplied to the utility grid, along with power directly generated by the wind turbines, during peak demand times. This system ensures a steady supply of power to the grid even during those periods when wind generation falls as a result of low wind speeds.

To control the transmission of power from the Rokkasho-Futamata Wind Farm to the national power grid, a smart grid monitoring and control system from Yokogawa Electric Corporation systems was installed. The system includes STARDOM network-based controllers and FA-M3 range-free controllers. The Yokogawa system factors in fluctuations in power production as a result of varying wind conditions and develops a power generation operation plan. With over 200 [MW] of NaS battery storage devices installed throughout the world, the company has leveraged their expertise with ceramics and electrical insulation systems to become a world leader in the development of sodium sulfur batteries. NGK NaS battery technology is typically used to stabilize intermittent renewable energy, provide emergency power, and as part of stand-alone power systems for small islands and isolated grids. NaS batteries have a high energy density, high efficiency of charge-discharge (89–92%), long cycle life, and they are fabricated from inexpensive materials. However, because of the operating temperatures of 300 to $350°C$ and the highly corrosive nature of sodium polysulfides, NaS battery technology is primarily suited for large-scale, non-mobile applications such as grid energy storage.

8.6 Energy and society

Industrialized societies run on energy, a tautological statement in the sense that it is obvious. Population, gross domestic product (GDP), consumption and production of energy, and production of pollution for the world and the United States are interrelated. The United States has less that 5% of the population of the world; however, in the world, the United States generates around 25% of the gross production and 22% of the carbon dioxide emissions, and is at 22% for energy consumption. Notice that countries such as China, India, Europe, the United States, Brazil, Russia, and Japan consume around 75% of the energy and produce 75% of the world's GDP and carbon dioxide emissions. The developed countries consume the most energy and produce the most pollution, primarily due to the increase in the amount of energy per person. On a person basis, the United States is the worst for energy consumed and carbon dioxide emitted.

The energy consumption in the United States increased from 32 [quad] in 1950 to 101 [quad] in 2009. One [quad] is equal to 1015 British thermal units

[Btu]. There was an increase in efficiency in the industrial sector, primarily due to the shock of the oil crisis of 1973. However, you must remember that correlation between GDP and energy consumption does not mean cause and effect. The oil crisis of 1973 showed that efficiency is a major component in gross national product and the use of energy.

A thought on energy and GDP: solar clothes drying (a clothesline) does not add to the GDP, but every electric and gas dryer contributes; however, they both do the same function. We may need to think in terms of results and efficient ways to accomplish a function or process and the actual life-cycle cost. The underdeveloped part of the world, primarily the two largest countries in terms of population (China 1.3×10^9 and India 1.1×10^9) are beginning to emulate resources and greenhouse gas emissions. One dilemma in the developing world is that a large number of villages and others in rural areas do not have electricity.

Just as there is growth in the population, it is necessary to improve the technology, not only because fossil fuels are running out but because actual technology is a contaminant and therefore not efficient. The Organisation for Economic Co-operation and Development (OECD) will support the development of electric cars, to contribute in the efficient use of energy and promote the use of renewable energy. It is believed that in 2030 half of the vehicles will be electric, so that by 2050 there will no longer be internal combustion engines.

8.6.1 Use of energy

It is a physical impossibility to have exponential growth of any product or exponential consumption of any physical resource in a finite system. There are numerous historical examples of growth: population, which increases 2–3% per year; gasoline consumption, 3%; world production of oil, 5–7%; electrical consumption, 7% per year. The U.N. projects over 9 billion people by 2050, with the assumption that the growth rate will decrease from 1.18% in 2008 to 0.34% in 2050.

However, even with different rates for growth, the final result is still the same. When consumption grows exponentially, enormous resources do not last long. This is the fundamental flaw in terms of ordinary economics and stating growth in terms of percentages. The theme is that all we need is economic development and the world's problems will be solved. However, the global economic crisis of 2008 and environmental problems have made some economists have second thoughts on continued growth.

8.7 Economics

Business entities always couch their concerns in terms of economics, such as claiming that we cannot have a clean environment because it is uneconomical. The thought here is that renewable energy is not economical in comparison to coal, oil, and natural gas. The different types of economics to consider are pecuniary, social, and physical. Pecuniary is what everybody thinks of as economics, or money. On that note, we look at life-cycle costs rather than our ordinary way of doing business, low initial costs. Life-cycle costs refer to all costs over the lifetime of the system.

Social economics are those costs borne by everybody, and many businesses want the general public to pay for their environmental costs. A good example is the use of coal in China, where there are laws for clean air, but they are not enforced. The cost will be paid in the future in terms of health problems. The estimates of pollution costs for generation of electricity by coal range from \$0.005 to \$0.1/$[kWh]$.

Physical economics are the energy costs and the efficiency of the process. There are fundamental limitations in nature due to physical laws. Energetics, which is the energy input versus energy in the final product for any source, should be positive. For example, production of ethanol from irrigated corn has close to zero energetics. So, physical economics is the final arbitrator in energy production and consumption. What each entity wants are subsidies for itself and penalties for its competitors. Penalties come in the form of taxes and environmental and other regulations, while incentives come in the form of subsidies, breaks on taxes, lack of social costs to pay on the product, and governmental funding of research and development. It is estimated that we use energy sources in direct proportion to the incentives that source has received in the past.

There are many examples of incentives and penalties for all types of energy production and use.

8.8 Solutions

We do not have an energy crisis, since we know that energy cannot be created or destroyed. We have an energy dilemma because of the finite amount of readily available fossil fuels, which are our main energy source today. The problem is the overpopulation and overconsumption. The world population is so large that we are doing an uncontrolled experiment on the environment of the earth. However, the developed countries were also major contributors to this uncontrolled experiment in terms of consumption. The solution depends on world, national, and local policies and what policies to implement, and even individual actions, like reducing consumption, having zero population growth, shifting to renewable energy, and reducing greenhouse gas emissions, environmental pollution, and military expenditures.

Continued exponential growth is a physical impossibility in a finite system, and the earth is a finite system. Previous calculations made about the future are just estimations, and possible solutions to our energy dilemma are as follows:

- Conservation and more efficient use of energy. Since the first energy crisis, this has been the most cost-effective mode of operation. It is much cheaper to save a barrel of oil than to discover new oil.
- Reducing demand, transitioning to zero population growth, and being a steady-state society.
- Redefine the size of the solar system and colonize the planets and space; however, this will not alleviate the problem on earth. From our present viewpoint, the resources of the solar system are infinite, and our galaxy contains over 100 trillion stars.

Because the earth is finite for population and our use of the earth's resources is also limited, a change to a sustainable society, which depends primarily on renewable energy, becomes imperative on a long timescale. In order of priority for the next 25 years:

1. Implement conservation and efficiency
2. Substantially increase the use of renewable energy
3. Reduce dependence on oil and natural gas
4. Use clean coal, which has to include all social costs
5. Make use of nuclear energy
6. Reduce environmental impact, especially greenhouse gases
7. Implement policies that emphasize items 1 and 2

State and local policies must be the same. Efficiency can be improved in all the major sectors: residential, commercial, industrial, transportation, and even the primary electrical utility industry. National, state, and even local building codes would improve energy efficiency in buildings. Finally, there are a number of things that you as an individual can do about conservation and energy efficiency.

For a few final comments, the possible future for human society involves conservation efficiency, with an orderly transition to sustainable energy and a steady state with no growth, catastrophe or with some revival. As overpopulation and overconsumption are affecting the earth, as an uncontrolled experiment, the most probable future for the population is catastrophe or catastrophe with some revival.

8.9 Theoretical problems

Problem 8.1. *Calculate the kinetic energy of a car that has a mass of* 1000 $[kg]$ *moving at* 10 $[m/s]$.

Solution 8.1. *The kinetic energy of moving objects is calculated by:*

$$KE = 0.5\ mv^2.$$

Thus, $KE = 5000\ [J]$.

Problem 8.2. *A* 12 $[V]$ *battery is rated at* 100 $[Ah]$. *It could deliver* 5 $[Ah]$ *for* 20 $[h]$, $E = 1 \times 0\ [Wh]$, *or* 1.2 $[kWh]$. *However, at a faster discharge rate, the values would be lower:* 85 $[Ah]$ *for* 10 $[h]$, 70 $[Ah]$ *for* 5 $[h]$. *Explain this behavior.*

Solution 8.2. *Decreased temperature results in less battery capacity, and for a lead-acid battery, storage capacity decreases around* 1% *for every* $1°C$ *drop in temperature. Explosions due to short circuits, and generation of hydrogen plus disposal of used batteries and toxic chemicals are problems.*

Problem 8.3. *High-speed flywheels are made from composite material. A flywheel is a ring with* 3 $[m]$ *radius, mass* = 50 $[kg]$, 20000 $[rpm]$. *How much energy is stored in the flywheel?*

Solution 8.3. *We have the following equations:*

$$I = 0.5(M)(R)^2,$$

$$I = 0.5(50)(3)^2\ [Kgm^2],$$

and

$$E = 0.5(I)(\omega)^2\ [J],$$

$$200\ [rpm] \approx 2094.395\ [rad/s],$$

$$E = 0.5(I)(2094.395)^2\ [J].$$

Problem 8.4. *Compare kinetic energy storage to batteries in terms of energy density and installed costs. Consider both systems to have the same amount of energy capacity.*

Solution 8.4. *There is no precise solution to this problem.*

Problem 8.5. *Find any commercial chemical capacitor storage system; note the source, company, and any other useful specifications. How much energy can it store?*

Solution 8.5. *There is no precise solution to this problem.*

8.10 Homework problems

Problem 8.6. *There is no single entirely "clean" energy generation method. Perform some brief research and find the drawbacks of the following systems regarding the environment and their byproducts during production/establishment, usage/operation, and disposal/cessation of the operational phases*

- *PV solar*
- *Wind*
- *Hydro*
- *Batteries*

Solution 8.6. *There are some key concepts any report regarding environmental issues around renewable energy should include:*

- *PV solar*
  - *Production and deposition imply the use and further treatment of silicon*
  - *Domestic usage of solar panels leads to serious issues of harmonic contents and phase unbalancing*
  - *Their efficiency is quite low, just like their lifespan. It is hard to weigh how effective they are as they are commonly used with batteries (and power electronics)*
  - *They are susceptible to cleansing issues regarding dust, salt (in the sea), and objects (like leaves), which makes them require maintenance*
- *Wind*
  - *The location of a wind farm needs to be entirely modified as affecting endogenous species and conditions*
  - *Off-shore windmills disturb the local habitat and demand complex transport and maintenance tasks that might be polluting*
  - *The intermittent nature of the wind makes batteries necessary, as well as power electronics*
- *Hydro*
  - *Their location implies the entire modification of the natural zone*
  - *Huge amounts of artificially-placed water can result in geological issues*
- *Batteries*

- *Their components are highly polluting*
- *Their deposition is normally incorrect*
- *They degrade quite easily; bad usage leads to early deposition*

Problem 8.7. *Suppose that a flywheel is about to be used to store excess energy from a power plant. Surpluses have been calculated to be about* 456 $[kW]$ *on average for* 30 $[min]$ *intervals. Calculate the dimensions of a low-speed steel flywheel* (180 $[rpm]$) *that is* 80% *efficient, and* 3 $[m]$ *tall.*

Solution 8.7. *In order to take the total amount of surpluses, an average amount of energy of*

$$E = 456\ [kW] \times \left(30\ [min]\frac{60\ [s]}{[min]}\right) = 821\ [MJ]$$

is required to be captured. We can use flywheel equations directly, by incorporating steel's density as follows:

$$E = 0.5(I)(\omega)^2\eta = 0.5(0.5\delta\pi r^4 h)\left(180\ [rpm]\left(\frac{[s]^{-1}}{60\ [rpm]}\right)\left(\frac{2\pi\ [rad/s]}{[s]^{-1}}\right)\right)^2\eta.$$

Steel density is $\delta = 8050\ \left[\frac{kg}{m^3}\right]$, *so*

$$821 \times 10^6\ [J] = 0.5(37.93 \times 10^3 r^4)(355.3)(0.8) = 5.39 \times 10^6 r^4.$$

The radius can be solved:

$$r = \sqrt[4]{\frac{821 \times 10^6}{5.39 \times 10^6}} = 3.51\ [m].$$

Problem 8.8. *Capacitors are also known as energy storing elements. Calculate the capacitance of a capacitor array that can hold* 1 $[MJ]$ @ 100 $[V]$. *Consider capacitor's stored energy to be represented by*

$$E = \frac{1}{2}CV^2.$$

Solution 8.8. *The capacitance value can be directly taken from the above equation:*

$$1\ [MJ] = \frac{1}{2}C(100)^2 = 200\ [F].$$

Problem 8.9. *Considering Problem 8.7 and taking its parameters, how fast should a fiberglass flywheel rotate to deliver the same amount of energy, having* 3.5 $[m]$ *radius? Consider its density to be* 0.055 $\left[\frac{lb}{in^3}\right]$.

Solution 8.9. *Fiberglass density is of about*

$$\delta = 0.055\ \left[\frac{lb}{in^3}\right]\frac{[in^3]}{(25.4 \times 10^{-3}[m])^3}\frac{0.4535\ [kg]}{[lb]} = 1522.1\ \left[\frac{kg}{m^3}\right].$$

The energy equations are as follows:

$$E = 456\ [kW] \times (30\ [min] \frac{60\ [s]}{[min]}) = 821\ [MJ],$$
$$E = 0.5(I)(\omega)^2\eta = 0.5(0.5(1522.1)\pi(3.5)^4 3)\omega^2(0.8),$$
$$E = 430.5 \times 10^3\ \omega^2.$$

So,

$$\omega = \sqrt{\frac{821 \times 10^6}{430 \times 10^3}} = 43.69 \left[\frac{rad}{s}\right],$$
$$\omega = 417\ [rpm].$$

Problem 8.10. *Perform some brief research regarding the latest lithium-based batteries. List their characteristics regarding energy density, power density, charge/discharge regimes, and lifespan.*

Solution 8.10. *There is no precise solution to this problem.*

8.11 Summary

The integration of different energy sources can be seen as complementary and should commonly obey a local energy plan, as shown in Figures 8.2 and 8.3. This means that not all the types of energy are suitable to being applied in a particular place due to their natural restrictions and storage requirements. Transmission lines and phase unbalancing also need to be considered so that the whole integration fulfills energy usage objectives.

Some types of generators can be seen as complementary in very specific senses. For example, an induction generator with a squirrel cage rotor will consume reactive power that can be provided by a synchronous machine. A capacitive bank can also be used to compensate the requested reactive power but will add complexity to the transients of the generating device. Broadly, the effects of unbalanced loads, faulty systems, and potential complementary generators must be measured and controlled. In the same spirit, a smart grid is required to manage complex networks, as it provides a means to deal with full-scale integration of electric devices.

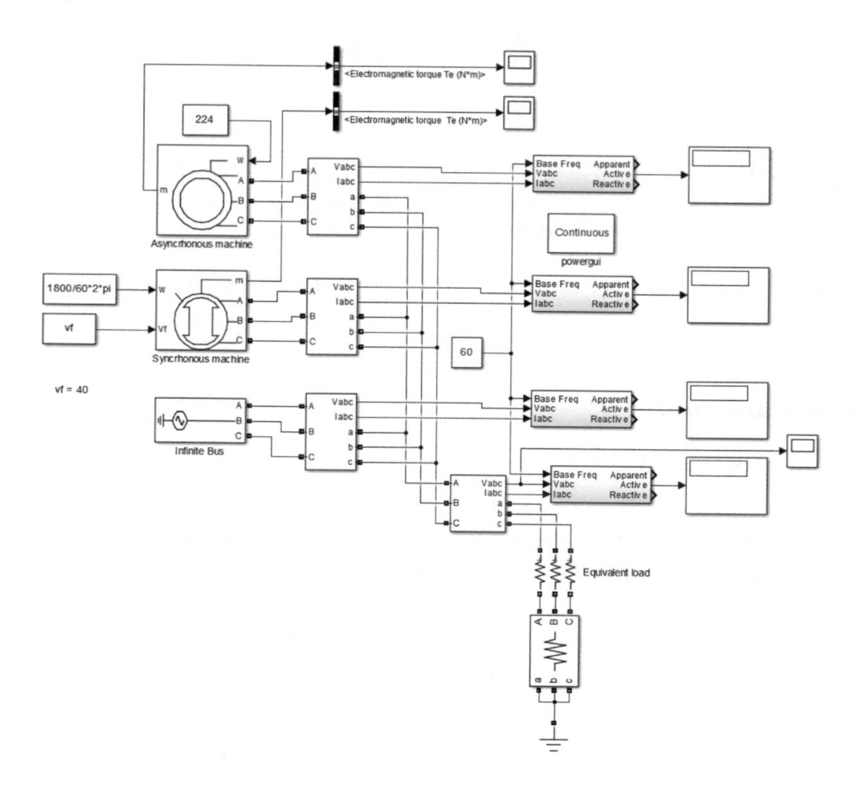

FIGURE 8.2
Integration of various sources 1/2

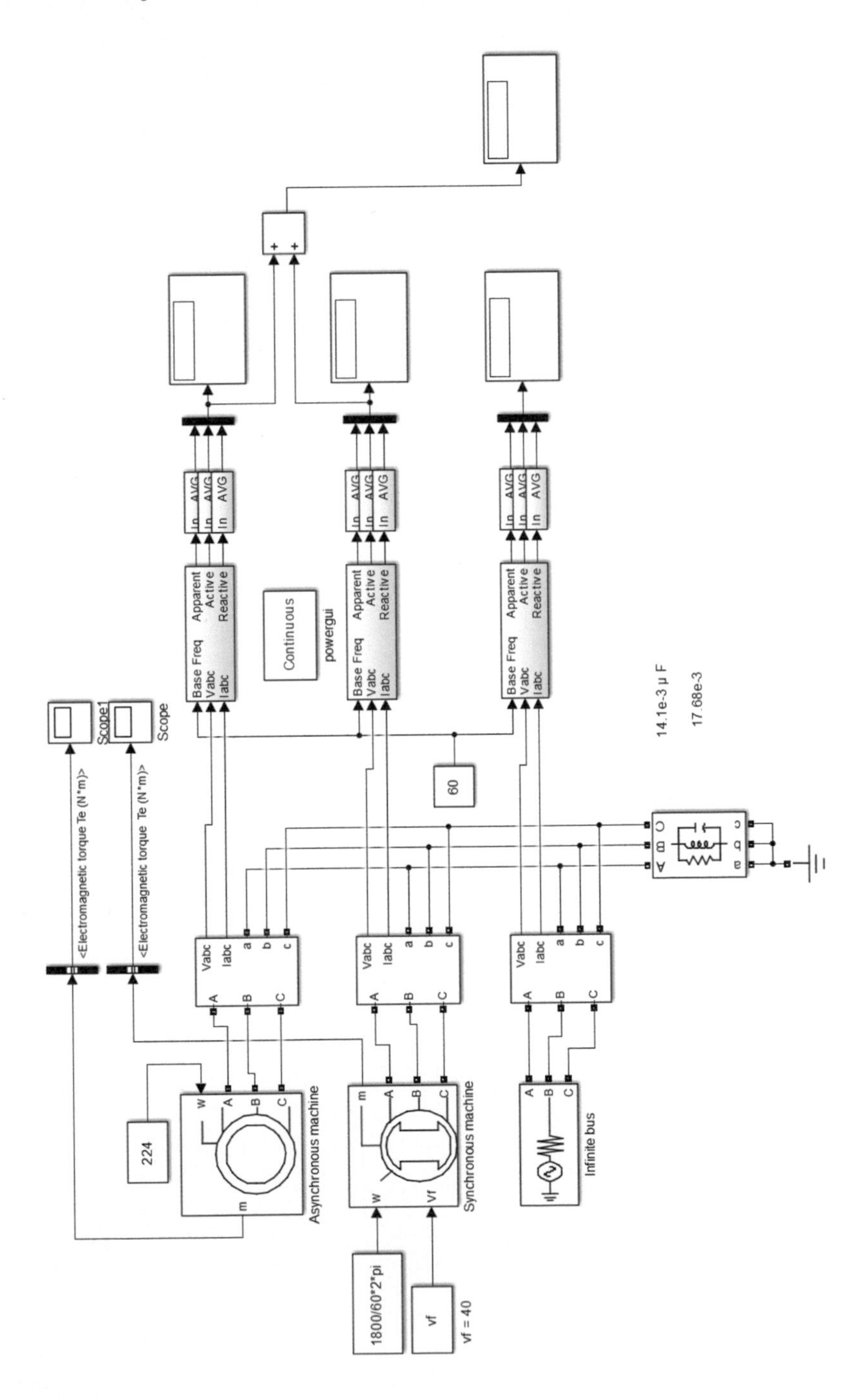

FIGURE 8.3
Integration of various sources 2/2

9

Smart Grid

The smart grid appeared as the response to human energy needs, and uses a communication system, that allows improvement of electric energy conditions.

A new professional curriculum is a vital element for the future electrical power industry. Different engineering topics need to be created to face the smart grid's technological obstacles. The development of educational programs requires skills and technologies outside of power engineering.

The electric system curriculum requires theory and experimental work in numerous areas, with a major focus on the following basic topics:

- Electric machines
- Power electronics
- Control systems
- Telecommunications
- Cybersecurity
- Automation
- Instrumentation
- Digital systems
- Economics
- Prediction
- Signal processing

Different authors have defined the smart grid system as [9]: Green Energy Act (Canada): "A nickname for an ever widening palette of utility applications that enhance and automate the monitoring and control of electrical distribution." "An automated, widely distributed energy delivery network characterized by a two-way flow of electricity and information, capable of monitoring and responding to changes in everything from power plants to customer preferences to individual appliances." "A smart grid is the electricity delivery system (from point of generation to point of consumption) integrated with communications and information technology." In fact, the smart grid concept

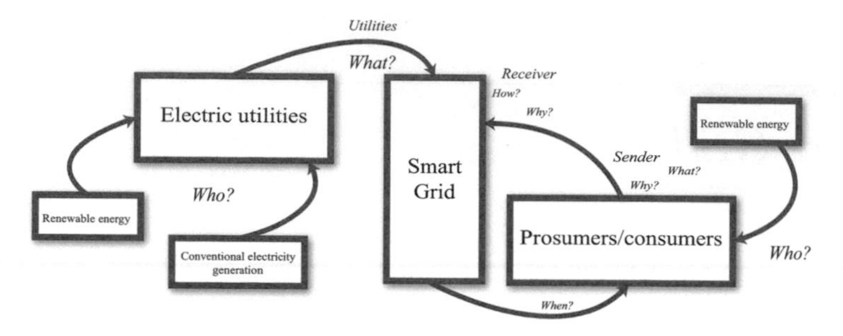

FIGURE 9.1
Supply chain for describing a smart grid

brings an active consumer participation in which the consumer is informed and the demand response and distribution energy are also reduced. The topology of conventional central generation has many problems when energy renewal resources are included in the interconnection system, so the smart grid distributes energy sources with plug and play conveniences' focus on renewable energy; thus, new services and products are in the market. In the case of natural disasters, severe damage is limited to relatively small areas, which gives the possibility of swift response.

For facing the areas that the new concept of smart grid has to solve, key factors have to be characterized in different models to get a complete idea; some of them are designed as per a well-known theory, such as the supply chain model proposed by [69]. When the electric grid is shown as a closed-loop supply chain model, presented in Figure 9.1, it is possible to recognize the main actors of the electric system answering the questions who, when, why, how, and what [66]. In this new topology, new areas will play a crucial role in the smart grid topology and it is clear that cybersecurity is in the top of the energy control system layers for supervising the grid.

A conventional electric grid is not able to handle all the electrical energy necessities of today; hence, some energy alternatives appeared to find new solutions. The traditional electric grid covers the central generation system, transmission, distribution, and end user. One of the main problems is storing energy from renewable sources of energy, like wind or solar energy, and the production of energy using renewable energy being prioritized in the distribution system. There are different technological conditions that have to be accomplished to construct a smart grid system. Figure 9.2 illustrates the main technological groups that form the smart grid concept.

When a smart grid is developed, it is expected that the smart grid can fulfill the following general attributes under operation:

- Evaluate grid health in real time
- Predict normal performance

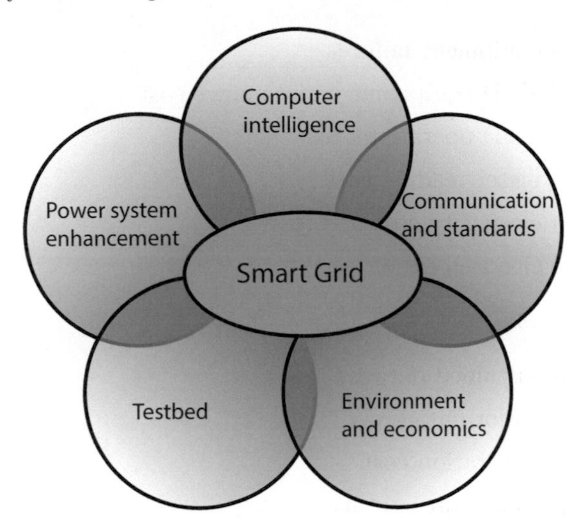

FIGURE 9.2
Key groups of the smart grid technology

- Anticipate energy emergencies and demands
- Adapt to new energy requirements
- Distribute resources and renewable energy
- Handle stochastic events
- Provide auto-correction and reconfiguration

9.1 Benefits of the smart grid

The U.S. Department of Energy provides a list of benefits regarding the smart grid:

- Economic factors
  - Optimized generator operation
  - Deferred generation capacity investments
  - Reduced ancillary service cost
  - Reduced congestion cost
  - Deferred transmission (T) capacity investments
  - Deferred distribution (D) capacity investments

  - Reduced equipment failures
  - Reduced T&D equipment maintenance cost
  - Reduced T&D operations cost
  - Reduced meter reading cost
  - Reduced electricity cost and losses
  - Reduced electricity theft
- Reliability factors
  - Reduced sustained outages
  - Reduced major outages
  - Reduced restoration cost
  - Reduced momentary outages
  - Reduced sags and swells
- Environmental
  - Reduced CO2 emissions
  - Reduced SOX, NOX, and PM-2.5 emissions
- Security
  - Reduced oil usage
  - Reduced wide-scale blackouts

Since the smart grid has shown its energy potential, it is important to make a prediction about the cost of upgrading the traditional electric grid presented in Figure 9.3 to a smart grid. As an example, the Electric Power Research Institute forecast that the cost to upgrade the U.S. electric grid to a smart grid could be between $338 billion and $476 billion, and could generate benefits of between $1,294 billion and $2,028 billion; as result, the upgrade of the electric grid is possible and beneficial for the utility system and end user. Some of the key drawbacks when a traditional grid is used are: centralized architecture and control, passive transmission and distribution, very extensive network (long paths and many components), lack of diversity.

As it could be observed, the communication systems inside a smart grid are very important and have to be restructured in order to get the communication system working properly. Figure 9.4 shows how the smart grid communication has to be adjusted.

The micro-grids are local and independent electrical grids that could integrate energy resources, transmission, distribution, and end-user. A local island that can operate as a stand-alone or grid-connected system is defined as a micro-grid. Normally, gas turbines or renewable resources power it and an

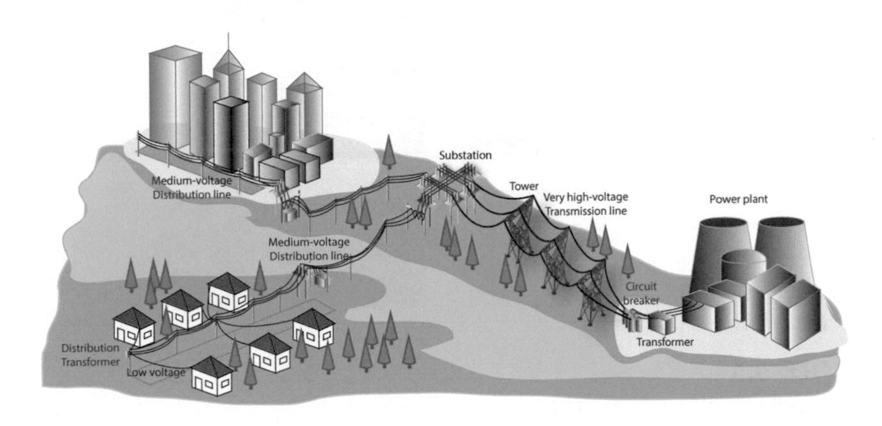

FIGURE 9.3
Conventional electric grid

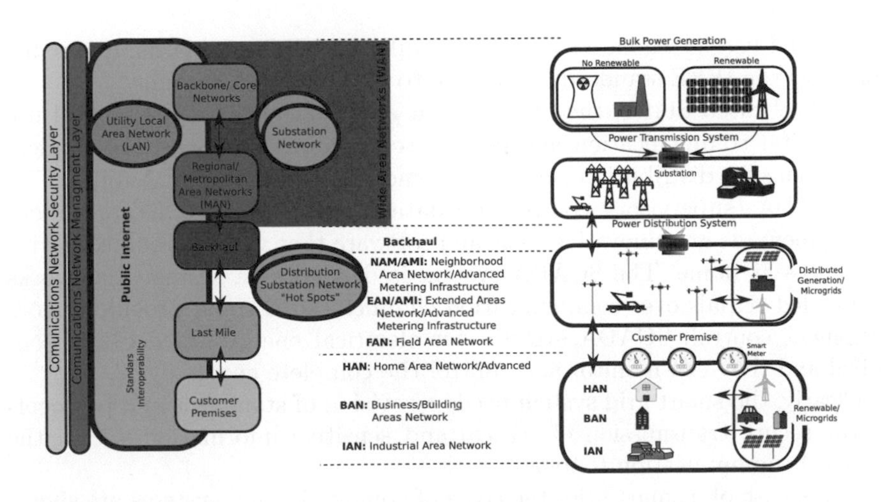

FIGURE 9.4
Communication in a smart grid system

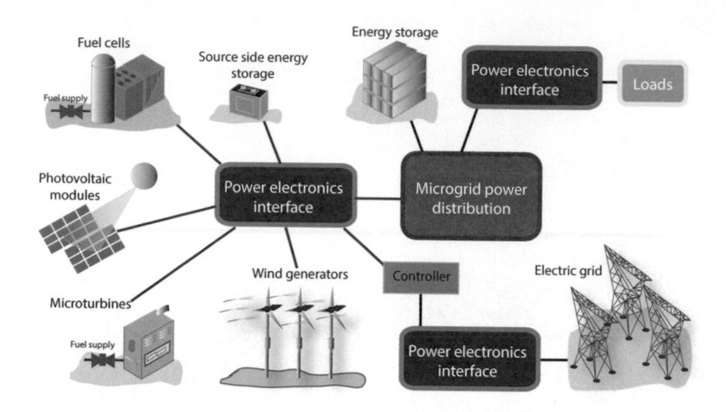

FIGURE 9.5
Basic topology of a micro-grid

inverter is included in order to connect it to the main grid. In addition, harmonics are filters that are used to improve the power quality. The micro-grid has advanced control systems and the smart grid includes complex decision support systems. Figure 9.5 shows a micro-grid. A conventional grid-tied system is not a micro-grid; one of the main limitations in a conventional grid-tie is the photovoltaic systems in which the operation at home is based on IEEE standard 1547, in which the inverter cannot power the home when the grid is disconnected.

The end-user is regulated at home by different IEEE standards. Figure 9.6 illustrates the IEEE standards that have to be taken into account.

In the United States, the transmission and distribution systems still use older digital communication and control technology. Advanced communication systems for distribution automation, such as remote terminal unit (RTU) and SCADA (supervisory control and data acquisition) systems, are under development, as are innovative tools and software that will communicate with appliances at home. The SCADA system is defined as a system that works with coded signals over communication channels, so it could provide remote equipment control. SCADA systems for electrical energy are digitally controlled and they can monitor and control the complete energy cycle.

Clearly, the smart grid system needs a revision of standards and protocols for the secure transmission of critical and sensitive information within the proposed communication topology.

In the case of a smart grid, the costs of communication systems are significant and sometimes an impediment, so an evaluation between the total cost and the overall performance is needed. In the communication system, maintaining the security between the control center and field devices is one of the most important problems that have to be faced.

Smart meters are a new kind of electrical meter (Figure 9.7) that measure

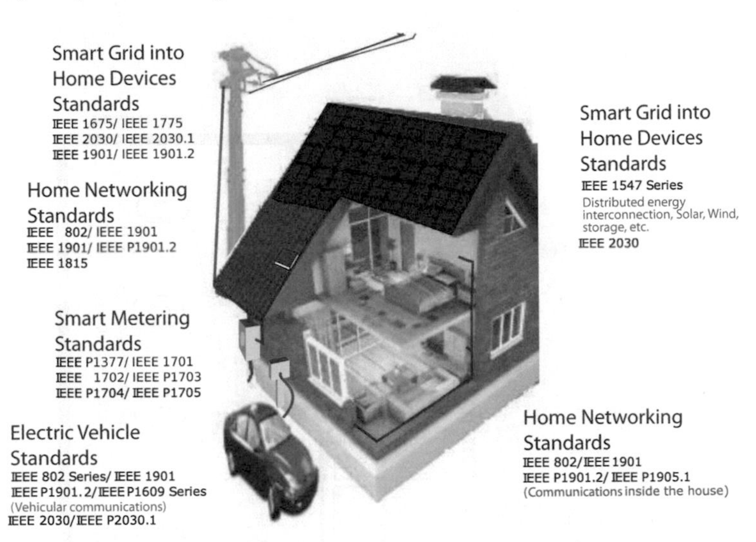

FIGURE 9.6
IEEE standards at home

energy usage in real-time, just as the old ones did. But, they also send information back to the utility by wireless signals. Instead of using a utility meter reader who goes to the property and manually does the periodical electric service reading, the information is send automatically to the utility. To improve the overall performance of the smart meters, a new group of appliances has been developed. Smart appliances like the toaster, the microwave, and even the refrigerator are based on micro-chips that allow your appliances to communicate with the smart meters to control the energy usage.

Smart meters are part of a global plan to upgrade the world electrical grid to a smart grid, which reduces energy consumption. The smart grid is being implemented in both Canada and Mexico in order to communicate with the United States. Designers are working on new standards that will integrate all of North America into a single, unified smart grid system. There is an initiative to generate a global smart grid to cover a continent, as presented in Figure 9.8. The Global Energy Network Institute (GENI) shows this Dymaxion (tm) Map of the world from the perspective of the North Pole, to reveal the global grid currently under construction. The lines are high-voltage electrical transmission lines that are capable of transferring large amounts of energy from continent to continent.

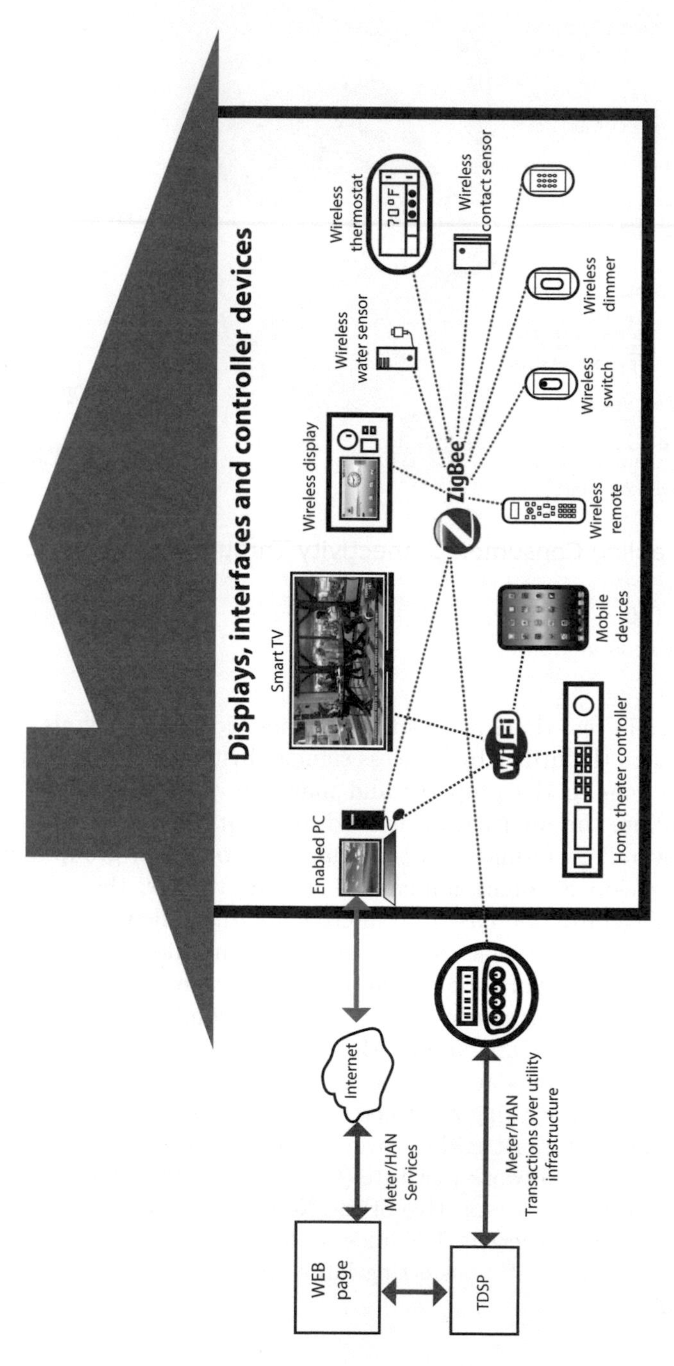

FIGURE 9.7
Smart meter and smart home

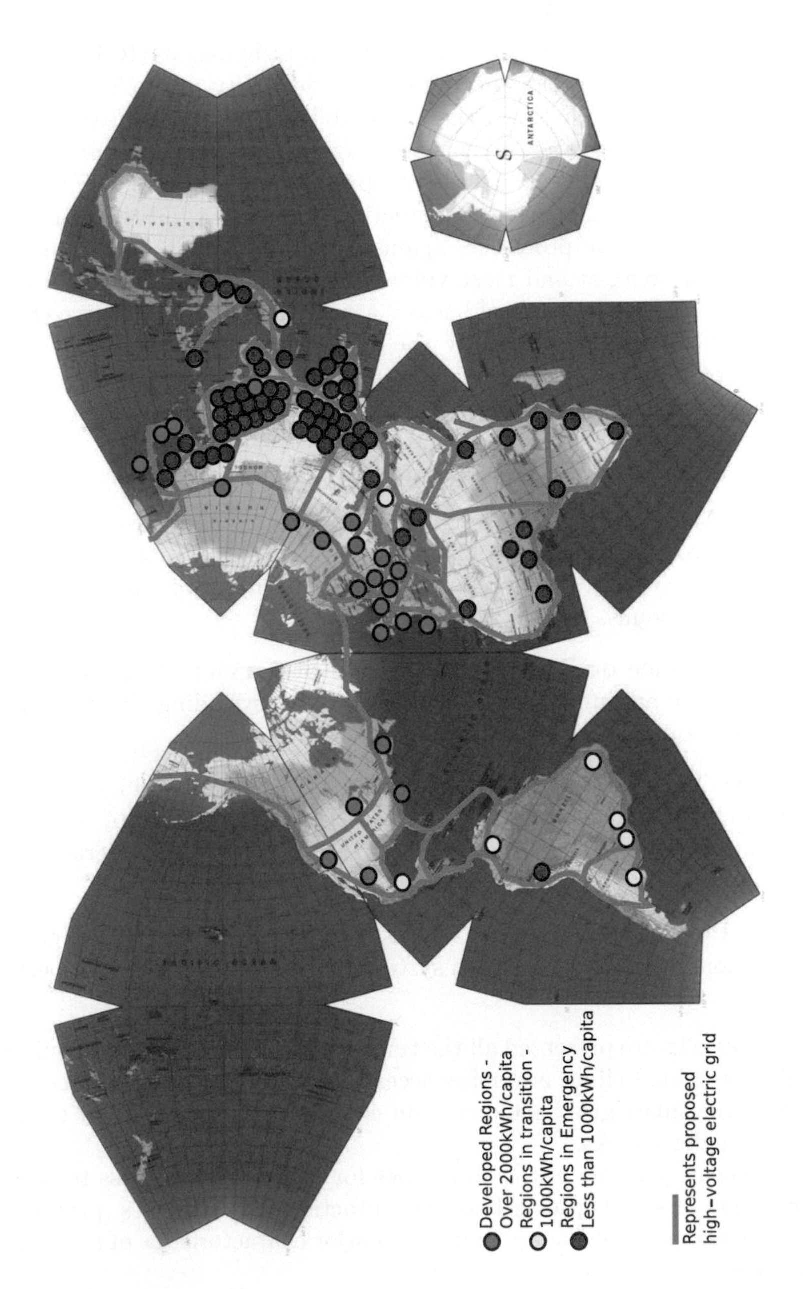

FIGURE 9.8
Global smart grid

9.2 Cyber-security

The presence of information and communication technologies (ICT), e.g., computers, networks, programs, and data, gives flexibility and efficiency to the smart grid operation. However, the vulnerability of ICT to threats must be considered [71].

Smart-grid cyber security focuses on protecting ICT from unintended or unauthorized access, changes, or destruction. By default, all the components in the smart grid have protection against intruders, but hackers are always looking for the weakest and most vulnerable link that designers overlooked. For example, in 2015 a couple of hackers attacked during a self-drive car test, and were able to remotely modify the a/c and radio stereo of the car. In fact, they could modify the acceleration, brakes, or steering, but these hackers didn't. The same attacks could be done on the smart grid elements, resulting in a big risk for both power distribution and people.

Cyberattacks can be classified as [25]:

Malware: Malicious software designed by hackers to run in a specific system. Viruses, worms, Trojan horses, ransomware, spyware, rootkit, etc. are examples of malware. Stuxnet is an example of malware affecting SCADA and PLC systems.

Denial-of-Service or Distributed Denial-of-Service: Makes a networked system busy and unavailable to its users by overloading the system with a huge amount of requests.

Zero-Day: This is an unpatched computer system vulnerability that is not yet disclosed to the public. The name refers to the vendor of the affected computer systems having zero days to fix it after the vulnerability has been disclosed.

Advance Persistent Threats: This is a generalized term for a group of professional hackers targeting a system continuously over a long period of time.

Cyberattacks are presented all the time, so a secure smart grid should have enough protection to limit adversary access. Furthermore, if the cyber-security is broken, the smart grid should contain elements that minimize or eliminate the effect of the attacks.

Cybersecurity is one of the 20 metrics for measuring progress toward implementation of smart-grid technologies, practices, and services, proposed by the U.S Department of Energy in [8]. Six major characteristics of a smart grid are proposed in [8]:

1. Enables informed participation by customers
2. Accommodates all generation and store options

3. Enables new products, services, and markets
4. Provides the power quality for the range of needs
5. Optimizes asset utilization and operating efficiency
6. Operates resiliently during disturbances, attacks, and natural disasters

Cybersecurity is defined as an emphasized metric of the smart grid operating resiliently during disturbances, attacks, and natural disasters. Resiliency refers to the ability of a system to react to events such that problematic consequences are isolated with minimal impact to the remaining system, and the overall system is restored to normal operation as soon as practical. This robustness reduces interruption of services, providing a direct benefit to consumers and service providers. Protection against all hazards, whether accidental or malicious, and the ability to span natural disasters, deliberate attacks, equipment failures, and human error, are included in resiliency. Operational resiliency has three essential properties: a) ability to change (adapt, expand, conform, contort) when a force is enacted; b) ability to perform adequately or minimally while the force is in effect; and c) ability to return to a predefined, expected normal state whenever the force relents or is rendered ineffective.

The study and research of vulnerabilities is necessary to quantify cybersecurity issues. Laboratories and test-bed environments are needed, considering the actual field conditions. Then, plenty of cyber-attacks should be characterized in order to understand or estimate the possible response of the smart grid when a cyber-attack is presented.

9.3 Exercises

9.3.1 Contribution of solar energy

Assume that there is demand for energy from a distant point and there is solar energy to be exploited. In this exercise, the student will intervene by reducing the consumption of energy from an older generation plant, using the surplus energy produced by solar photovoltaic systems.

The reduction of even a minimum of absorbed energy will certainly have an impact on environmental pollution produced by a plant of the older generation.

Set the resistive load to the second value and close the R2 relay to supply energy coming from the coal plan (see Figure 9.9).

Close R4 relay to transfer energy coming from the plant to the load and observe the power consumption in the maximum demand meter.

In this situation you can see active power required from the resistive load

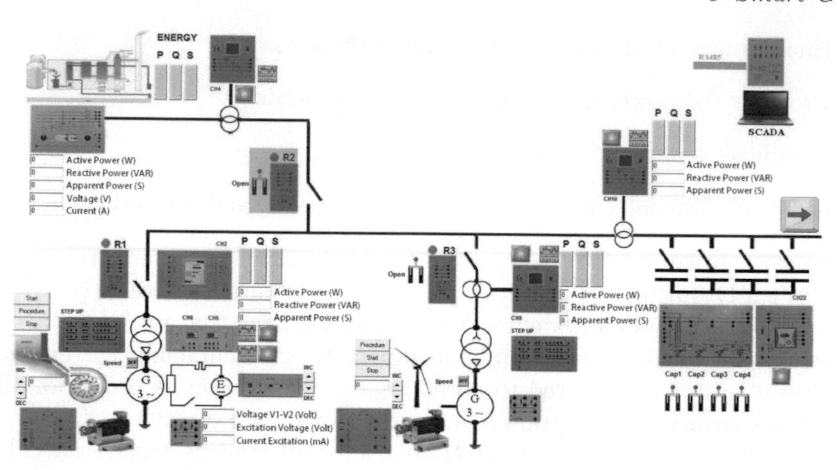

FIGURE 9.9
Solar energy global panel

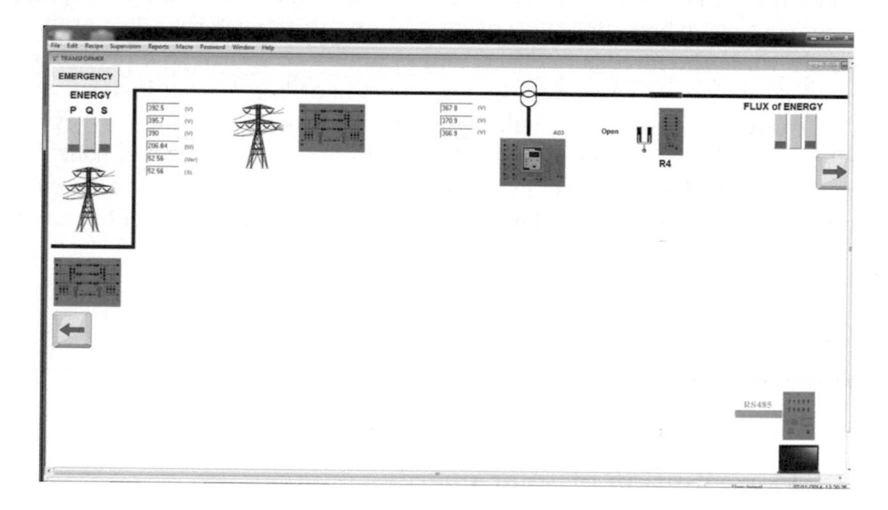

FIGURE 9.10
Transformer panel

Resistive load	Active power [W]	Reactive power [VAr]
2	207.64	52.32
3	310	101

and a minimum of reactive power required from the primary of the step-down transformer (see Figure 9.10).

Insert the hydroelectrical energy to the network using the hydroelectrical plant simulator.

Observe the brushless motor starting, and after that, enter in the procedure for the automatic synchronization connection (see Figure 9.11).

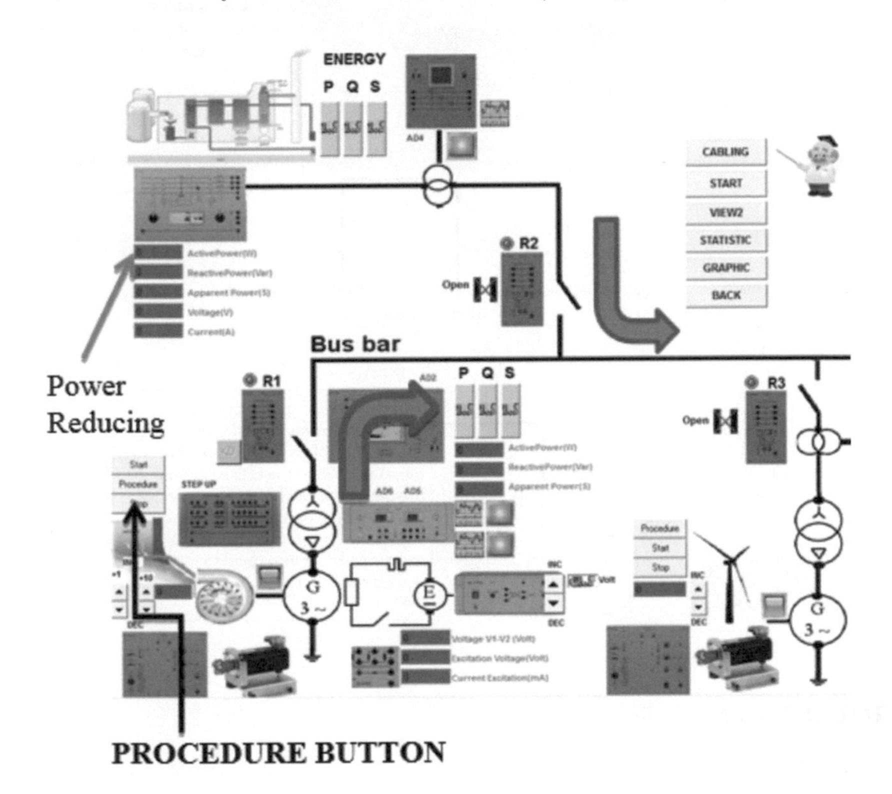

FIGURE 9.11
Automatic synchronization connection

Increment the speed to 1800 [rpm] and then increment the excitation field until the delta output of the generator goes up to 230 $[V]$.

Look at the AC display phase-to-phase voltage in the electrical power digital measuring unit. Check, at the same time, either the secondary voltage or frequency on the software; the level has to be equal to that of the network 400 $[V]$.

Look to the display of the generator synchronizing relay and reduce the speed angle shift and the angle at under 20 degrees (see Figure 9.12).

When the hydroelectrical part is connected to the grid, you have to pay attention to stabilizing the system, incrementing the active power consumption from the load, and stabilize the system regulating the speed and excitation. Check the LED near the turbine; it has to be green and stable (see Figure 9.13).

Observe the reducing of power coming from the coal plant when you introduce active power to the network.

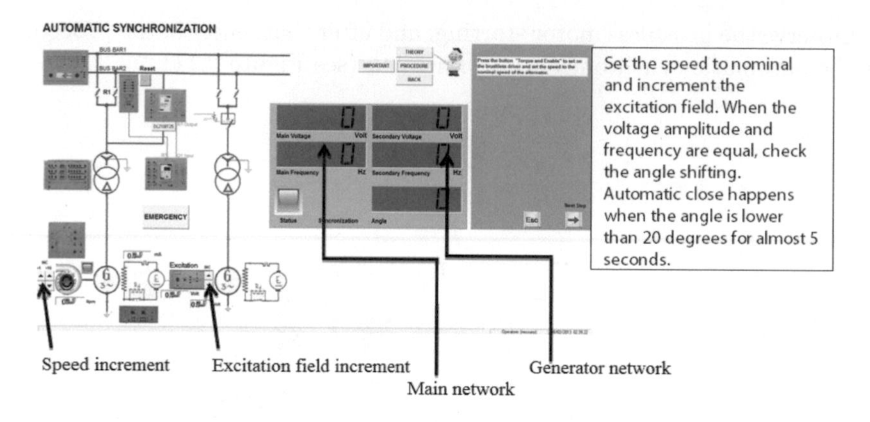

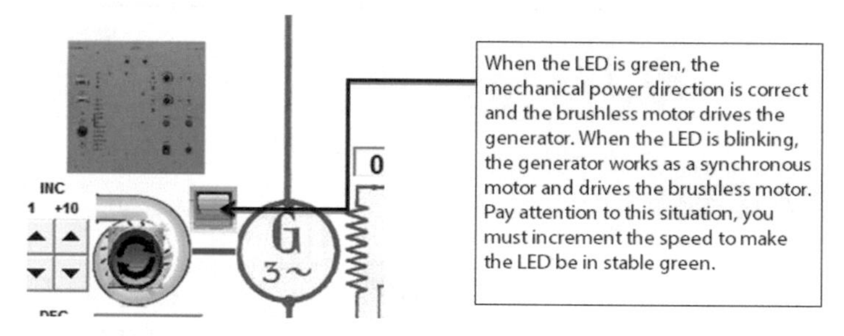

FIGURE 9.12
Display panel

The hydroelectrical contribution, in this case, also permits you to reduce energy supplied from old plant generators. Observe the meter near the coal plant simulator.

	Grid Side		Synchronous Gen.		Load	
Resistive Load	Active Power [W]	Reactive Power [VAr]	Active Power [W]	Reactive Power [VAr]	Active Power [W]	Reactive Power [VAr]
2	111	-540	145.4	593	-235	-68
4	281	-350	160	550	-400	-180
ω=1818 [rpm], I= 1.76 [A]						

9.3.2 Contribution of wind plant

Assume that there is demand for energy from a distant point and there is a wind power plant near that point.

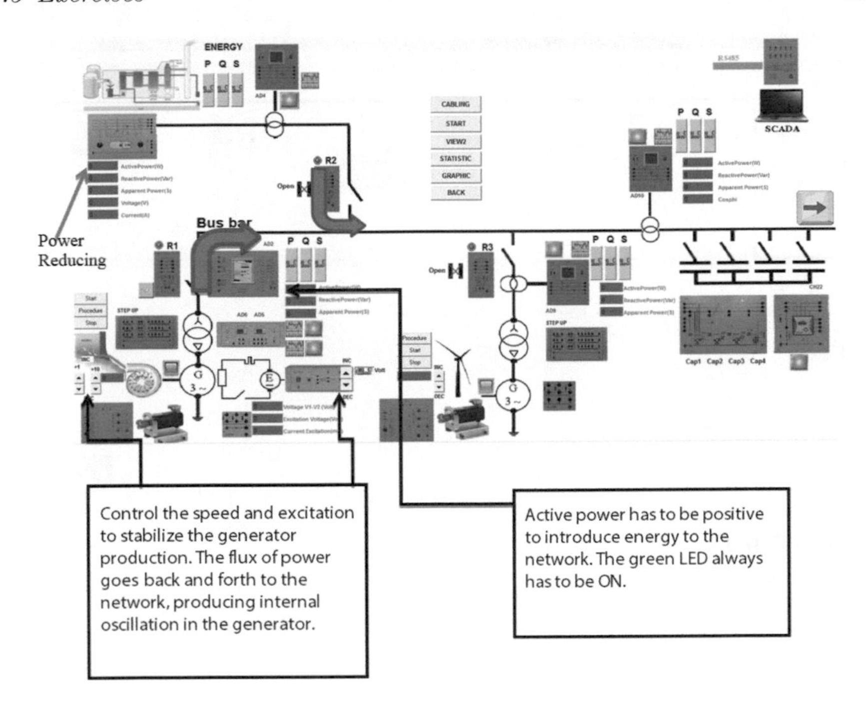

FIGURE 9.13
Stabilizing the system

A wind power plant is going to be connected to a network oriented to reducing coal plant power consumption.

Set the resistive load to the second value and close the R2 relay to supply energy coming from the coal plant. Close the R4 relay to transfer energy coming from the coal plant to load and observe the power consumption in the maximum demand meter (see Figure 9.14).

In this situation you can see the active power required from the resistive load and a bit of reactive power required from the primary of the step-down transformer. Insert the energy to the network using the wind-system plant simulator.

Check to see if the brushless motor is starting (see Figure 9.15).

Increment the speed to 1800 [rpm] and press the button "Procedure" (see Figure 9.16).

Close the relay R3 only when the speed is 1800 [rpm]. Increment the speed of the brushless wind generator to increment the active power contribution (see Figure 9.17).

Observe the reducing of power coming from the coal plant when you introduce active power to the network.

The wind power plant's electrical contribution also, in this case, permit

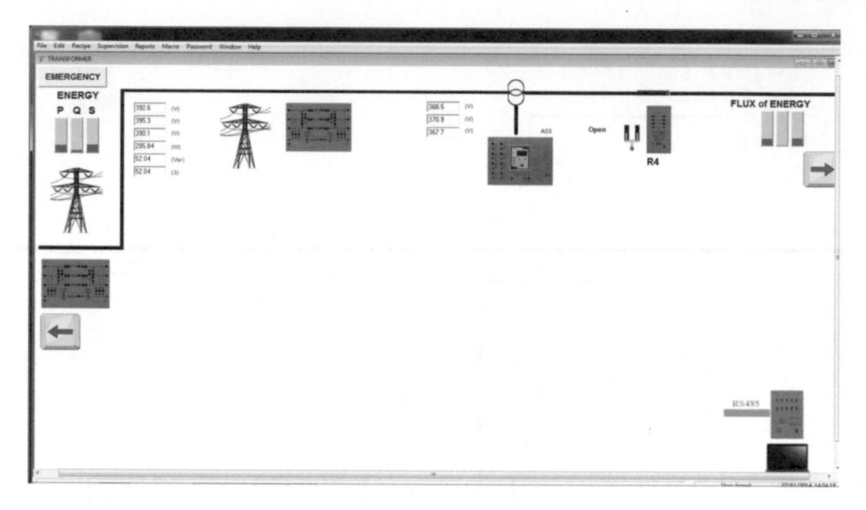

FIGURE 9.14
Transfer power panel

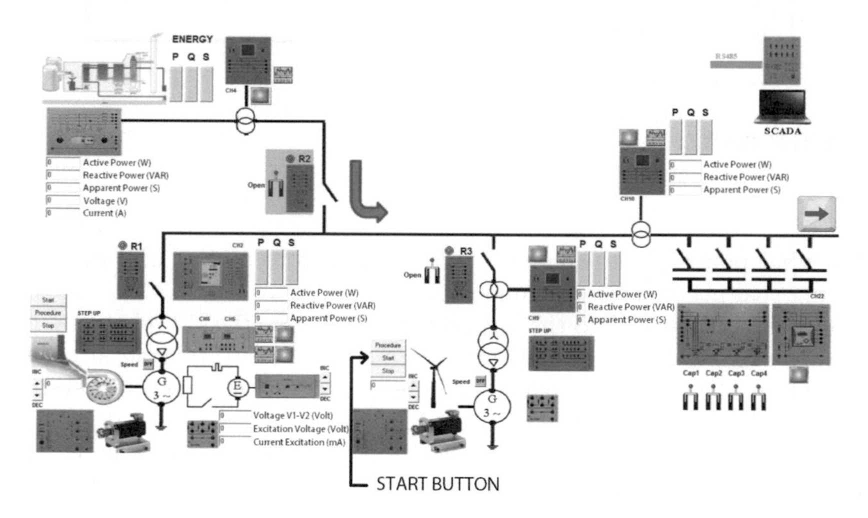

FIGURE 9.15
Starting the motor

the reduction of energy coming from old plant generators. Observe the meter near the coal plat simulator.

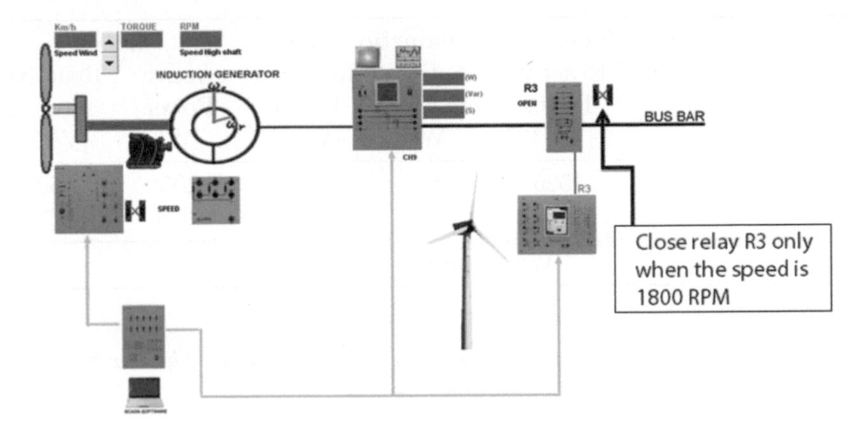

FIGURE 9.16
Speed increment

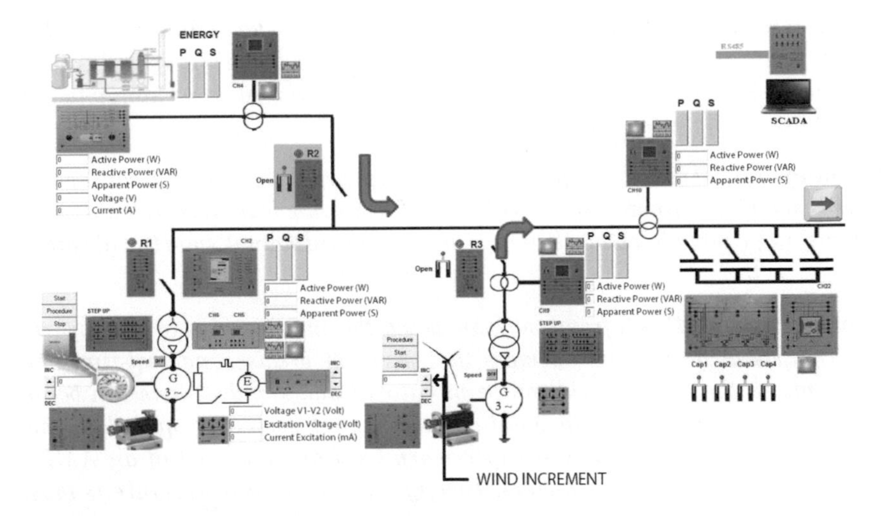

FIGURE 9.17
Wind generator increment

9.4 Homework problems

Problem 9.1. *Navigate to the carbon footprint calculator from myclimate.org and perform a calculation of your personal footprint.* ht tp s: // co 2. my cl im at e. or g/ en/f oo tp ri nt _c al cu la to rs /n ew ?g cl id =C jw KE Aj wu tX IB RD V7-S Dv di Ns Uo SJ AC Il Tq li h61Z 9b JP4c rT Bt qQ 1o PF Lf W6p AU t2m az sl pj U3s Vx oC -A Tw_w cB

Resistive Load	Grid Side Active Power [W]	Grid Side Reactive Power [VAR]	Induction Gen. Active Power [W]	Induction Gen. Reactive Power [VAR]	Load Active Power [W]	Load Reactive Power [VAR]
2	187	1526	48	-1455	205	-52
4	384	1653	35	-1467	-389	-179
Wind speed = 3000						

Solution 9.1. *There is no specific solution to this problem. The teacher would ask the student to write about resources usage.*

Problem 9.2. *Visit the Smartrid.gov site and find at least 5 reported smart-grid or micro-grid projects throughout the available documents. Report them, taking into account the funding required and the expected outcomes.*

Solution 9.2. *There is no specific solution to this problem. However, students can be asked to be as specific as desired. Projects at the proposed site are thoroughly explained and presented, so their technical issues, economic factors, and follow-up inquiries are also available.*

Problem 9.3. *The smart grid is mostly supported by one key concept: advanced metering infrastructure (AMI). Perform some brief research about such systems and conclude about their pertinence toward actual smart-grid integration.*

Solution 9.3. *AMI is a framework on which the smart grid is to be supported. It is composed of electronic devices that perform precise and real-time measurements so the electric grid can be monitored and key decisions can become possible during operation. In addition, the digital systems required to make their information available at a supervisory level are also part of an AMI.*

AMI goes beyond smart meters. Briefly, the whole infrastructure is mostly oriented to data analysis and its translation into information.

Problem 9.4. *Another big concept behind smart-grid integration is the energy management system (EMS). Perform some brief research about such systems and conclude about their pertinence toward actual smart-grid development.*

Solution 9.4. *EMS is not a new concept and is commonly related to SCADA systems. As the smart grid can easily move beyond a simple human-machine interface conceptually, the EMS is a concept that makes smart-grid definition easier, as it states the foundations where the smart-grid is to be implemented.*

Problem 9.5. *The smart grid approach to energy management has not been thoroughly developed. There are many issues precluding its final realization. Perform some brief research and determine the top five issues that hinder smart grid realization. Provide an explanation for each concept.*

Solution 9.5. *There are many obstacles regarding smart-grid integration to the current grid. Some of them are listed here so they can be used as a reference:*

- *Even though there are some related projects, how the technology is to be integrated is not clear yet*
- *Most parts of the world are grid-unautomated, e.g., North America is below 40% automated*
- *Corporations are already developing grid products even if the smart-grid has not been technically defined*
- *Smart-grid projects are substantially more expensive than common electric projects*
- *The electric grid is not available for testing, so R&D efforts have been obscured*

9.5 Simulation

It is important to consider that most assumptions made regarding system interconnections depend on the infinite bus stiffness, which is considered to be infinite indeed. This implies that it can give/take any amount of active or reactive power while holding the voltage characteristics of the network. Although this is known to be unrealistic, it will be accurate as long as the bus surpasses other sources in terms of installed capacity.

In addition, the incorporation of new sources should normally follow from local needs and must be calculated so their power capacity is as required. If a generator is not fully used, its efficiency will drop. On the other hand, the incorporation of many generators will decimate the efficiency of the infinite bus as its generators will require less energy (see Figure 9.18).

Energy projects must then be preceded by exhaustive analysis before their application. Undoubtedly, the incorporation of new "domestic power" sources to the grid should be regulated and measured, so the already installed systems and the new ones fit a known consumption profile.

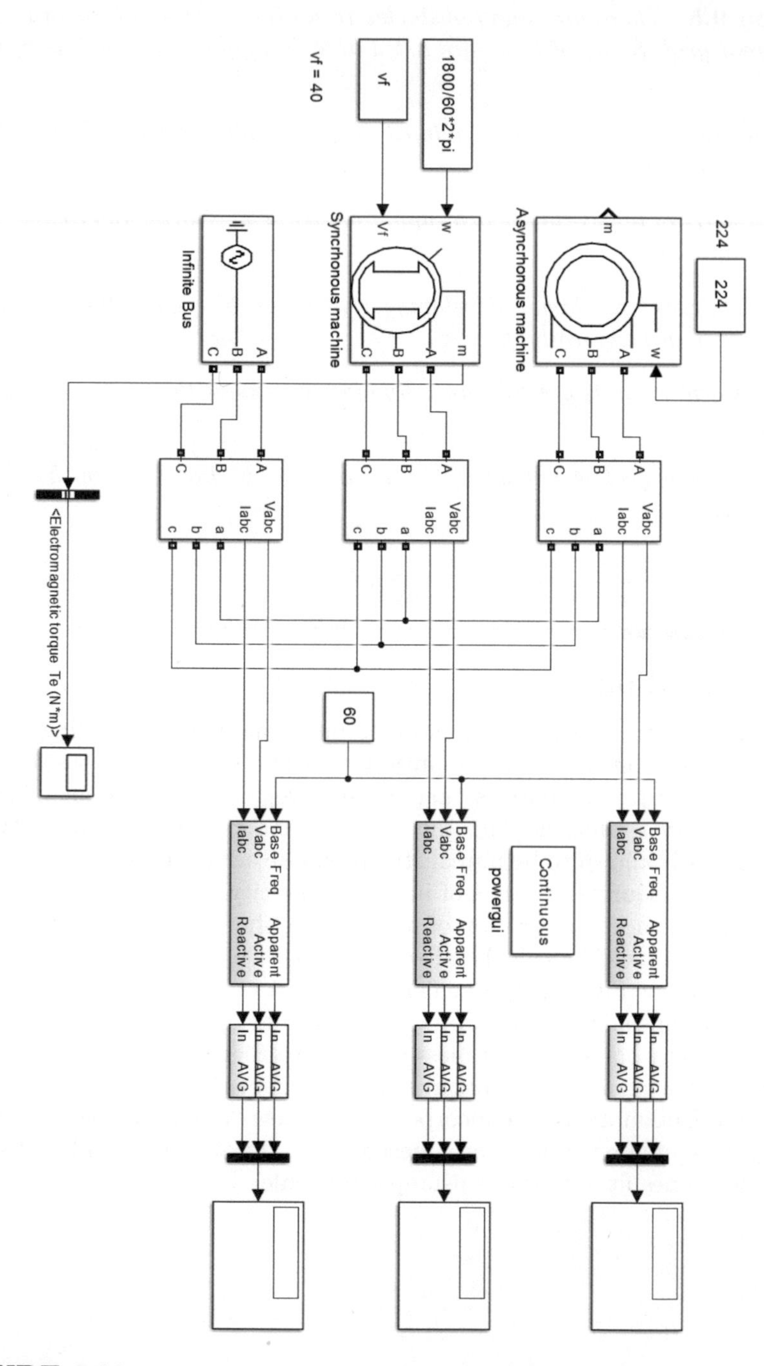

FIGURE 9.18
Power contribution of different power sources

10

Power Electronics in Power Systems Using LabView, LabView-FPGA, and Multisim

10.1 Introduction

"Today, approximately 30 percent of all power generation utilizes power electronics between the point of generation and consumption. By 2030, it is expected that up to 80 percent of all generated electricity will utilize power electronics" [59]. Moreover, the power electronic stages have reduced their losses (see Figure 10.1), so they are very attractive for controlling power systems. Hence, it is very important to design power electronic stages using software tools such as LabVIEW.

Not only are National Instruments customers completing more projects on time (`http://forums.ni.com/t5/Power-Electronics-Development/gp-p/grp-1891`), but they're doing it in less time with fewer resources than those who are following the traditional design approaches. This is made possible by using a platform-based approach that can fusion the best of build and the best of buy, as well as leveraging high-level software with flexible off-the-shelf hardware. These are integrated together, so that you get the opportunity to focus on innovation, not your implementation. Graphical system design and the LabVIEW reconfigurable inputs and outputs (RIO) architecture provide quality products and the confidence to innovate, while getting your project done in less time. The RIO relationship between price-performance and number of systems deployed is presented in Figure 10.2.

The two main goals of the power electronics design using simulation tools are the following. A traditional power electronics design is done using a V process that is shown in Figure 10.3. This process starts with co-simulation that allows us to move to prototype in a fast manner because the design V process is highly improved, such that co-simulation provides real results that are much closer to the experimental design (see Figure 10.4).

Embedded systems are used in the V process for designing power electronics systems. Thus, the availability of heterogeneous field programmable gate array (FPGA) hardware contains an array of integrated digital signal processor (DSP) cores capable of efficiently executing floating-point math operations, so that the co-simulation and the deployment process can be improved. On the other hand, these modern FPGAs are actually hybrid devices containing

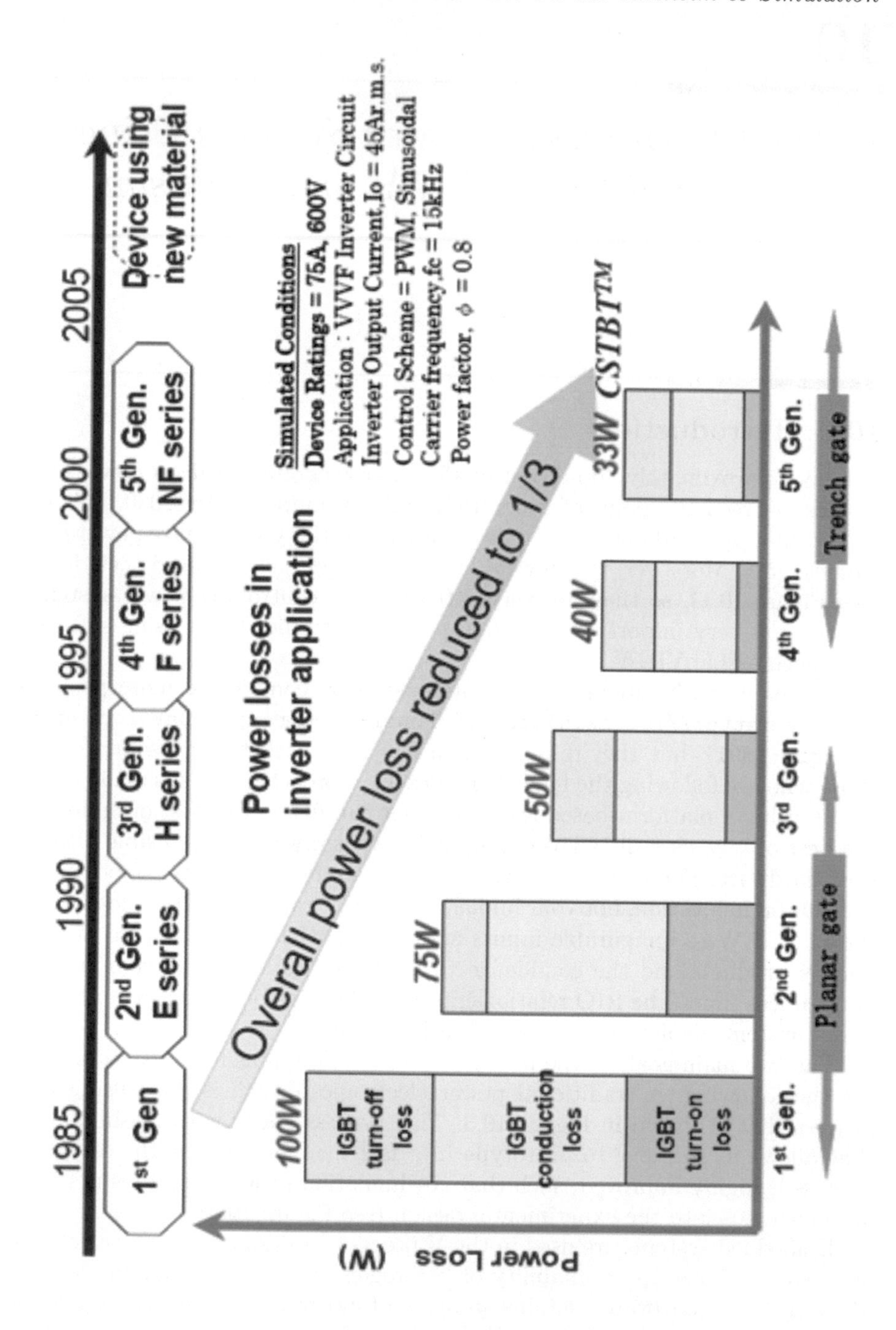

FIGURE 10.1
Generations of power electronics and power losses

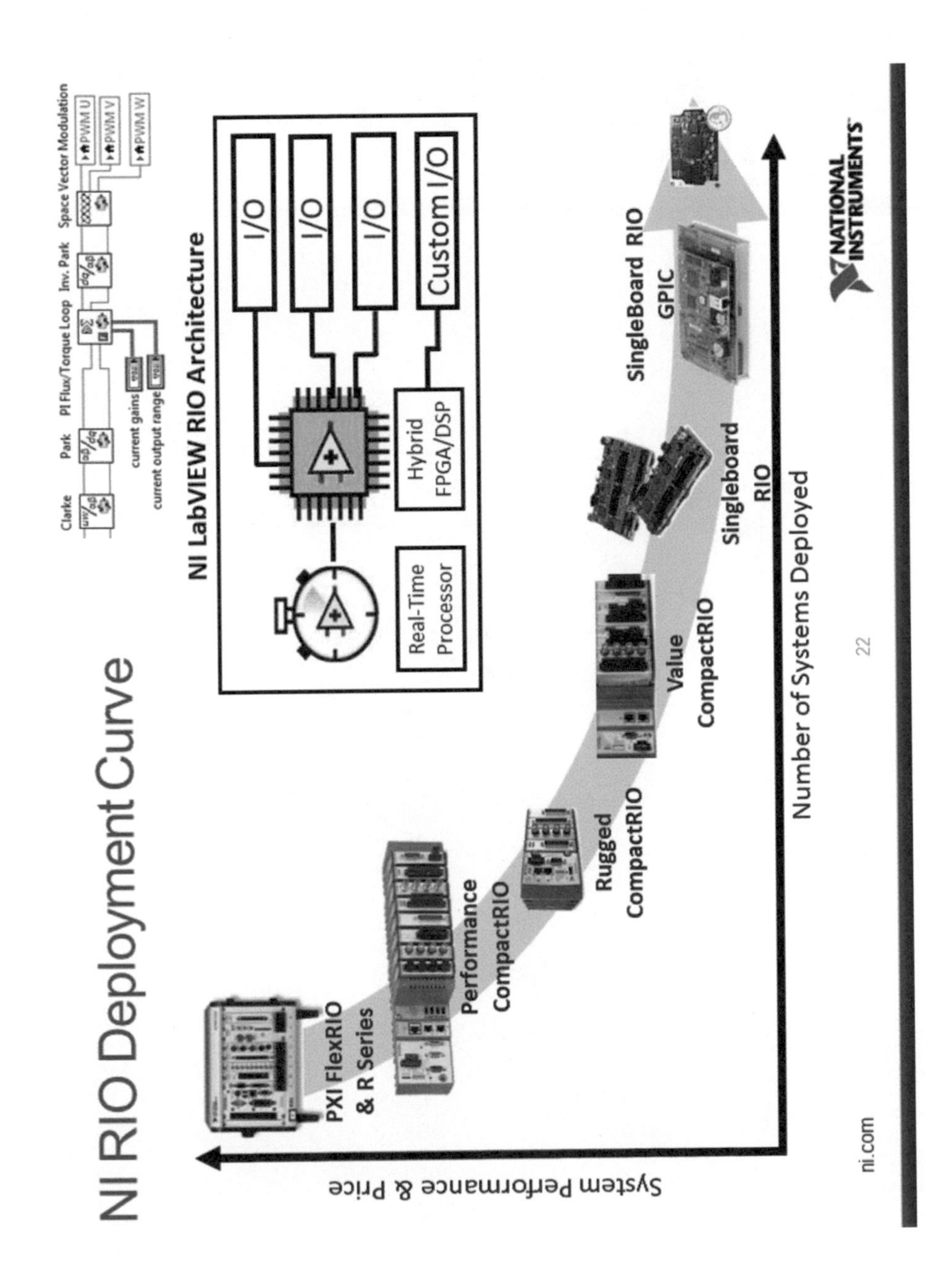

FIGURE 10.2
NI-RIO deployment curve

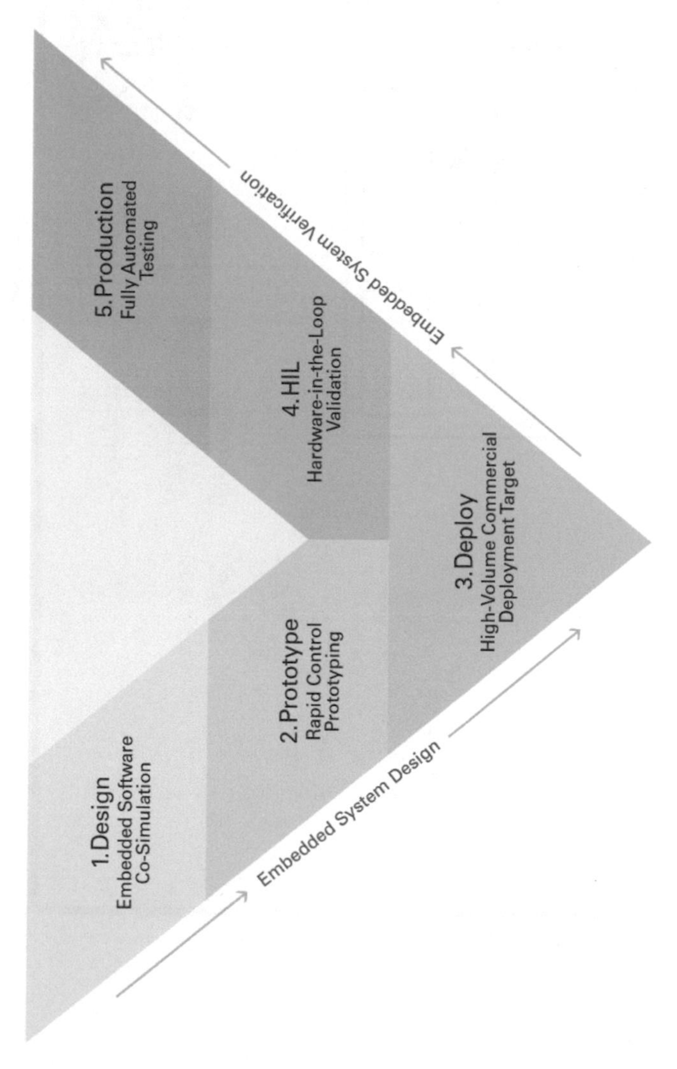

FIGURE 10.3
Design V process

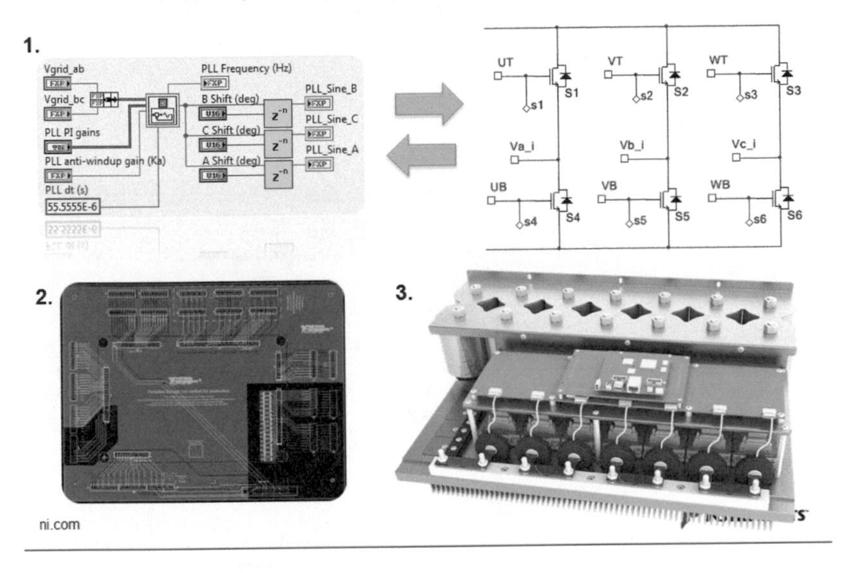

FIGURE 10.4
Co-simulation, interface board, and commercial deployment

hard-core DSP processing elements, integrated within the reconfigurable computing fabric, and capable of achieving MHz speeds for simulation and control, so the design process requires less time. This FPGAs can be combined with technology to analyze electric circuit, power electronics, and power system models in order to get results that allow us to deploy controllers in power electronic stages (see Figure 10.5).

10.2 Co-simulation LabVIEW and Multisim

As power electronics is one of the most important topics in power systems (`http://www.ni.com/power/electronics/`), it is enormously important to understand and to design it [3]. Besides, there are several simulations tools that can improve the design of power electronics in power systems. As a result, LabVIEW (`http://www.ni.com/manuals/`) has several tools that allow us to create simulations, co-simulations, and hardware in the loop systems. In addition, the conventional power electronics performance can be improved when a software design is done. Power electronics has become an essential part of power systems (see Figure 10.6). The basic elements in power electronics are the semiconductors that are electronic switches (IGBTs, GTO, MOSFET,

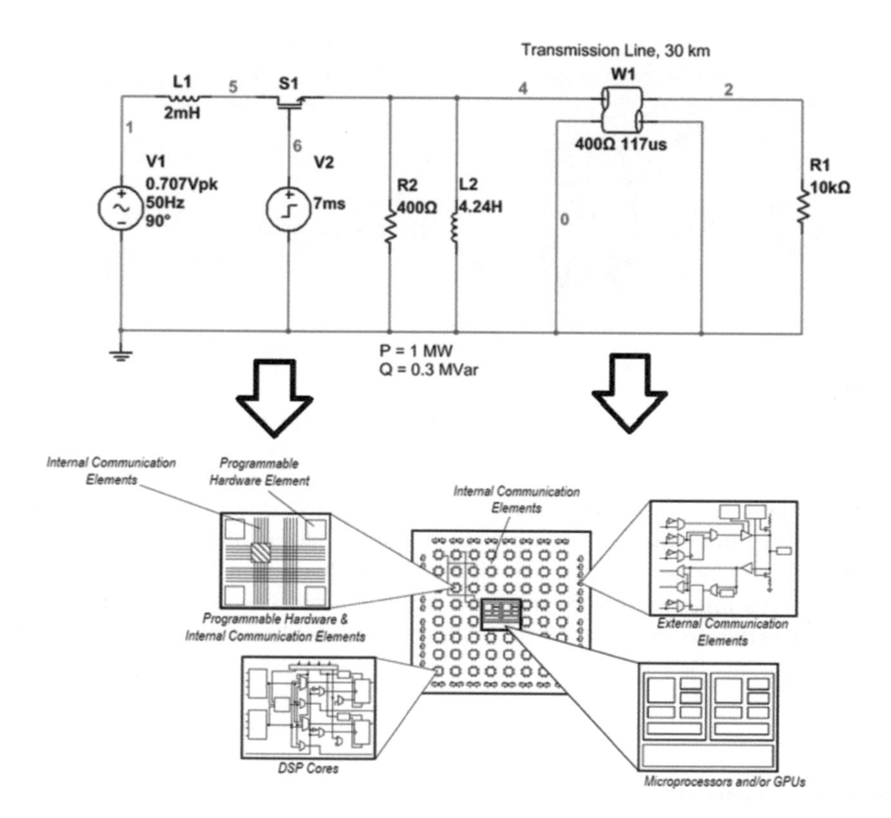

FIGURE 10.5
Automatic Multisim to LabVIEW FPGA conversion enables this power system model to be simulated at 2 MHz speeds using heterogeneous FPGA hardware

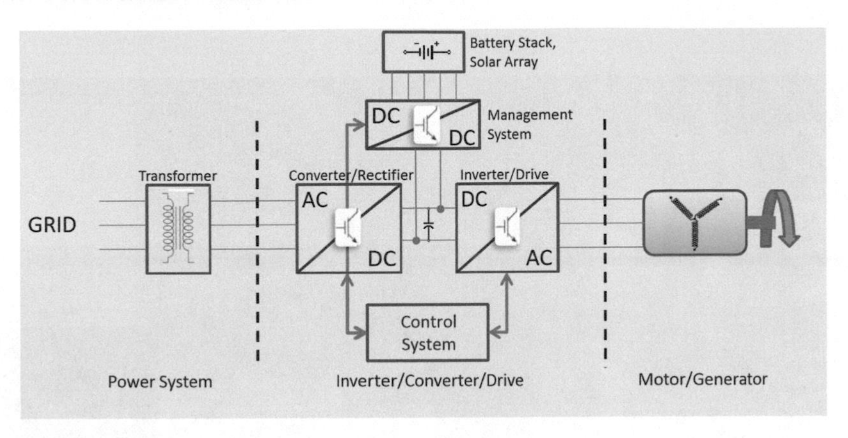

FIGURE 10.6
Power system with power electronics elements

etc.). Those components can either conduct or block the electric current in the power system when a control signal is provided. They are very effective for controlling power systems since the time to change their conductivity condition is in the range of microseconds. Thus, a power electric system can be controlled using power electronic stages. When the power electronic stages are designed for power systems, it is important to use simulation tools that help to design the power electronic stage in a precise manner and to save time and economic resources. As a result, the faster way to move from simulations to commercial deployments, and product generation, is the co-simulation in which the subsystems are linked as a coupled problem that is simulated in a distributed manner. One of the most important industrial and academic tools is the co-simulation that is a closer representation of the real system in which the power electronic stage can be analyzed in a comprehensive manner (see Figure 10.6). This chapter deals with the LabVIEW FPGA and Multisim co-simulation tools to understand the power electronics design and fundamentals for power systems. In this chapter, you will gain a basic understanding of how the LabVIEW graphical system design tool chain can be used to allow for power electronics control. Figure 10.7 illustrates a block diagram about designing power electronics using LabVIEW FPGA and Multisim as the first step in the power electronic design [54]. This diagram can be determined by the design V process (see Figure 10.3).

One of the most important advantages of using Multisim (Multisim Manual) and LabVIEW in co-simulation is designing and analyzing a complete power electronic system, including applications such as wind turbines. Thus, co-simulation helps to effectively design across both the analog and digital domains. Traditional platforms are unable to accurately simulate both analog and digital together, so design errors cascade to the physical prototype and lead to an ineffective and lengthy design approach. With the new co-simulation

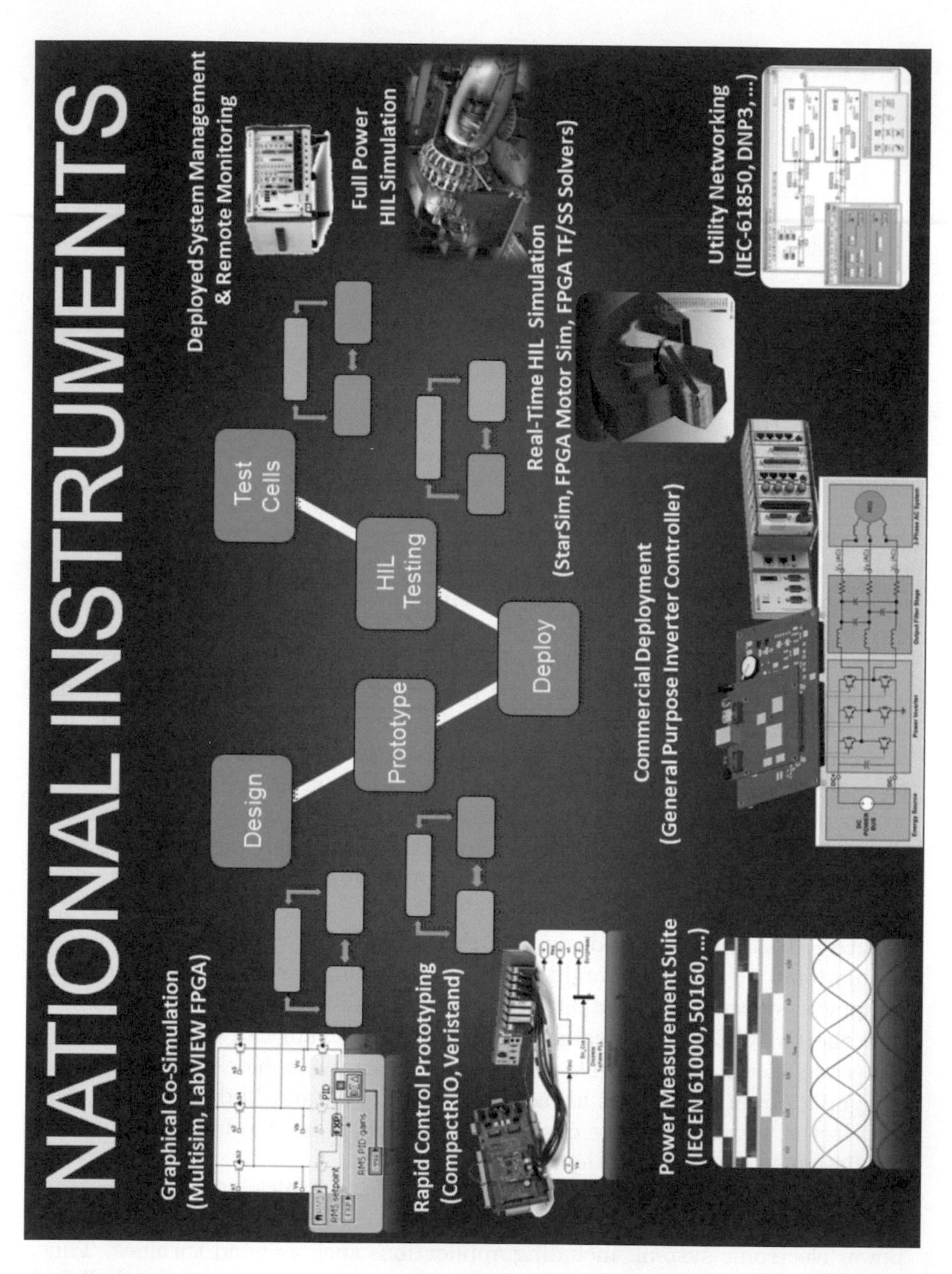

FIGURE 10.7
Design block diagram for power electronics based on LabVIEW FPGA, Multisim, and LabVIEW

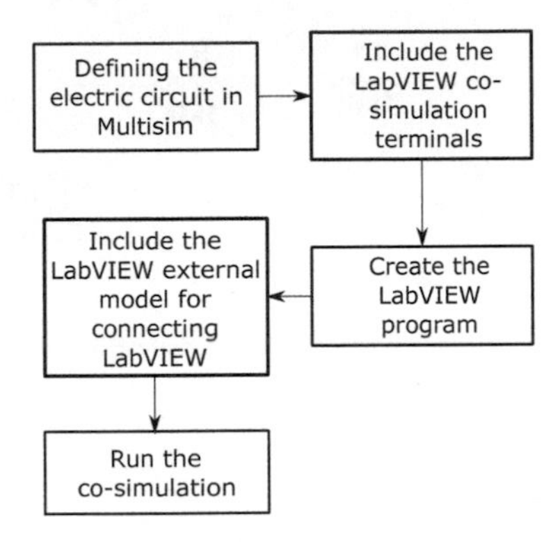

FIGURE 10.8
Basic block diagram for running a co-simulation

capabilities between Multisim and LabVIEW, a power electronic circuit with accurate, closed-loop point-by-point simulation and an entire analog and digital system can be designed. Hence, LabVIEW and Multisim take advantage of two distinct simulation engines [54]: Multisim simulation for the accuracy of analog and mixed-signal circuitry, and LabVIEW for the effective design and implementation of control logic. This is made possible by a unique time-step negotiation between both simulators using the external model interface in LabVIEW. The following programs have to be installed for running a power electronics co-simulation.

1. Install LabVIEW Full or Professional version
2. LabVIEW Control Design and Simulation Module
3. Multisim

Figure 10.8 shows a general flow diagram for designing a co-simulation program of an R-L-C electric circuit between LabVIEW and Multisim. As observed, the process is very easy and friendly.

The following example shows an R-L-C circuit that is modeled using co-simulation between LabVIEW and Multisim [2]. The steps necessary to complete a co-simulation are presented below.

Step 1: Place the R-L-C electric circuit and the voltage source in the topology in Multisim, as shown in Figure 10.9.

Step 2: This circuit will be able to send and receive data to and from the LabVIEW simulation engine when the communication terminals are placed. These terminals in Multisim are hierarchical block or sub-circuit (HB/SC) terminals. Right-click and select Place on the schematic ≫ HB/SC connector from the shortcut menu, or simply type <**Ctrl-I**>. Place one HB/SC connector above and to the left of the schematic and another connector

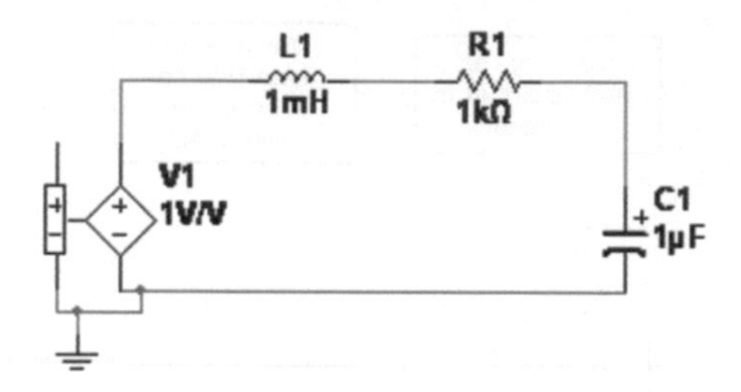

FIGURE 10.9
Components placed in Multisim

above and to the right of the schematic. Rotate the second connector 180 degrees by holding **Ctrl** and tapping **R** twice. Connect the HB connectors to the circuit by selecting them from the context menu (see Figure 10.10.)

To configure the HB/SC connector as an input or output to LabVIEW, the **LabVIEW Co-simulation Terminals** window must be opened. Navigate to **View≫LabVIEW Co-simulation** Terminals. Notice that the HB/SC connectors placed earlier appear in this window. To configure each connector as an input or output, select the desired choice from the **Mode** settings. Then configure each connector as a voltage or current input/output by selecting the **Type**. Finally, if it has placed two input or output terminals that are needed to function as a differential pair, a **Negative Connection** can be selected. Configure IO1 as an **Input** and IO2 as an **Output**. You can immediately see the shape of the connector changing on the schematic based on whether it's an input or an output connector (see Figure 10.11). The final electric circuit in Multisim can be seen in Figure 10.12.

Step 3: Creating a Digital Controller in LabVIEW is the last step for connecting LabVIEW and Multisim. To pass data back and forth between LabVIEW and Multisim, you must use a Control & Simulation Loop in LabVIEW. Navigate to the block diagram in LabVIEW (white window), right-click to open the **Functions Palette**, and navigate to **Control Design & Simulation≫Simulation≫Control & Simulation Loop**. Left-click and drag the loop onto the block diagram (see Figure 10.13). Now add the Halt Simulation function to your VI to stop the Control & Simulation Loop. Right-click the block diagram and **navigate to Control Design & Simulation ≫Simulation ≫Utilities ≫Halt Simulation**. Left-click to place the Halt Simulation function on the block diagram and then right-click the Boolean input of the Halt Simulation VI and select **Create ≫Control**. This creates a Boolean control on the front panel that you can use to halt, or stop, the simulation while it is running. Next, place the Multisim Design VI that facilitates the communication between the LabVIEW and Multisim simulation engines. Right-click the block di-

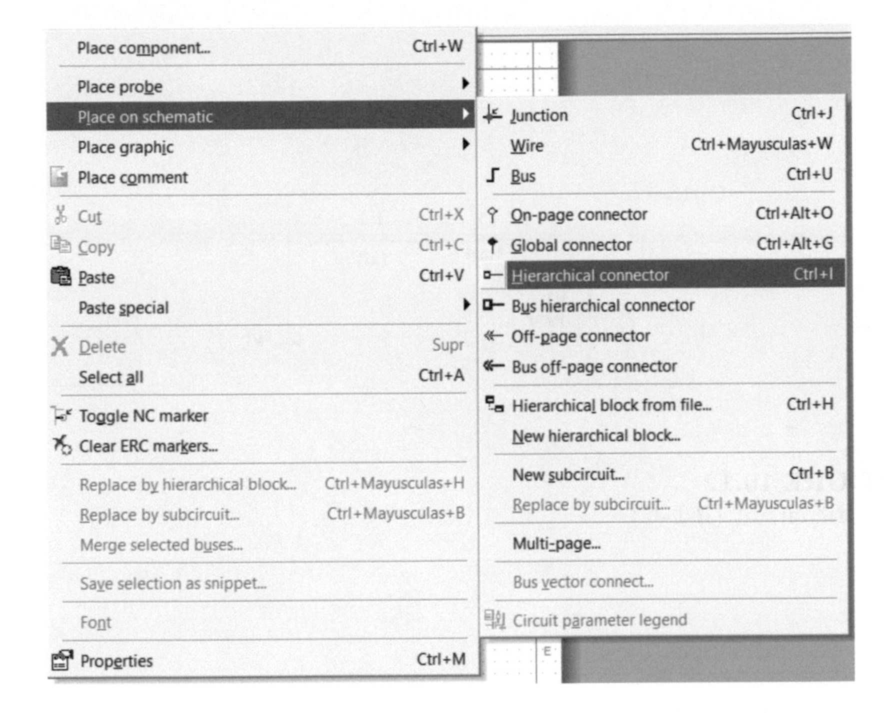

FIGURE 10.10
HB connectors in Multisim

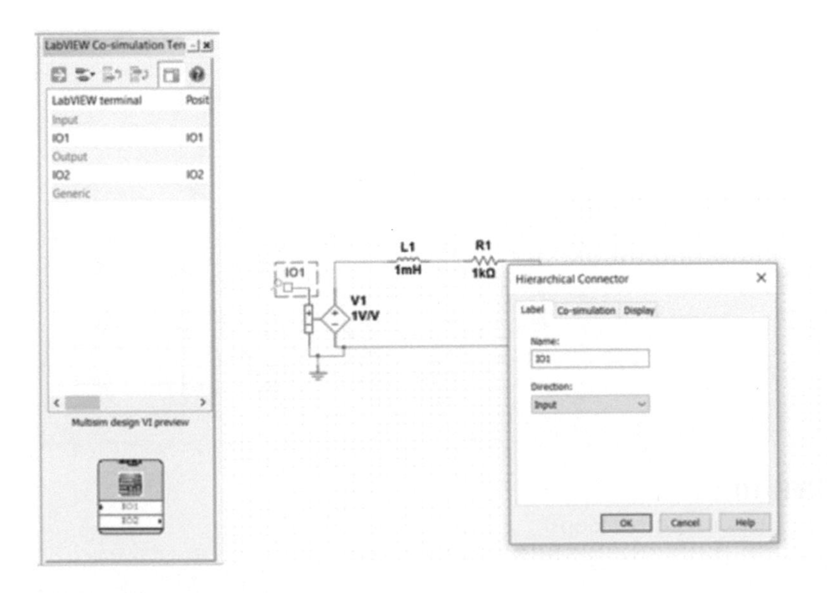

FIGURE 10.11
Input or output section

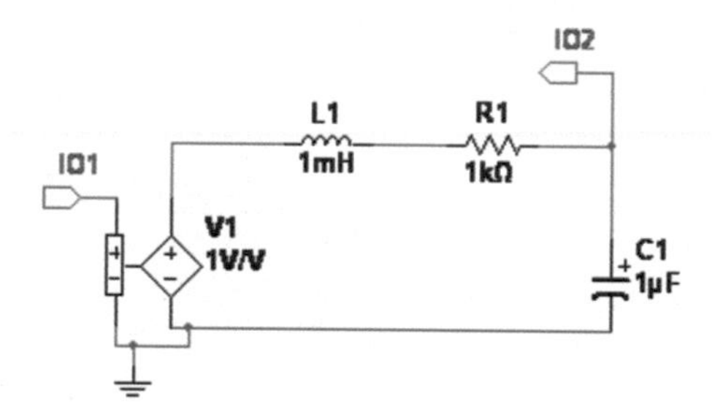

FIGURE 10.12
Electric circuit (R-L-C)

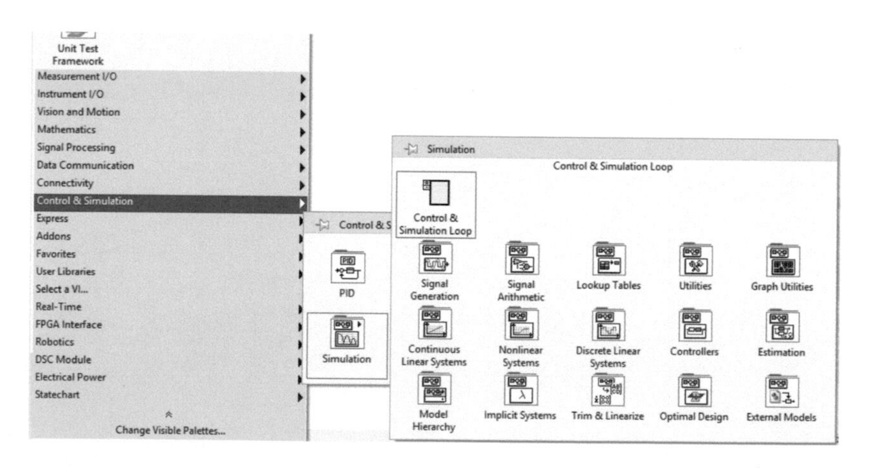

FIGURE 10.13
Selecting the simulation loop

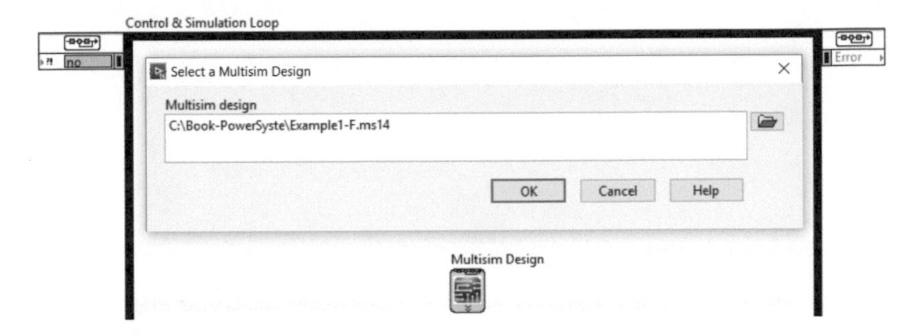

FIGURE 10.14
Selecting the simulation block diagram in Multisim

agram and navigate to **Control Design & Simulation ≫Simulation ≫External Models ≫Multisim ≫Multisim Design**, and left-click to place the VI within the Control & Simulation Loop; it must be placed within the loop. The **Select a Multisim Design** dialog box appears after you position the Multisim Design VI. You now need to either specify the file path or browse to the location by clicking the **Browse** button and locating the file on your hard disk.The Multisim Design VI now populates the terminals, identically to the Multisim Design VI Preview within the Multisim environment, with the proper inputs and outputs. Left-click on the double down-arrow and expand the terminals if they are not visible. Figure 10.14 shows the selection of a Multisim file in LabVIEW. To pass data to the Multisim design, you must first create a numeric control on the front panel. To easily accomplish this, right-click on the input, $Voltage_{In}$, and select **Create≫Control**. This places a numeric control terminal on the block diagram that is already wired to the input of the Multisim VI. The block diagram terminal for the control has a corresponding control on the front panel, which is the user interface in LabVIEW. To quickly navigate between the block diagram and front panel, simply press <**Ctrl-E**> (see Figure 10.15).

Using the same design process, it is possible to generate more complex co-simulations such as DC-DC converters. One of the most well-known DC-DC converters is the Buck converter, which is designed to supply electrical energy from a power source with a higher voltage V_{DC} to the load with a lower voltage V_{out} (see Figure 10.16). When the semiconductor switch Q1 is on, the electrical energy is delivered from the source V_{DC} through switch Q1 and inductor L2 to the load. When the output voltage (V_{out}) is high enough, this energy link will be shut down by turning off Q1. Energies stored in L2 and C2 will maintain the load voltage V_{out}. The electrical current in the circuit is sensed using XCP1. Figure 10.17 shows the co-simulation diagram in which a conventional PID controller is included in order to reach the set point of voltage. Figure 10.18 depicts

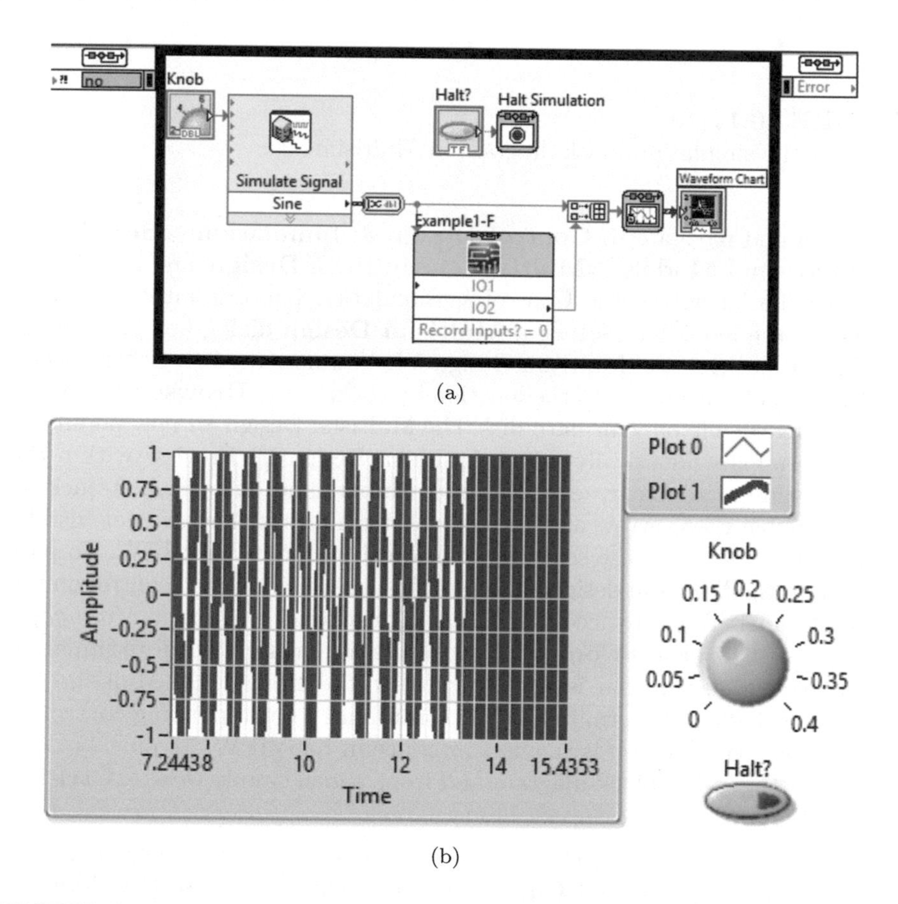

FIGURE 10.15
Block diagram LabVIEW simulation (a) and frontal panel (b)

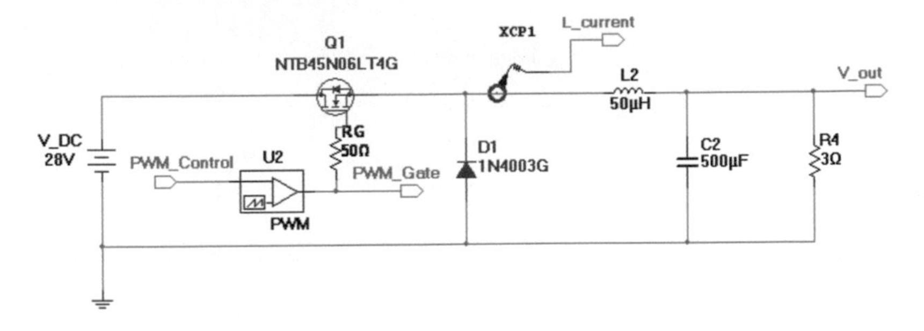

FIGURE 10.16
Buck converter designed in Multisim

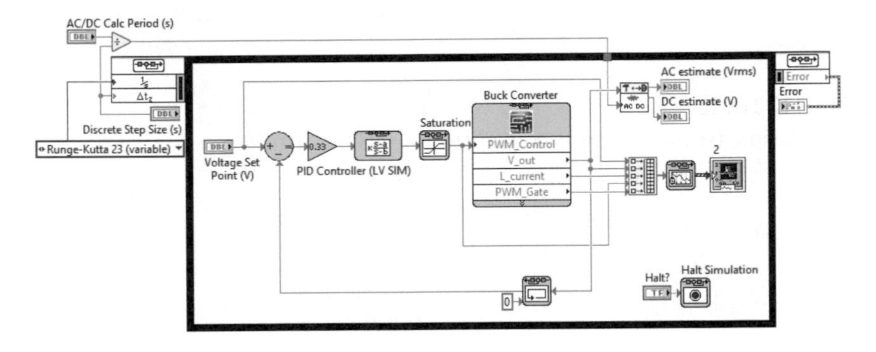

FIGURE 10.17
Co-simulation diagram between LabVIEW and Multisim

the results when the controller is adjusting the PWM command and the output voltage reaches the set point of voltage.

A Boos DC-DC converter can also be analyzed using co-simulation. The topology of a Boost converter is shown in Figure 10.19, and the co-simulation is presented in Figure 10.20. The input voltage V_{in} is lower than the output voltage V_{out}. When Q1 is on, the diode (D2A) is reversed-biased and the output stage is isolated. Hence, the inductor receives energy from V_{in}. On the other hand, when Q1 is off, the output stage receives energy from the inductor as well as V_{in}; thus, the output voltage (V_{out}) is increased. The general relationship between the duty ratio (d), V_{in}, and V_{out} is defined by Equation 10.1, when the DC-DC converter is in continuous-conduction mode. When the PWM command is changed, the output voltage is directly affected (see Figure 10.21).

$$V_{out} = \frac{V_{in}}{1-d}. \tag{10.1}$$

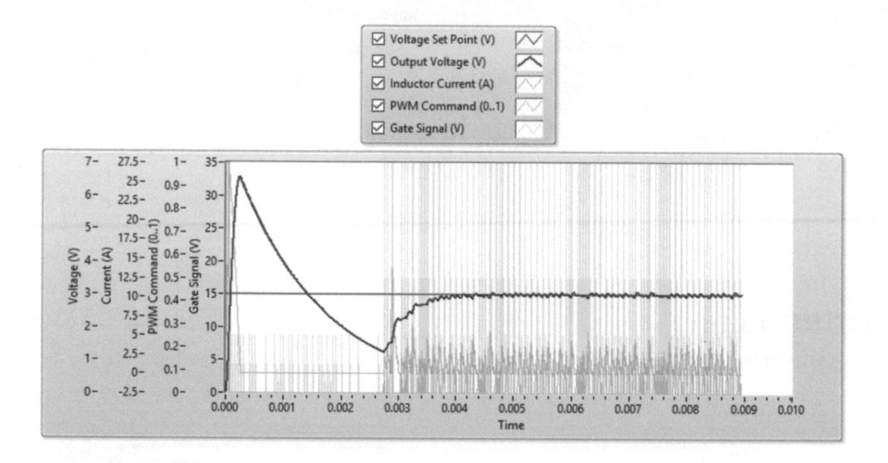

FIGURE 10.18
Co-simulation results using a conventional PID controller

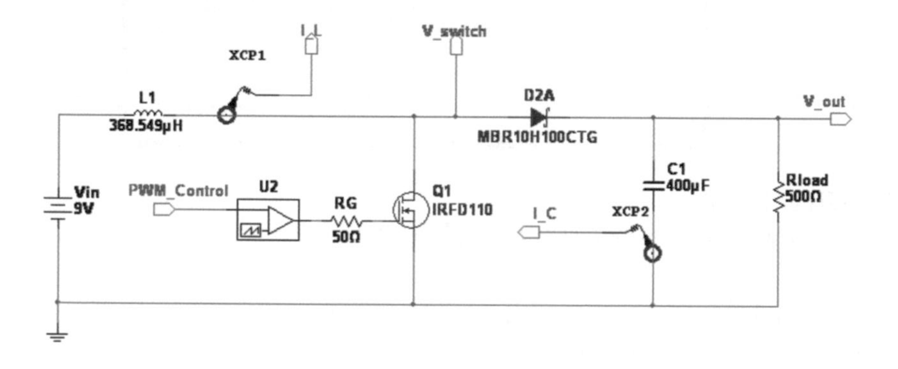

FIGURE 10.19
DC-DC boost converter in Multisim

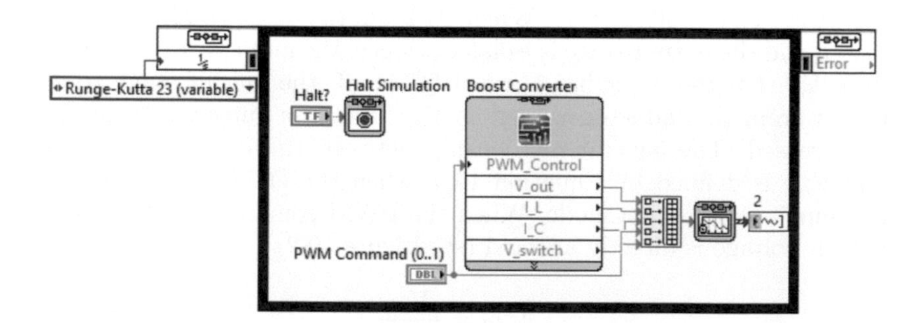

FIGURE 10.20
Co-simulation block diagram

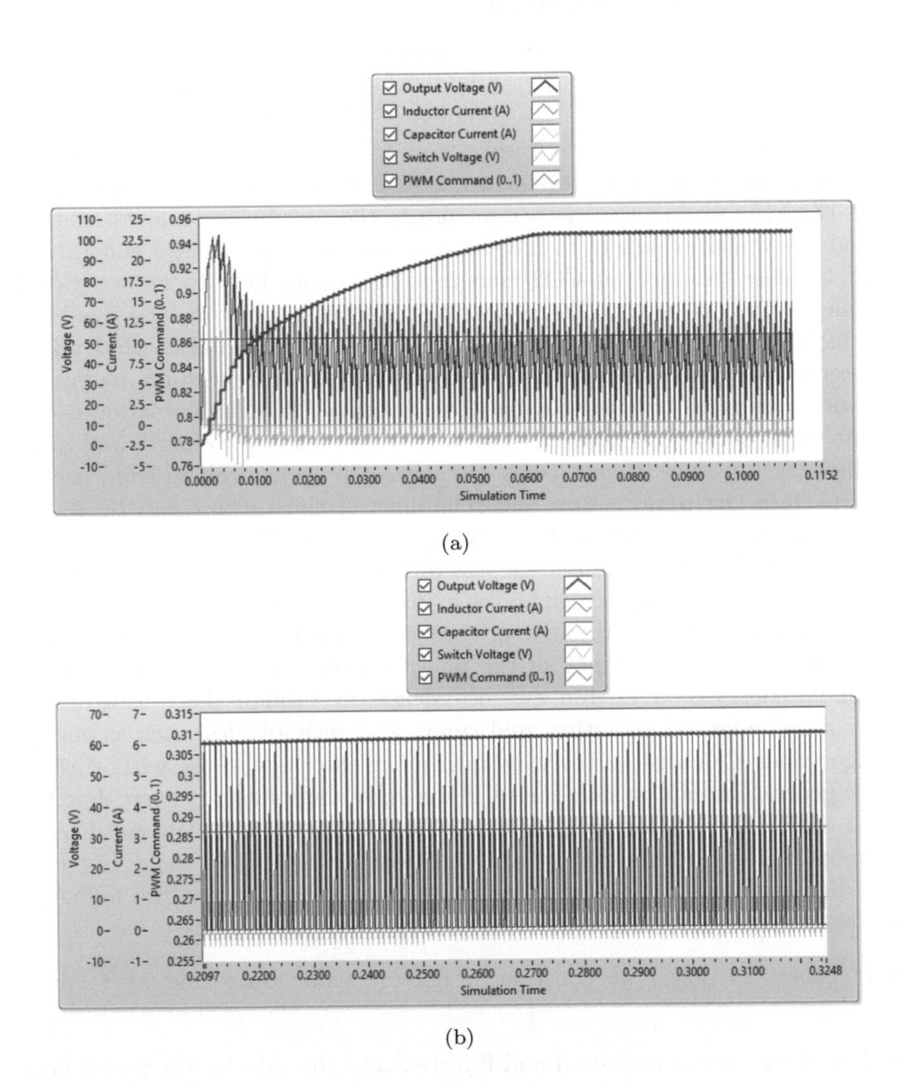

FIGURE 10.21
Results of Boost converter for a PWM command of 0.89 (a) and 0.3 (b)

10.3 Case study: Intelligent wind turbine control with LabVIEW simulation

Nowadays, wind energy only supplies a fraction of the total power demand in the world, but it is growing very fast, while the cost per watt of the electricity produced falls proportionally. The progress of wind power in recent years has exceeded all expectations, with Europe leading the global market [17, 20], and global installed capacity increasing 14, fold from 2000 to 2011, sufficient to cover 3% of the world's electricity demand. In the near future, many countries around the world are likely to increase the level of wind energy generation. Wind turbines produce no CO2 emissions and could help to reduce global greenhouse gas emissions. The cost of the electricity provided by wind energy facilities has been dropping since the 1980s, which makes wind power an interesting economical alternative energy source for developing regions of the world compared to fossil fuel-based generation. The cost reduction is due to new technologies and higher production scales, leading to larger and more reliable wind turbines. Hence, control systems have a key role in wind energy systems since the performance and reliability of the turbine can be significantly enhanced by intelligent control systems [18]. However, the inherent variability of the wind resource creates difficulties in forecasting the energy production of wind farms, which makes management of the renewable resource more challenging. This variability of the wind energy resource production is in contrast to the consistency of the power output from conventional fossil fuel-based energy sources, but can be mitigated through the use of intelligent control, energy forecasting, and smart grid technologies, such as energy storage systems located by the wind farm. The efficiency of the wind turbine power generation can be maximized if the rotational speed is such that the ratio between the blades' tip speed and wind speed, called the tip-speed ratio, is the optimal one at any time (λ_{opt}). One popular turbine topology in recent years has been the doubly-fed induction generator, which has the following features:

- It generates power at constant frequency while operating at variable speeds.
- The slip power is recovered and injected into the grid by the power converter.
- It can generate power with good efficiency over a reasonable speed range.
- Since only a fraction of the power goes through the switched-mode power converter, its rating is significantly smaller than in variable speed/variable frequency schemes in which all of the turbine power is transferred through a back-to-back AC-to-DC converter and DC-to-AC inverter.

Classical wind turbines are characterized by lower inertia than classical power plants. In the case of back-to-back converter/inverter topologies, common with permanent magnet synchronous machine (PMSM) type turbines, the intermediate DC-voltage bus creates an electrical decoupling between machine and grid. This decoupling can lead to a reduced amount of stored kinetic energy in the turbine, but the effect can be compensated for through suitable implementations of machine control. In the case of variable-speed wind generators, [42] suggested using the blade and machine inertia as a kinetic energy storage system to participate in regulation of primary frequency control. However, releasing or storing kinetic energy can only be considered as a portion of the primary control scheme. Since wind persistence is limited, the power reserve cannot be guaranteed for the long term.

It is also suggested in [17] to maintain a power reserve with the help of the pitch control when the wind generator works close to rated power. It is also possible to obtain a power reserve with the help of generator torque control but at a cost with regards to overall turbine power efficiency. In the case of the doubly-fed induction machine (DFIG), the power captured by the wind turbine is converted into electrical power by the induction generator and it is transmitted to the grid by both the stator and the rotor windings. The control system generates the pitch angle command and the voltage command signals at the rotor in order to control the power of the wind turbine, the DC bus voltage, and the voltage and frequency at the grid terminals. This section shows an intelligent control system that is proposed for DFIG wind turbines, which is intended to be cost effective and provide pitch control while reducing the mechanical stresses. In the near future the power rating of the wind turbines will increase further, especially in offshore applications. Doubly fed induction generator machines can be designed to work over a reasonable speed range for improving overall power quality and system efficiency.

An important factor in the design of variable-speed wind turbine controls is the random excitation of the system caused by wind turbulence. Obtaining an optimal control design that balances all of these requirements, constraints, and challenges requires high-performance digital control systems.

Intelligent control (IC) systems have the ability to work well when the operation point of the system is changing, because the IC system can implement its derived control laws from a number of different sources, including human knowledge. Fuzzy logic systems, for example, are able to synthesize information from human experts. Also, artificial neural networks can implement complex non-linear control laws that are automatically derived from the desired input-to-output response mapping of the overall wind turbine system.

10.3.1 Wind turbine operation

According to [17], the double-fed induction generator can be explained by the power flow shown in Figure 10.22. The input is the mechanical power from the spinning blades and the main power output is the electrical power flowing from the stator windings. Since the generator is an asynchronous machine, the value of the slip, s, defined as the ratio $\frac{\omega_r}{\omega_s}$ between the slip and synchronous angular frequency, is lower than 1 and, consequently, the rotor power, P_r, is only a fraction of the stator power, P_s. The sign of Pr is a function of the slip sign when the mechanical torque applied to the rotor is positive. The synchronous speed, ω_s, is normally a positive constant for generation with a constant grid voltage. Thus, the slip power can flow in both directions: to the rotor from the supply, and from the supply to the rotor. The speed of the machine can be controlled from either the rotor- or stator-side converter in both super- and sub-synchronous speed ranges. Thus, the machine can be controlled as a generator or a motor in both super- and sub-synchronous operating modes, yielding all four quadrants of operation [17]. Below the synchronous speed in the motoring mode and above the synchronous speed in the generating mode, a rotor-side converter operates as a rectifier and a stator-side converter as an inverter, whereby slip power is returned to the stator. Below the synchronous speed in the generating mode and above the synchronous speed in the motoring mode, the rotor-side converter operates as an inverter and stator-side converter as a rectifier, whereby slip power is supplied to the rotor. At the synchronous speed, slip power is taken from the supply to excite the rotor windings, and in this case the machine behaves as a synchronous machine.

The following conditions can occur in the induction generator:

1. P_r is positive for negative slip (speed greater than synchronous speed).
2. P_r is negative for positive slip (speed lower than synchronous speed).
3. P_r is transmitted to the DC bus capacitor and raises the DC voltage for super-synchronous speed operation.
4. P_r is taken out of the DC bus capacitor and decreases the DC voltage for sub-synchronous speed operation.

The grid-tied power converter, C_{grid}, is used to generate or absorb the power flowing to the grid, P_{gc}, in order to keep the DC voltage constant. In steady-state for a theoretical lossless AC/DC/AC converter, P_{gc} is equal to P_r and the speed of the wind turbine is determined by the power, P_r, absorbed or generated by the rotor-tied converter, C_{rotor}. The phase-sequence of the AC voltage generated by C_{rotor} is positive for sub-synchronous speed and negative for super-synchronous speed. The frequency of this voltage is equal to the product of the grid frequency and the absolute value of the slip. A beneficial feature for utility grid operation is that C_{rotor} and C_{grid} have the capability for generating or absorbing reactive power and can be used to control the reactive power or the voltage at the grid terminals [37].

10.3.2 Wind turbine per unit (P. U.) model

Figure 10.23 shows the simulation model of the wind, control system, and wind generator. The turbine simulation model was designed using the National

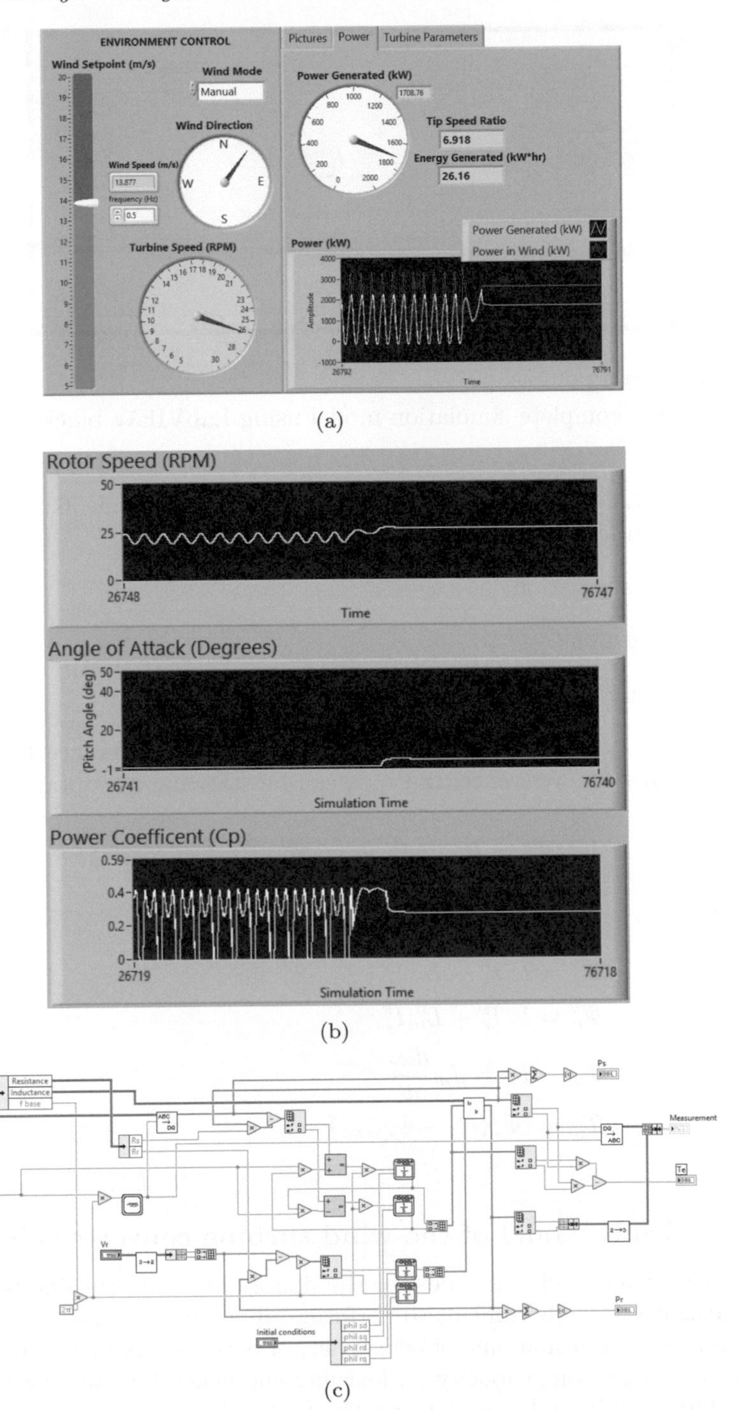

FIGURE 10.22
Wind turbine simulation results (a and b) and wind turbine model in LabVIEW (d)

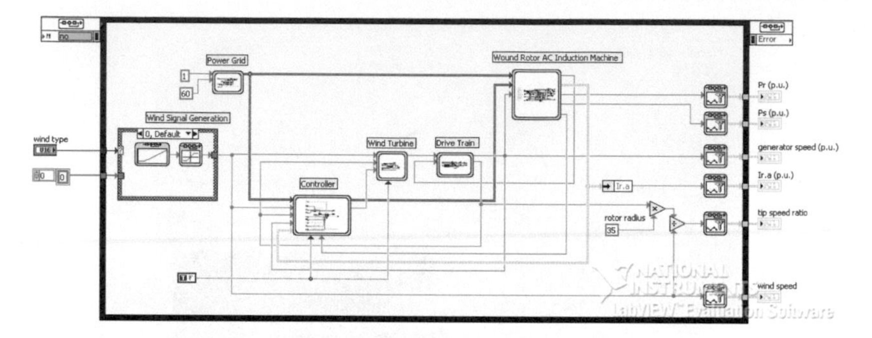

FIGURE 10.23
Wind turbine complete simulation model using LabVIEW block diagram

Instruments LabVIEW control design and simulation module [62]. This model does not include the following effects:

- Magnetic saturation
- Airgap harmonics
- Skin effects

The dynamic equations for the induction machine in reference frame k are defined below. The reference frame, k, can be stationary or rotating.

$$
\begin{aligned}
V_s^k e^{j\theta_k} &= I_s^k e^{j\theta_k} R_s + j\omega_k \Psi_s^k e^{j\theta_k} + e^{j\theta_k} \frac{d}{di} \Psi_s^k \\
V_r^k &= I_r^k R_r + j(\omega_k - \omega_r)\Psi_r^k + \frac{d}{di}\Psi_r^k \\
\Psi_s^k &= L_s I_s^k + L_m I_r^k \\
\Psi_r^k &= L_r I_r^k + L_m I_s^k \\
T_m - T_L &= J_m \frac{d\omega_r}{dt} \\
T_m &= (\Psi_{s\alpha} i_{s\beta} - \Psi_{s\beta} i_{s\alpha})
\end{aligned}
$$

10.3.3 Fault model of the wind turbine converter circuit

Since the converter circuit model of the doubly fed wind turbine is strongly nonlinear, it is usually difficult to perform online fault diagnostics. However, the input-to-output mapping of neural networks is also nonlinear and features the benefit of a strong capacity for learning and generalization of error symptoms. The mapping relation between the fault information, the fault type, or the cause of the fault in the wind turbine converter circuit can be learned and saved by a neural network. This enables the automatic online diagnosis of faults locally within the turbine control system itself. Taking, for example, the power converter of Figure 10.1, a possible fault condition is the disconnection of one converter bridge arm. In this scenario, the fault status of the converter can be divided into five basic types.

1. Primary fault condition 001: There are no faults detected in the electric power electronics components. The converter is operating normally.
2. Primary fault condition 010: A fault is detected in only one converter. This fault can be sub-classified into six types:
   a. Fault detected in component 1 (001)
   b. Fault detected in component 2 (010)
   c. Fault detected in component 3 (011)
   d. Fault detected in component 4 (100)
   e. Fault detected in component 5 (101)
   f. Fault detected in component 6 (110)
3. Primary fault condition 011: A fault is detected in the same phase of two converters. These faults can be sub-classified into three types:
   a. Fault detected in components 1 and 4 (001)
   b. Fault detected in components 2 and 5 (010)
   c. Fault detected in components 3 and 6 (011)
4. Primary fault condition 100: A fault is detected in two components of a half-bridge. These faults can be sub-classified into six types:
   a. Fault detected in components 1 and 3 (001)
   b. Fault detected in components 2 and 4 (010)
   c. Fault detected in components 3 and 5 (011)
   d. Fault detected in components 4 and 6 (100)
   e. Fault detected in components 5 and 1 (101)
   f. Fault detected in components 6 and 2 (110)
5. Primary fault condition 101: A fault is detected in two crossing components. These faults can be sub-classified into six types:
   a. Fault detected in components 1 and 2 (001)
   b. Fault detected in components 2 and 3 (010)
   c. Fault detected in components 3 and 4 (011)
   d. Fault detected in components 4 and 5 (100)
   e. Fault detected in components 5 and 6 (101)

TABLE 10.1
Tasks commonly solved by artificial neural networks (ANNs)

Task	Description
Function approximation	Linear and nonlinear functions can be approximated by neural networks. The ANN performs a model or curve fitting function.
Classification	1) Data classification: The ANN assigns data to a specific class or subclass. Useful for identifying patterns. 2) Signal classification: Time-series data is classified into subsets or classes. Useful for identifying the signal source.
Unsupervised clustering	Identifies order in data. Creates clusters of data of unknown classes.
Forecasting	Predict the next values in a time-series waveform.
Control systems	ANN commonly performs function approximation, classification, unsupervised clustering, and forecasting. Also, ANNs are used in model identification and analysis of open and closed loop systems.

f. Fault detected in components 6 and 1 (110)

An 8-bit coding scheme is used to describe the fault whereby the eighth bit distinguishes converter 1 from converter 2 (a value of 0 indicates converter 1), the primary fault type is encoded from the seventh bit to the fifth bit, the sub-classification type code is indicated with the fourth bit to the second bit, and an odd or even checksum is located in the first bit.

10.3.4 Artificial neural networks for fault detection

Artificial neural networks (ANNs) are applied in several problems [63], but there are typically around five main problems in which the performance of artificial neural networks stands out. Those problems are listed in Table 10.1.

Hebbian neural networks are an unsupervised and competitive ANN and they can be applied to different associative problems such as fault detection. As unsupervised networks, these ANNs only have information about the input space. Their training is based on the fact that neuron weights store the

TABLE 10.2
Hebbian learning procedure

Algorithm	Hebbian learning procedure
Step 1	Determine the input space. Specify the number of iterations $iterNum$ and initialize $t = 0$ Generate small random values of weights w_i
Step 2	Evaluate the Hebbian neural network and obtain the outputs x_i
Step 3	Apply the updating rule (4)
Step 4	If $t = iterNum$ then STOP Otherwise, go to Step 2

fault detection information. The weights stored in the ANN can only be reinforced if the input stimulus provides sufficient output response values. In this way, weights change proportionally to just the output signals. By this fact, neurons compete to be a dedicated reaction to a part of the input. Hebbian neural networks are thus considered the first self-organizing nets. The learning procedure is based on the following statement pronounced by Hebb: As A becomes more efficient at stimulating B during training, A sensitizes B to its stimulus, and the weight on the connection from A to B increases during training as B becomes sensitized to A.

Then, Steven Grossberg developed a mathematical model for this sentence [63]:

$$w_{AB}^{new} = w_{AB}^{old} + \beta x_B x_A,$$

where w_{AB} is the weight between the interaction of two neurons A and B, x_i is the output signal of the i neuron. $x_B x_A$ is so called the Hebbian learning term. The next algorithm gives the Hebbian learning procedure (Table 10.2).

Figure 10.24 shows a Hebbian network for a classification problem of 8-bit input signal based on LabVIEW. In the same way, a key point signal can be supplied in order to get its classification process about the fault diagnosis.

In fact, one of the main signals to observe in fault diagnosis is the direct current pulse voltage V_d of the three phases commute circuit and look at the output port of inductance load that contains the information about whether the transistor is fault or not, according to the fault model. If the fault type is different, the wave of V_d is different, and the fault type can only be analyzed by the six spectrum quantity, containing the direct current a0 quantity of voltage, the 1,2,3-harmonic amplitude value (A1 A2,A3), and the 1,2-harmonic phase (P1, P2); as a result, V_d is the main point signal in the wave method. Since the phase information is needed in fault diagnosis, the time $\frac{\pi}{6}$ with phase A voltage passing the zero acts as a reference point at every diagnosis. For the

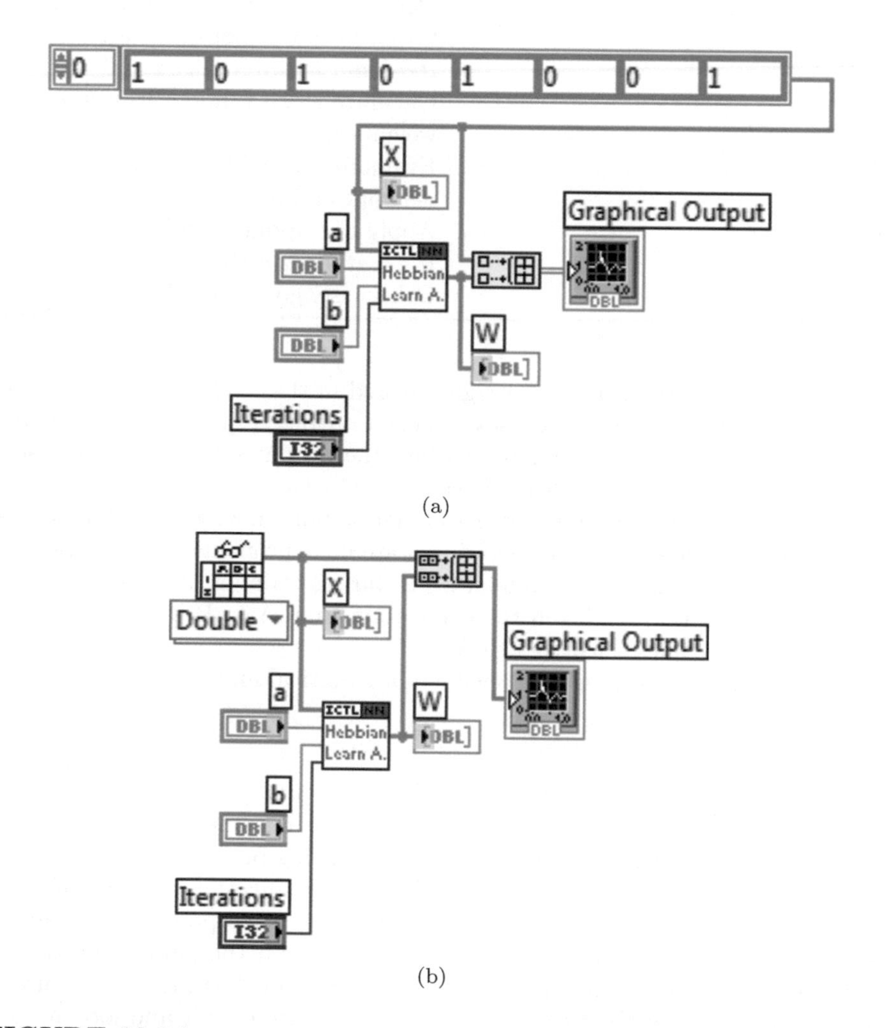

FIGURE 10.24
(a) Hebbian network 8-bit classification and classification using generated fault signals (b)

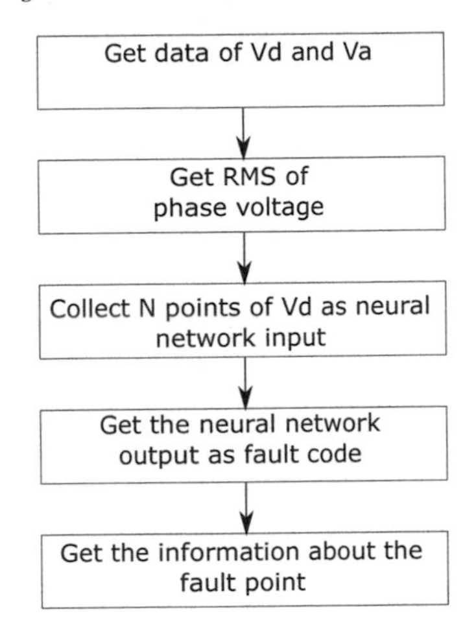

FIGURE 10.25
Fault diagnosis process using neural networks

FIGURE 10.26
Results using Hebbian artificial neural network

wave method, the N pieces of unitary data collected in a period are directly input into the neural networks. A simple flow chart of the fault diagnostic process is presented in Figure 10.25.

The results obtained using Hebbian artificial neural networks when a fault signal was generated are presented in Figure 10.26. The angle β acts like a sampler at fixed intervals (8°), the result presented was getting the angle $\beta =$ 45° for a big-type fault. The results show a good classification and recognition system. So, the use of neural networks for detecting all the fault signals of the converter presented below gets zero error.

Using a control topology based on intelligent control systems, the performance of the wind turbine can be improved. The proposed control system [62] includes a fuzzy logic controller for the pitch angle, a grey system for predicting the wind profile, and two ANFIS for controlling the generator currents. Figure 10.27 shows the proposed topology in which the fuzzy logic controller is a PI fuzzy logic controller [63]). The ANFIS was tuned with constant outputs; for tuning the ANFIS controller, the hybrid training was applied. The simulation results are presented in Figures 10.27, 10.28, and 10.29.

The above results were obtained without the power electronics stage (inverter) that is one of the main stages in the simulation of a wind generator.

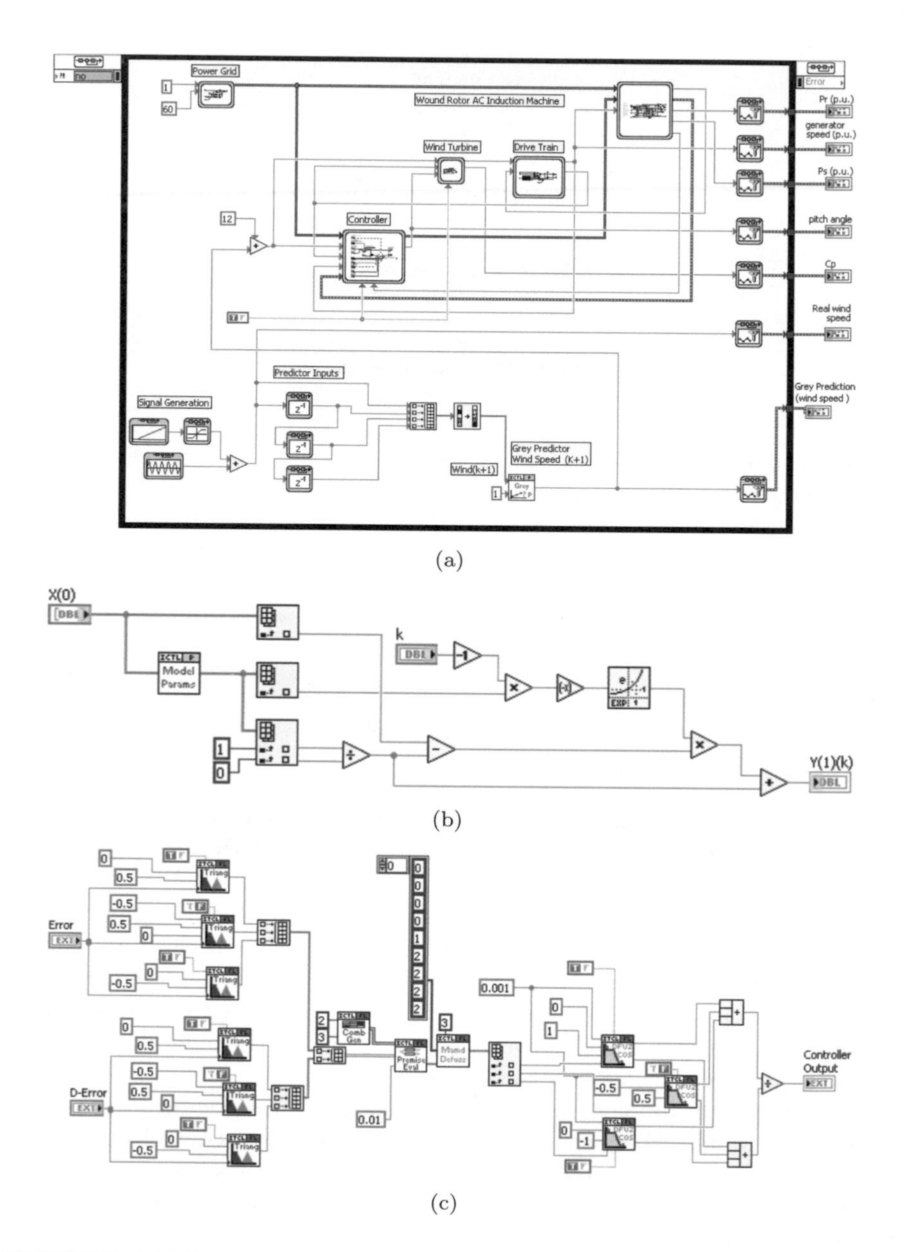

(a)

(b)

(c)

FIGURE 10.27
(a) Main control and simulation loop, (b) model estimation process, and (c) detailed view of the fuzzy controller

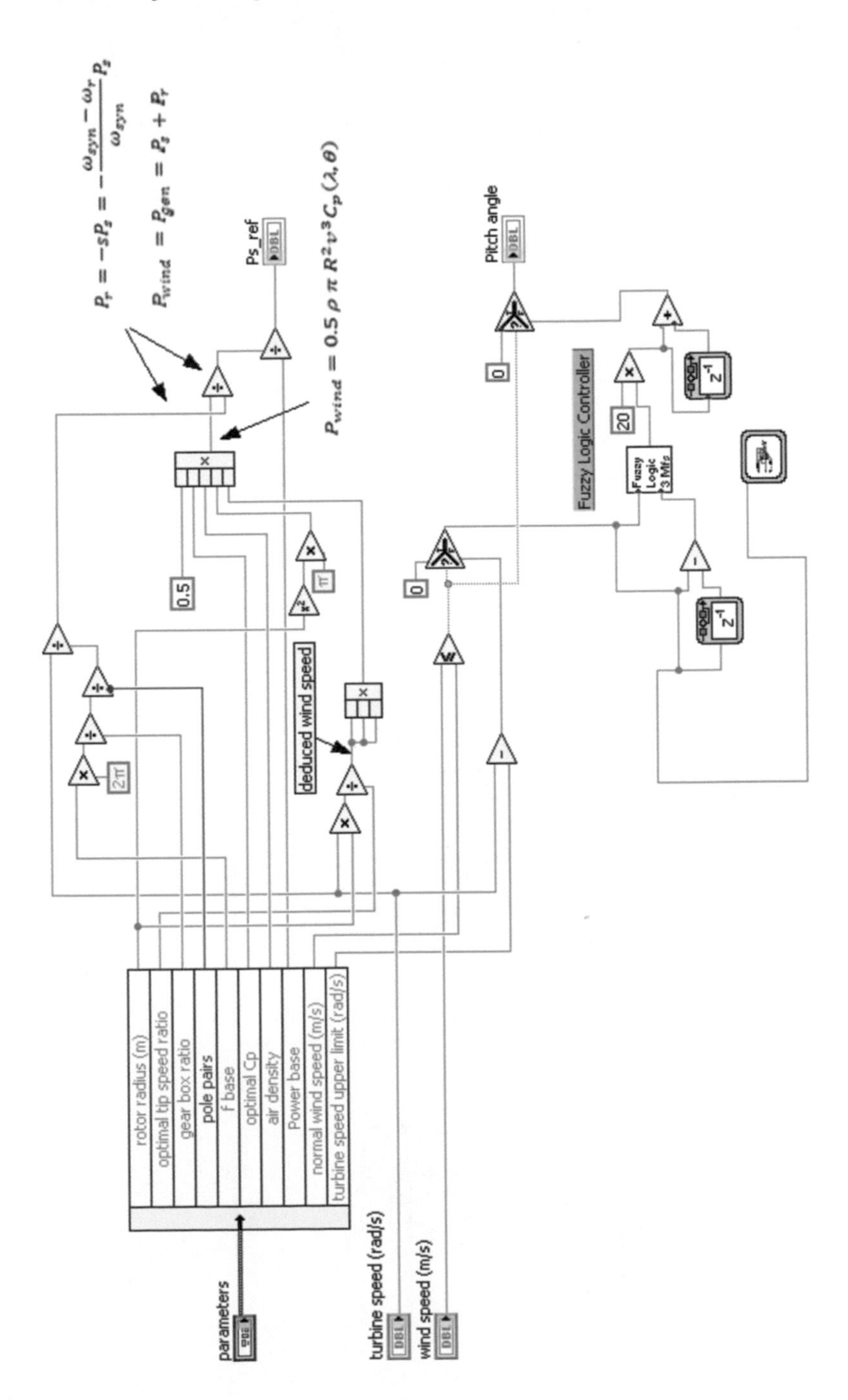

FIGURE 10.28
Wind speed controller

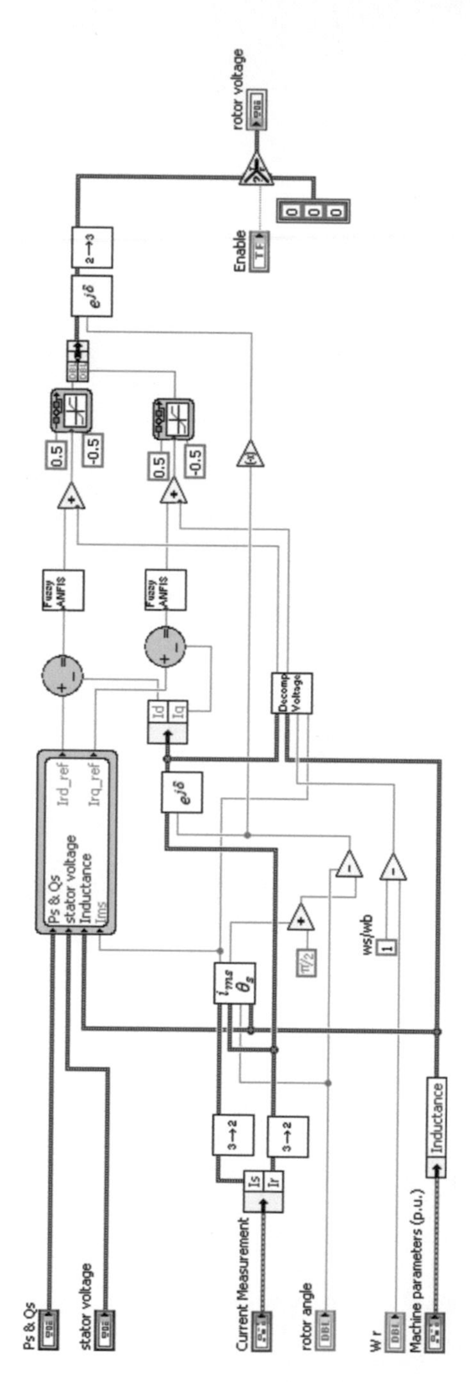

FIGURE 10.29
ANFIS incorporation to rotor voltage computation

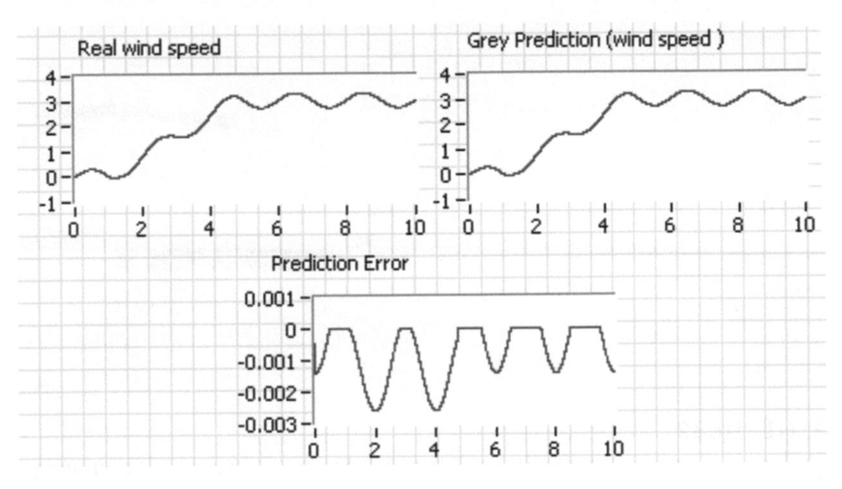

FIGURE 10.30
Frontal panel of grey predictor results (real wind, grey predictor, and prediction error)

Thus, the following test was done, including the power electronic stage, using a space vector modulation and ideal switches. The parameters of the controller were kept constant, using the previous tuning (without the inverter). As the results show, the controllers can deal with the pulse distortion generated for the power electronic stage. The selection of SPWM helps to decrease the harmonics and also increase the DC bus energy that can be used. Figure 10.31 presents the rotor voltage using the space pulse width modulation (SPWM), and Figure 10.32 shows the performance of the wind turbine using intelligent control systems.

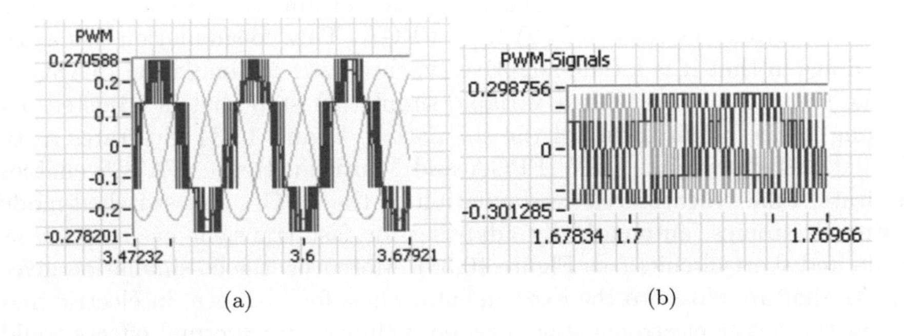

FIGURE 10.31
Rotor voltage: (a) reference signals and rotor voltage and (b) three-rotor voltage using the SPWM

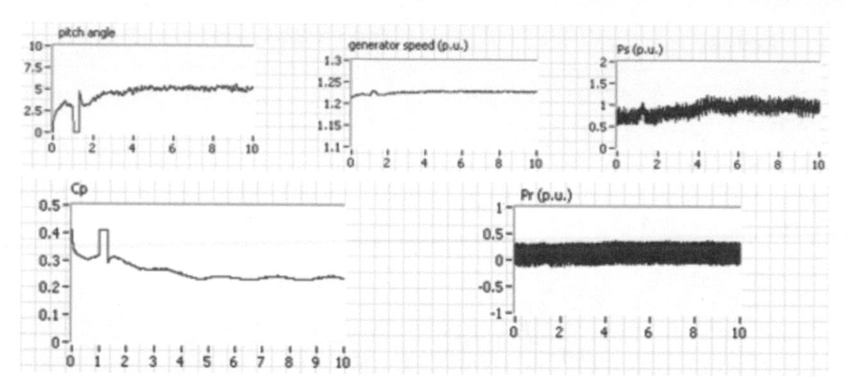

FIGURE 10.32
Controller performance including the power electronic stage using space pulse width modulation

10.4 Co-simulation LabVIEW FPGA and Multisim

When a designer is using a conventional simulation as a tool for making power electronic stages, there are multiple problems that have to be addressed before a final prototype is achieved. Thus, the time and design effort are large and the simulation can be far from the experimental prototype. The flow diagram for designing has one direction when conventional simulation is used. On the other hand, when a co-simulation between LabVIEW FPGA and Multisim is implemented, a bidirectional way is established (see Figure 10.33). Hence, the validation of the design is more accurate [54].

The co-simulation between LabVIEW FPGA and Multisim is divided in two parts. The first one is modeled in Multsim and the second one is the control program designed in LabVIEW FPGA. This co-simulation has a big advantage in that it is a time-adaptive synchronized co-simulation (TASCS), so the simulation runs with a variable simulation step that is adjusted according to the requirements of the simulation. Figure 10.34 illustrates a co-simulation between LabVIEW FPGA and Multisim. Hence, TASCS enables accurate, time-synchronized co-simulation of the FPGA and switched-mode power electronics, automatically adapting for fast transient events such as faults and short-circuits (see Figure 10.35). Moreover, the co-simulation gives results that are closer to the experimental ones; for instance, in electric machines the power electronic stage can be included and thermal effects could be analyzed (see Figure 10.36).

The next exercise is a basic co-simulation between LabVIEW FPGA and Multisim. It gives a basic introduction about co-simulation [54]. The first part of the exercise is to show off the project explorer window hierarchy, now with two new targets, the RT controller and FPGA, and then jump into the VI

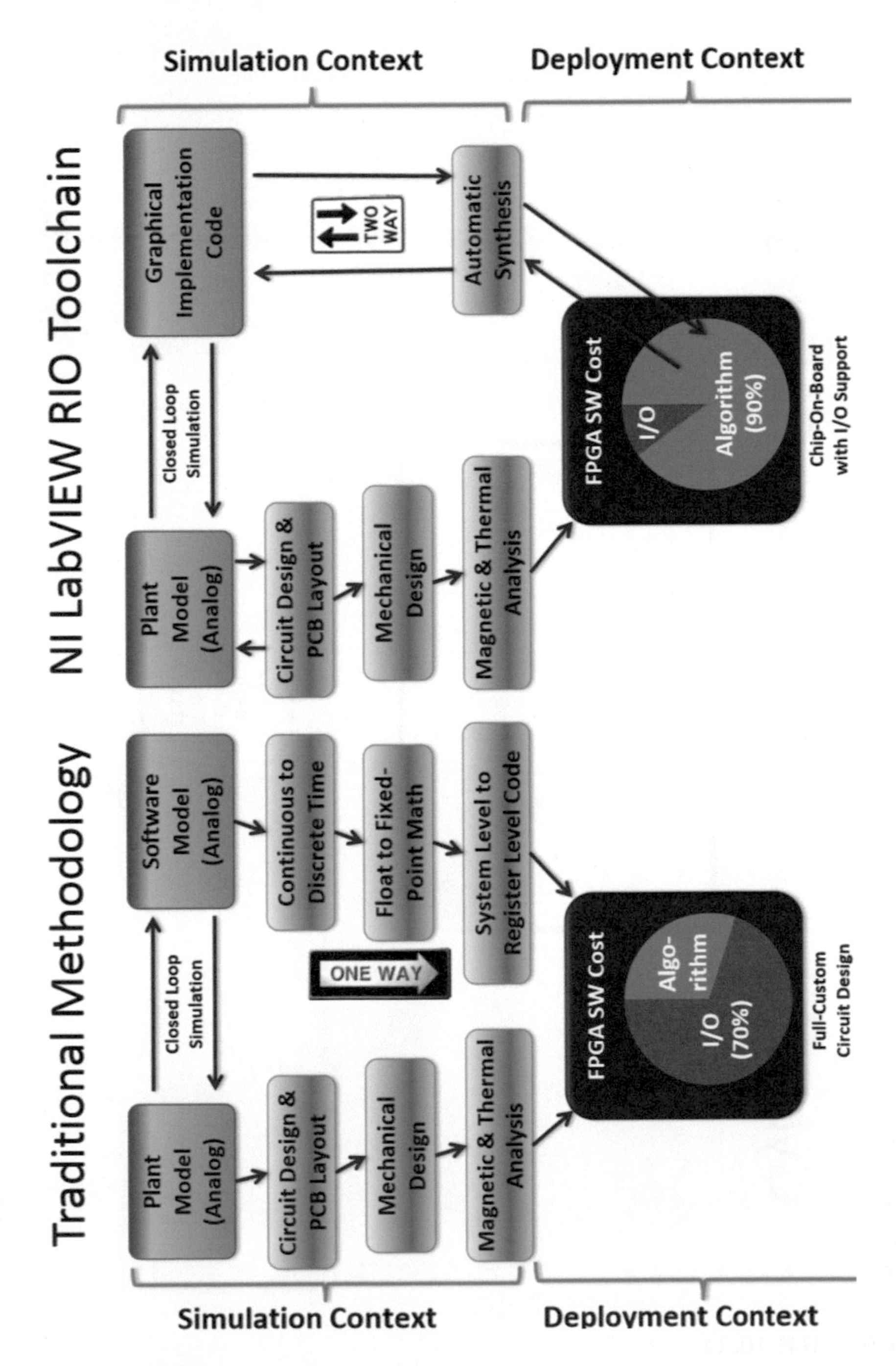

FIGURE 10.33
Reducing the simulation to deployment cycle from weeks to hours

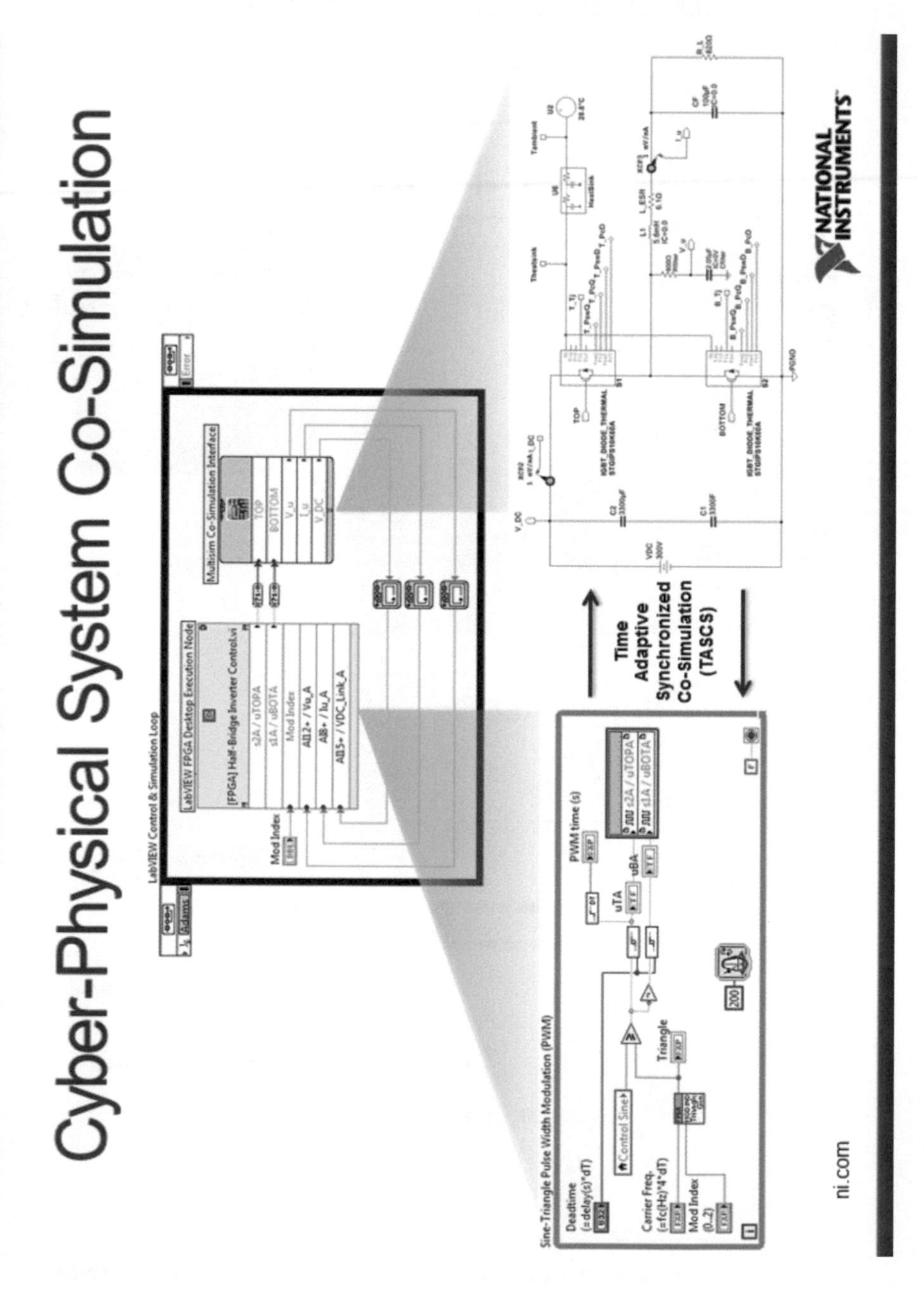

FIGURE 10.34
Co-simulation between LabVIEW FPGA and Multisim

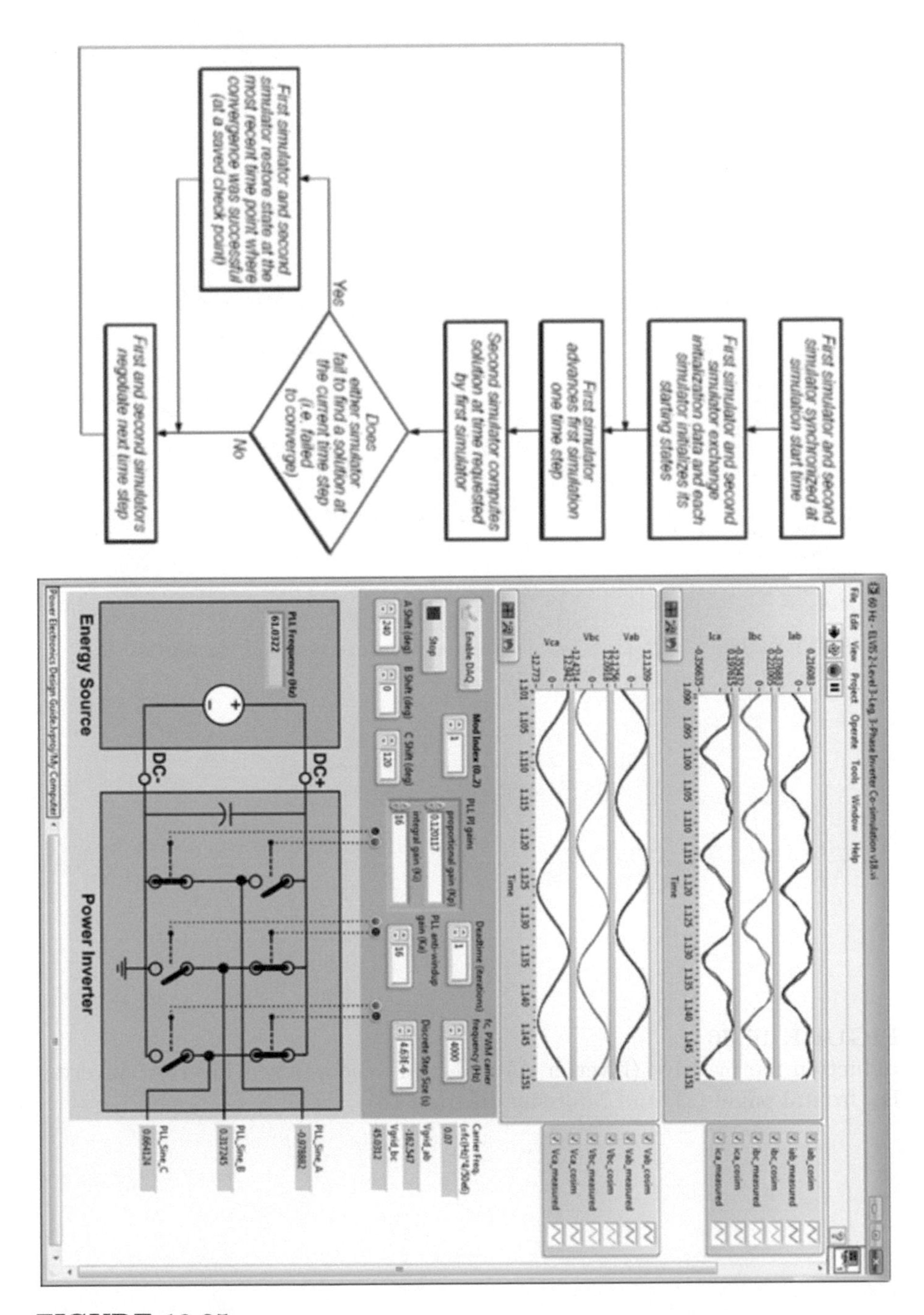

FIGURE 10.35
Co-Simulation flow diagram

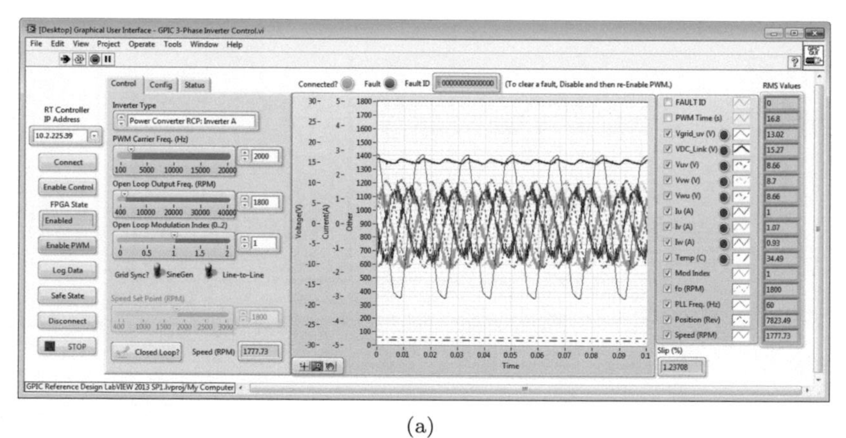

(a)

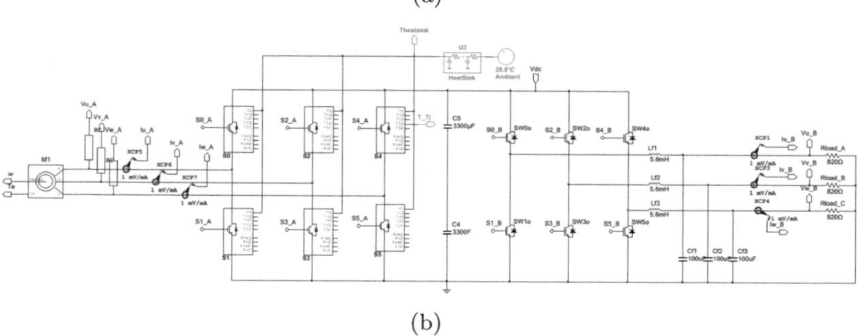

(b)

FIGURE 10.36
Induction motor drive (inverter A) and grid synchronized inverter (inverter B), frontal panel (a), and Multisim diagram (b)

front panel and block diagram, highlighting the differences from LabVIEW programming on Windows (see Figure 10.37). Follow the next basic steps in order to create and connect the co-simulation. Install the required software using the Measurements and Automation Explorer (see Figure 10.38) and connect the hardware (see Figure 10.39). Finally, the co-simulation file using LabVIEW FPGA can be created (see Figure 10.40).

1. Disable WiFi
2. Launch NI MAX
3. Refresh remote systems
4. Format disk
5. Install firmware: LabVIEW RT Add-Ons
6. LabVIEW PID and Fuzzy Logic Toolkit
7. NI system configuration
8. System State Publisher / Network I/O
9. Network Streams/ Protocols and Buses
10. NI-Watchdog
11. Set LV Project IP address
12. Connect

10.5 Case study: A half-bridge inverter control

This example will present how an application running on the FPGA using the desktop execution node is deployed. The application switches two IGBTs in order to produce an analog signal.

10.5.1 Background

The field-programmable gate array (FPGA) allows a user to perform a variety of tasks, including signal processing, at higher speeds than most processors. The FPGA desktop execution node allows a user to run an FPGA application on the development computer with simulated I/O. By specifying the desired FPGA clock speed, as well as the number of ticks before inputs are read and outputs are written, a user can determine the functionality of their FPGA application without needing to compile and deploy the application to

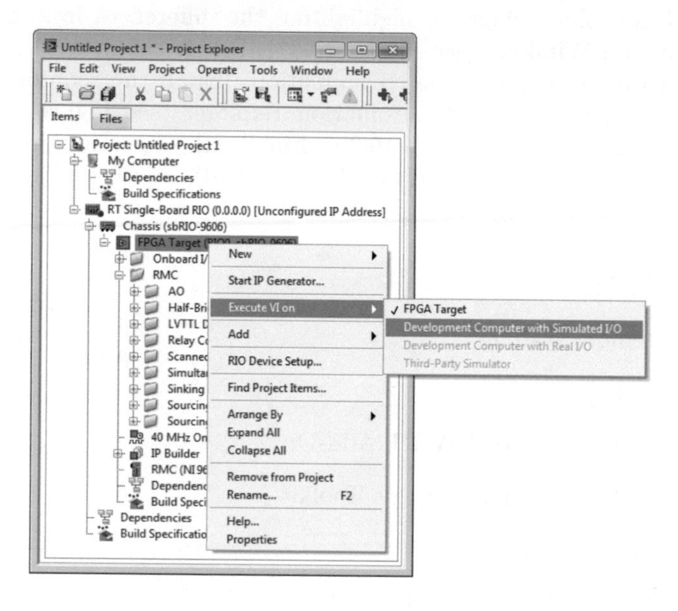

FIGURE 10.37
Project explorer

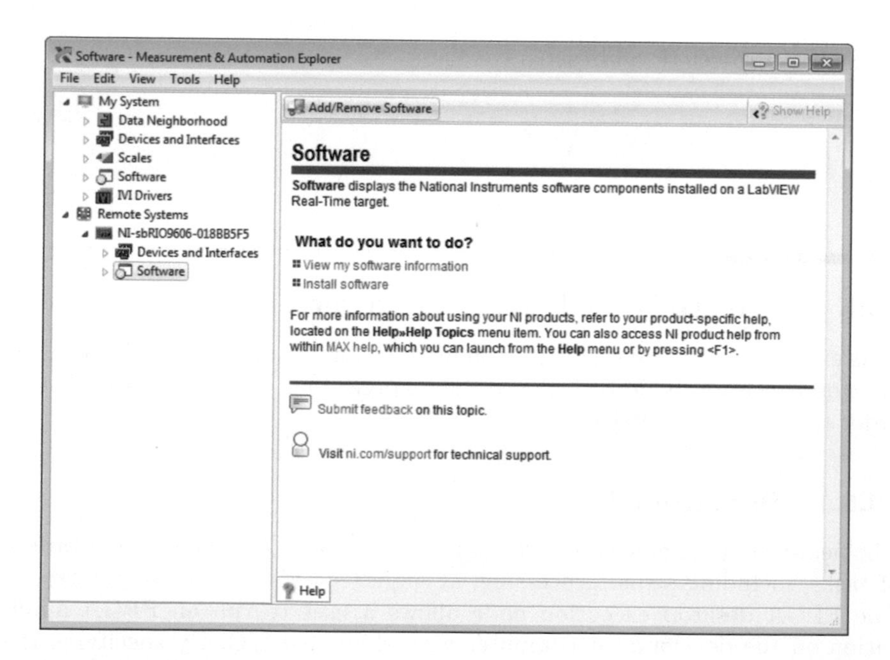

FIGURE 10.38
Measurements and automation explorer

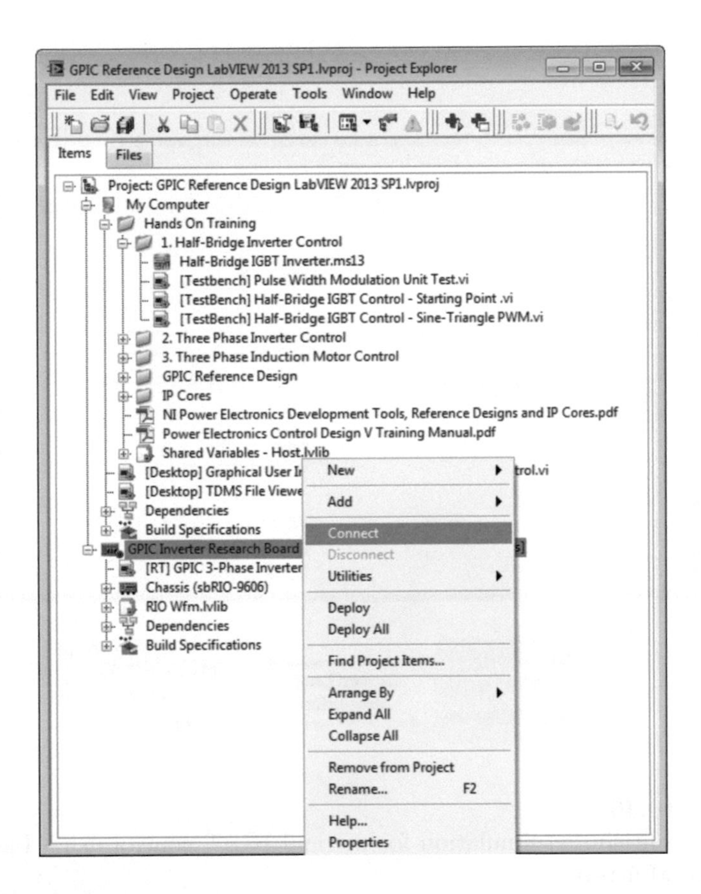

FIGURE 10.39
Connect the GPIC inverter board

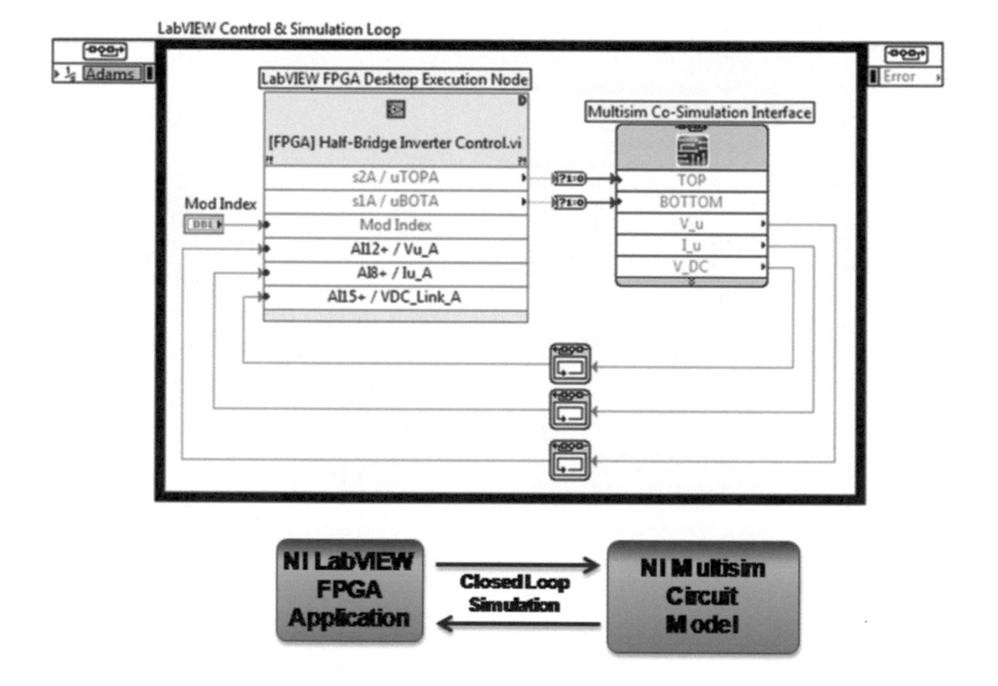

FIGURE 10.40
FPGA — Multisim co-simulation for manual IGBT control using LabVIEW FPGA and Multisim

the FPGA hardware [54]. Figure 10.41 shows the FPGA co-simulation. To effectively simulate an FPGA design using the FPGA desktop execution node, you must understand two time paradigms, real-world time and simulated time. Real-world time is the physical amount of time that elapses while something occurs. Because an FPGA is a programmable circuit, it takes a fixed amount of real-world time to execute such a circuit. To simulate a design, you can create a model that reflects the functional behavior of the FPGA and execute this model on a computer processor. However, such a model does not have the same real-world timing as the FPGA. To create a model that has the same real-world timing as the FPGA, you must use simulated time. Simulated time is an event-driven model of real-world time. When LabVIEW executes block diagram nodes, the execution causes simulated time to advance a certain number of steps. Clock ticks are the unit of simulated time where one clock tick is representative of a single cycle of the referenced FPGA clock [6].

For this exercise, you will need to use the LabVIEW FPGA in order to control the power inverter. A power inverter consists of a voltage source, two insulated-gate bipolar transistors (IGBTS), an inductor, a capacitor, and a resistor (see Figure 10.42). The IGBTs switch in order to change the current flowing through the resistive load while the inductor and capacitor filter and smooth the output signal. The power inverter performs DC-to-AC conversion.

10.5.2 Procedure

To run the half-bridge inverter control, a Multisim file and two TestBench VIs are needed. As shown in Figure 10.43, a project was designed. On the other hand, the Multisim file is used for studying the thermal conditions in the semiconductors, as shown in Figure 10.44. This Multisim file has to be constructed in order to create the co-simulation. In the Multisim editor environment, if you do not see the **LabVIEW Co-Simulation Terminals** pane (red box in the screenshot above), go to the **View** menu, select it, and then position it in the desired location (see Figure 10.45).

It is necessary to define a LabVIEW project. This file is the front panel of the half-bridge IGBT controller. In the LabVIEW project (see Figure 10.46), double-click on **[TestBench] Half-Bridge IGBT Control - Starting Point.vi**.

The goal of this example is to control the IGBTs (switches) in such a way as to avoid damaging the hardware. In addition to the switches, other controls include a slider for the modulation index and a button for stopping the program. Important indicators include the IGBT temperatures, a dial for the output voltage, both instant and RMS, as well as a chart that plots various waveforms from the circuit. Open the block diagram by pressing **CTRL+E** or by navigating to **Window≫Show Block Diagram** and observe the layout. Press **CTRL+H** to open the **Context Help** window and place the mouse over the blue box that reads **Desktop Execution Node**. This will cause the **Context Help** window to display information about the desktop execution

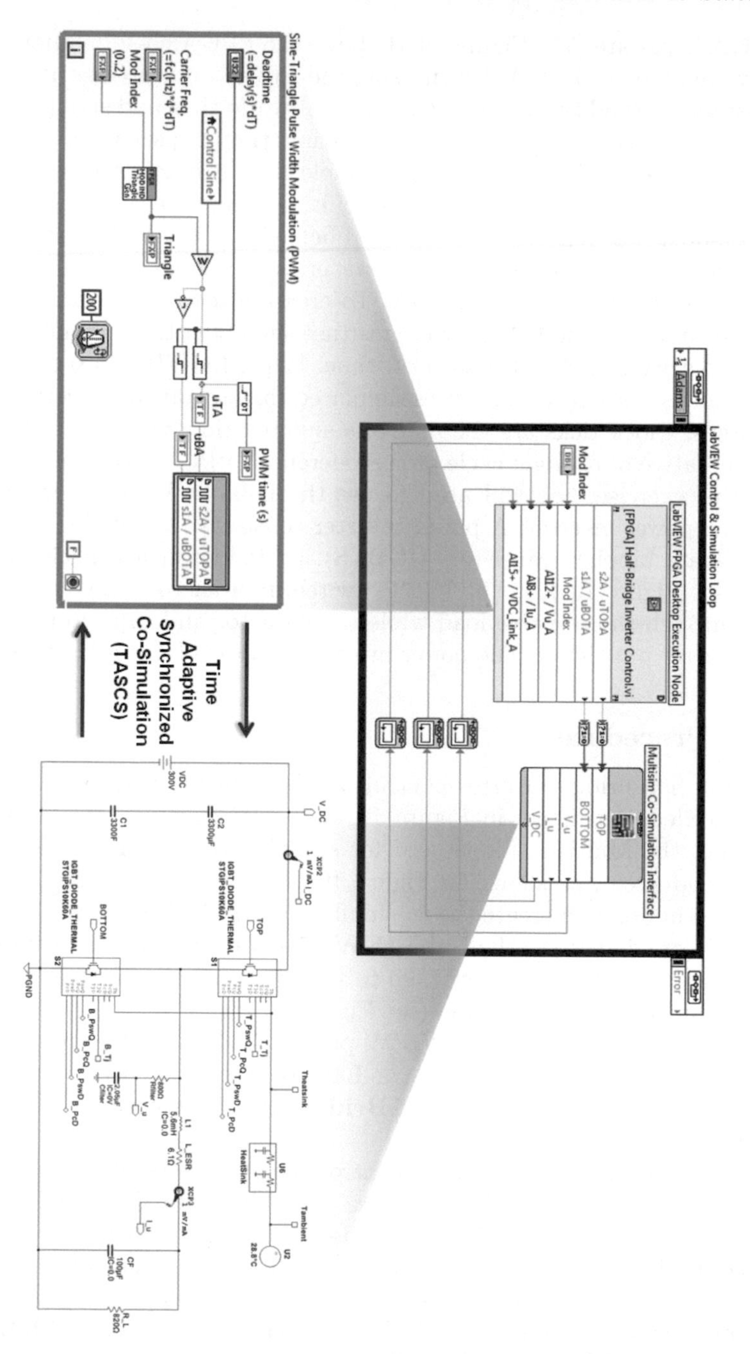

FIGURE 10.41
To enable high fidelity co-simulation, the LabVIEW FPGA desktop execution node interfaces to the LabVIEW FPGA application while the Multisim co-simulation interface connects to the power electronics circuit model. A desktop testbench application is shown.

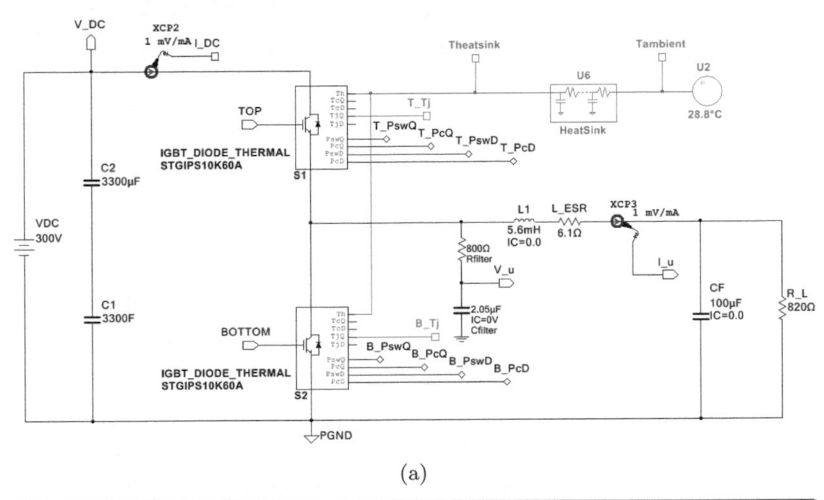

(a)

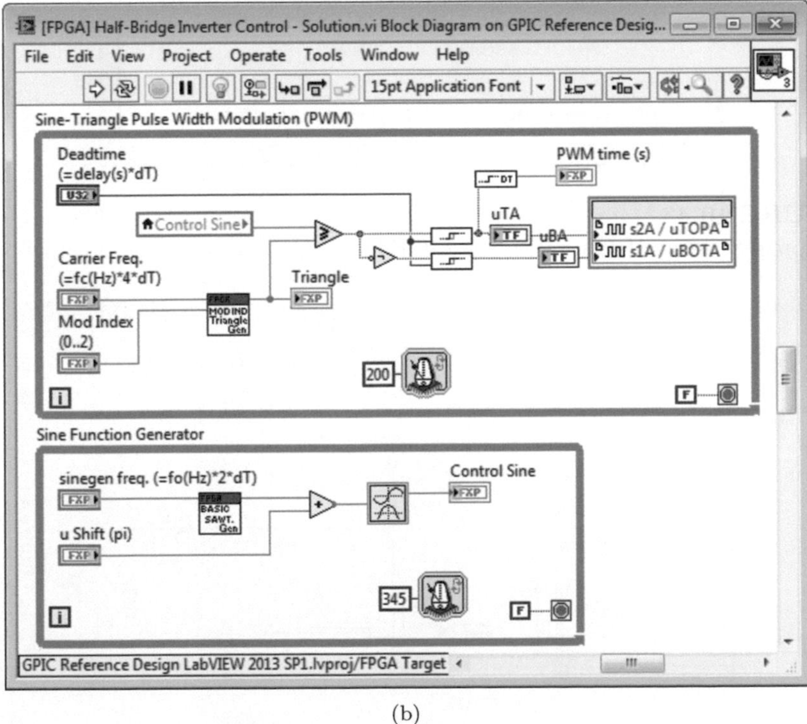

(b)

FIGURE 10.42
The Multisim power electronics circuit (a) and LabVIEW FPGA control application (b) are co-simulated with a common sense of simulated time

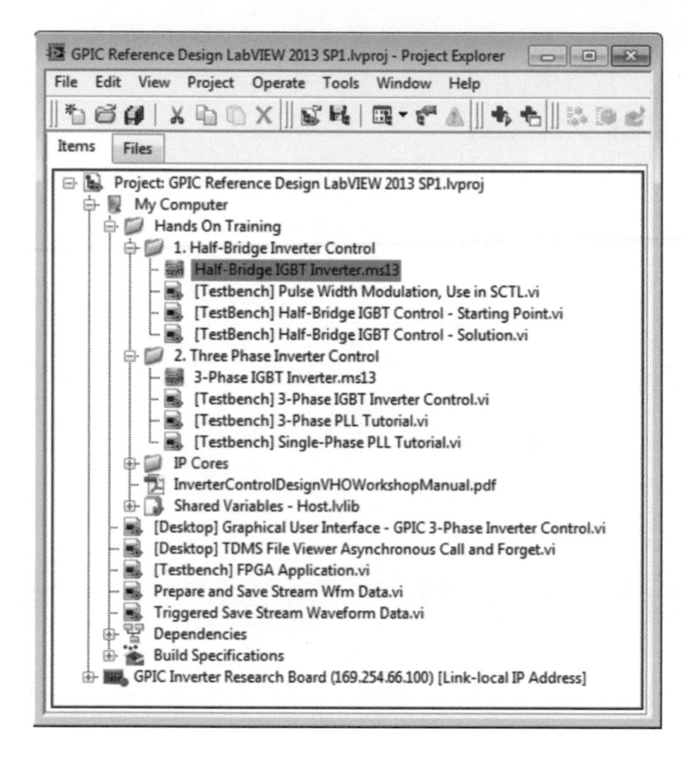

FIGURE 10.43
The project design for a half-bridge IGBT

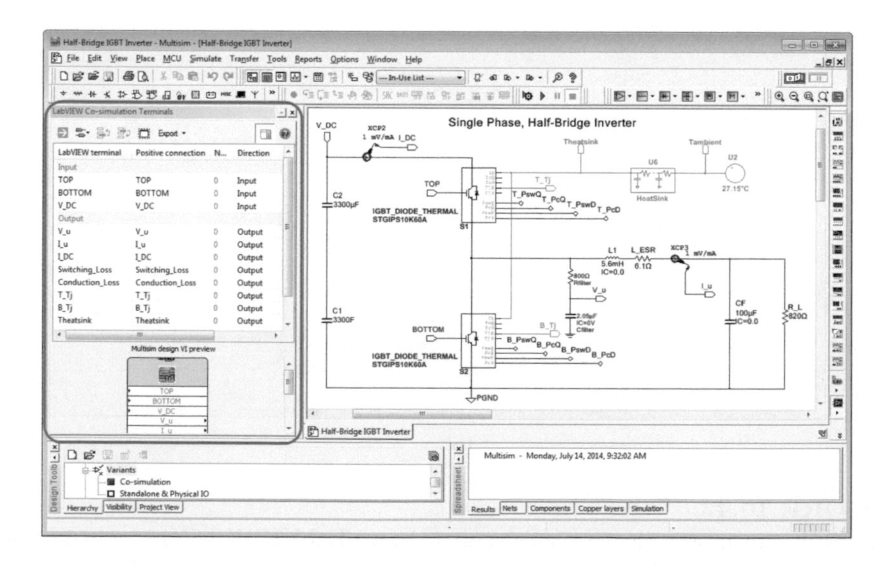

FIGURE 10.44
Thermal analysis of IGBTs using Multisim

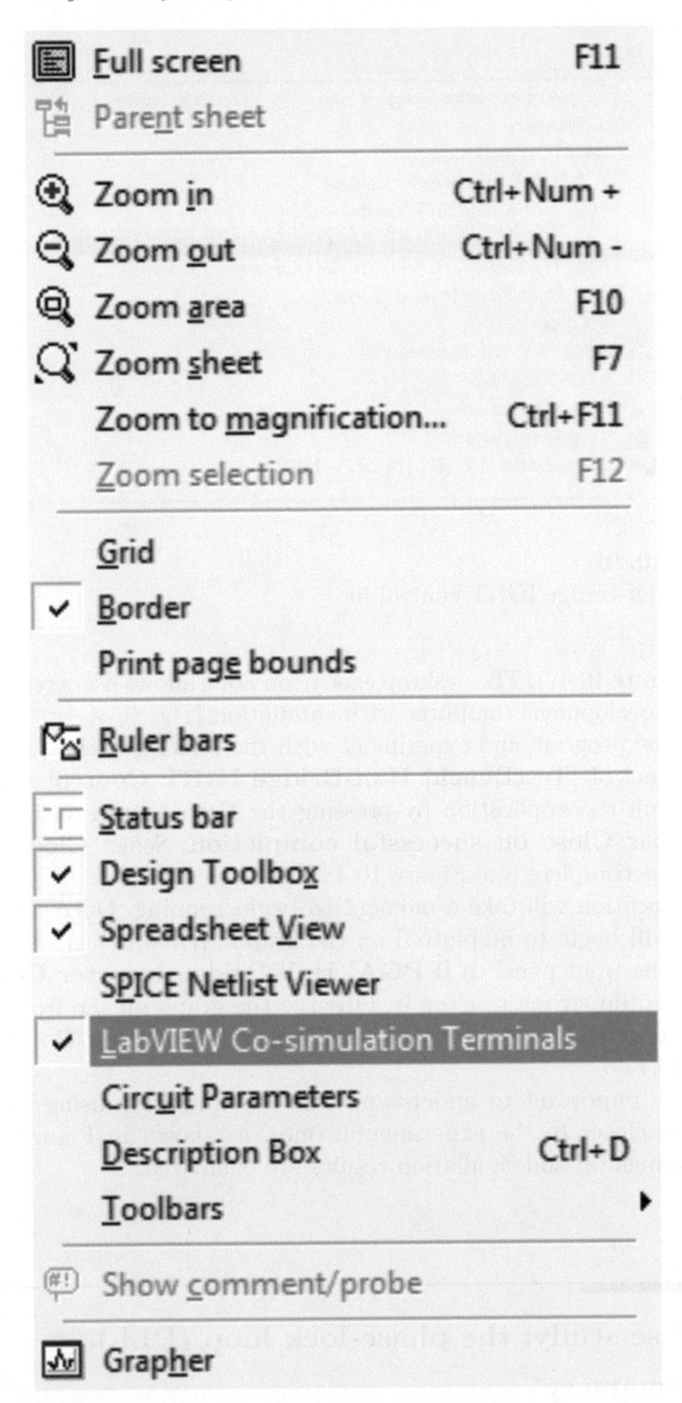

FIGURE 10.45
LabVIEW co-simulation terminals

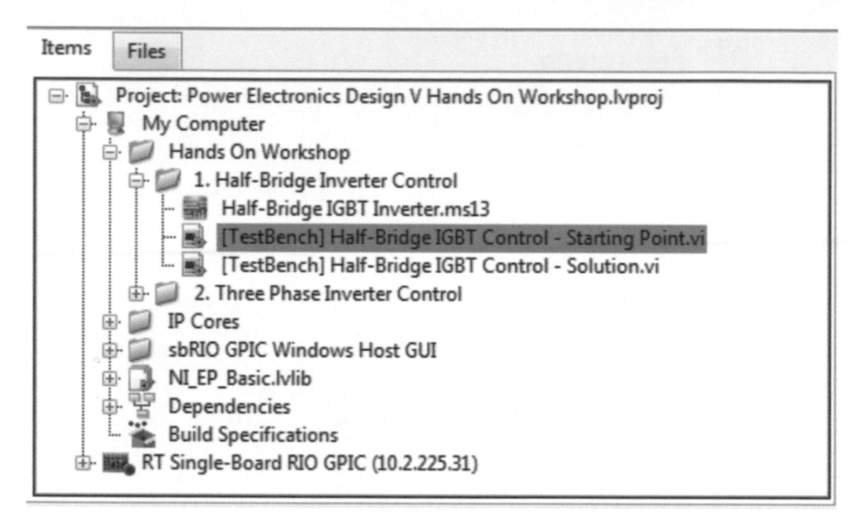

FIGURE 10.46
LabVIEW half-bridge IGBT control file

node (see Figure 10.47). The desktop execution node allows users to run FPGA VIs on the development computer with simulation I/O.

To run the program and experiment with the IGBT switches, navigate to the front panel of **[TestBench] Half-Bridge IGBT Control - Starting Point.vi**. Run the application by pressing the **Run** button on the toolbar. Check the box **Close on successful completion**. Select **Close** once the deployment is complete (see Figure 10.48.)

The application will take a moment to begin running. Once running, the waveforms will begin to be plotted on the graph. While the program is running, open the front panel to **[FPGA] Half-Bridge Inverter Control.vi**. Left-click the top switch to close it. Observe the graph on the front panel of the **[TestBench] Half-Bridge IGBT Control - Starting Point.vi**. (See Figure 10.49.)

It is very important to understand that co-simulation using FPGA can gives results closer to the experimental ones, as shown in Figure 10.50, in which experimental and simulation results are compared.

10.6 Case study: the phase-lock loop (PLL)

10.6.1 Summary

In this example, you will be creating your own 3-phase phase-locked loop (PLL) in order to understand how one might be implemented in a 3-phase

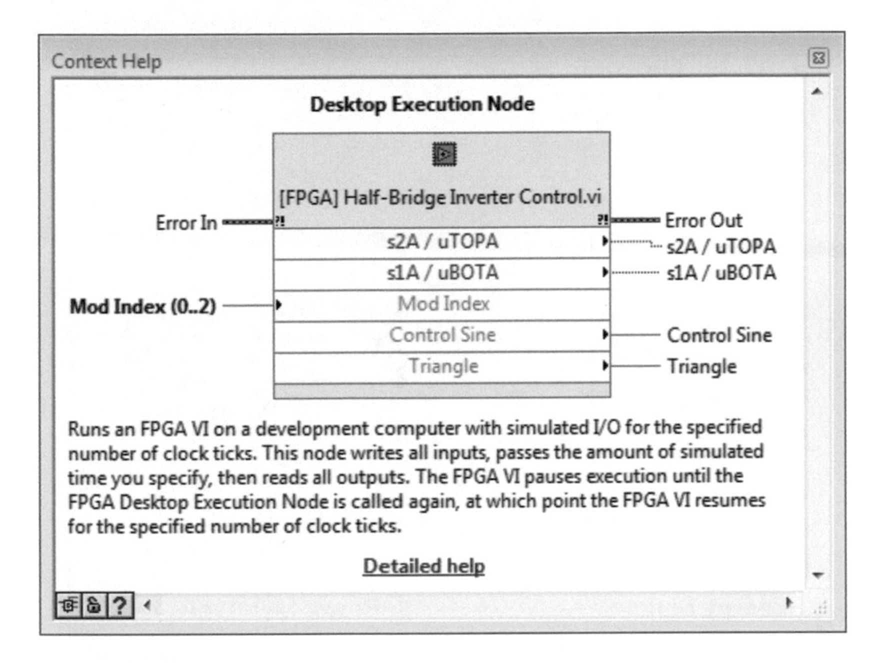

FIGURE 10.47
Desktop execution node

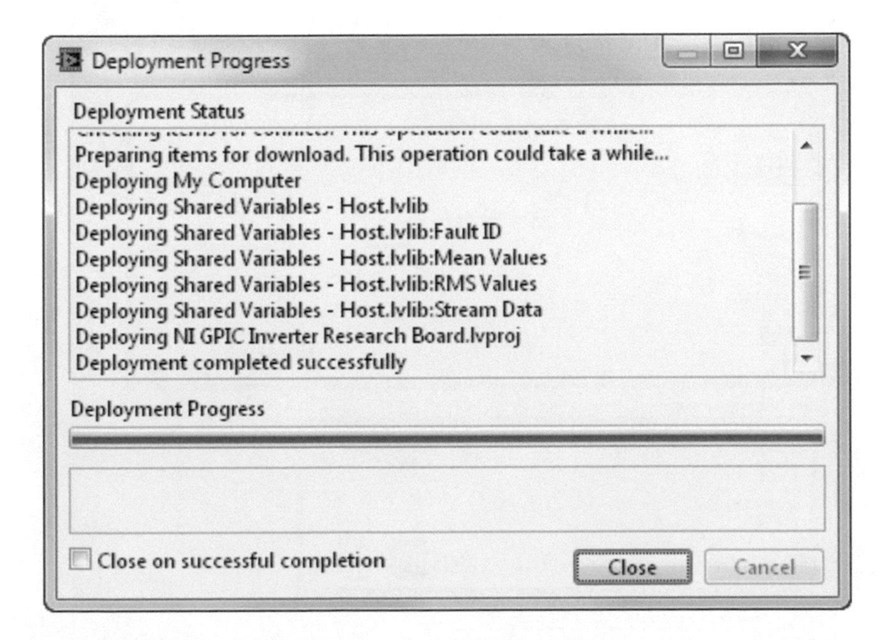

FIGURE 10.48
Deployment process

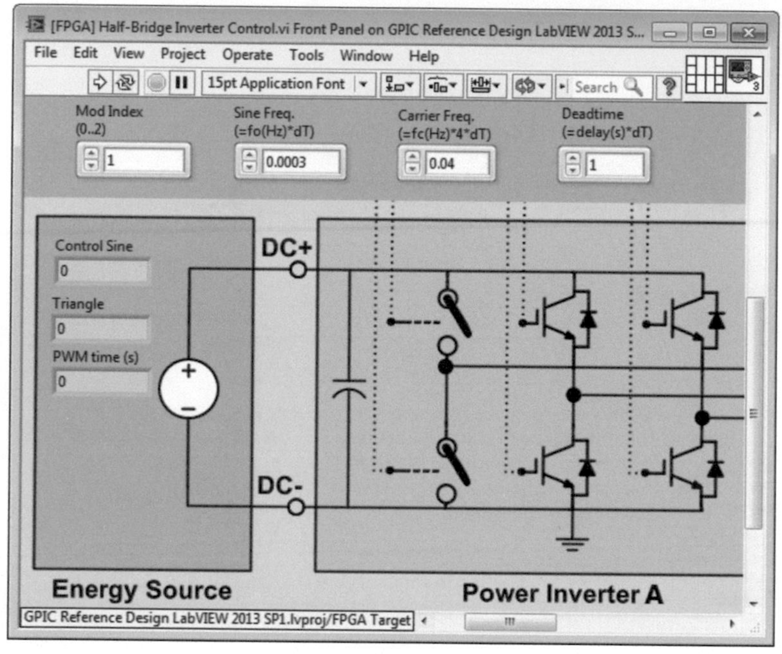

(a)

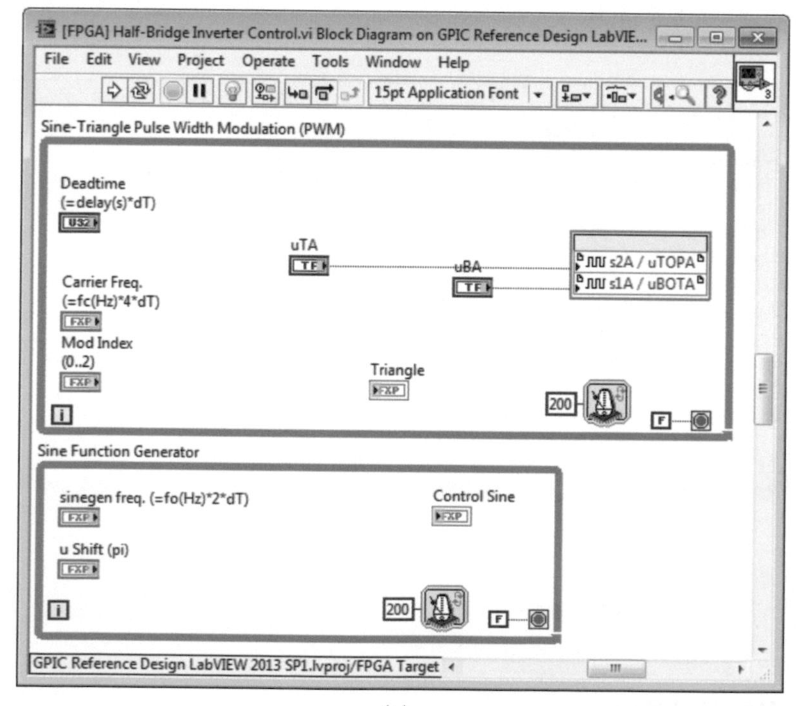

(b)

FIGURE 10.49
Frontal panel (a) and block diagram (b) of the co-simulation

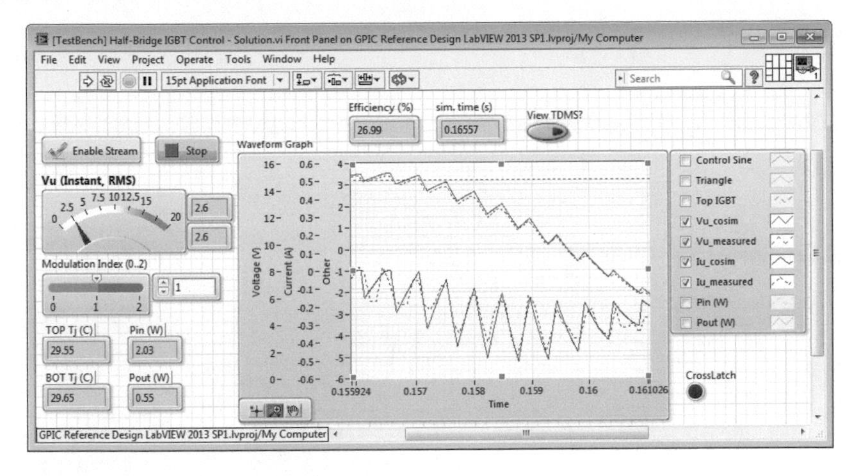

FIGURE 10.50
Comparing simulated against experimental results

inverter. Phase-locked loops are important to grid-tied power because in order to supply power to a grid, the inverter output must be phase locked to the grid frequency. Tasks you will do to complete this exercise include [54]):

- Implementing a 3-phase PLL
- Simulating a PLL application using the desktop execution node
- Experimenting by varying the input and outputs to your PLL application

10.6.2 Background

Phase-locked loops are a form of feedback control. A PLL produces an oscillating output, which begins operating with a set frequency. An input is then passed to the PLL to serve as a reference point. The PLL adjusts the output to then closely match the input phase. The longer the PLL runs, the more closely the PLL output is able to track the input, given that the input is stable. PLLs are useful in a variety of synchronous applications. Such applications include clock multipliers, demodulation, and power-grid phase locking. In order to connect an inverter to the grid, the inverter output's phase must be within a few degrees of the grid signal phase. This is where the PLL will be applied. By using the grid signal as a reference, the PLL is capable of generating a waveform that will be phase-locked to the grid signal. This output can then be used to drive PWM in order to generate the inverter output, as shown in Figure 10.51.

For creating a three-phase PLL, a project has to be created. There are a Multisim file and three VIs in the project, as shown in Figure 10.52.

The VI includes a front panel of the PLL application in which a vertical

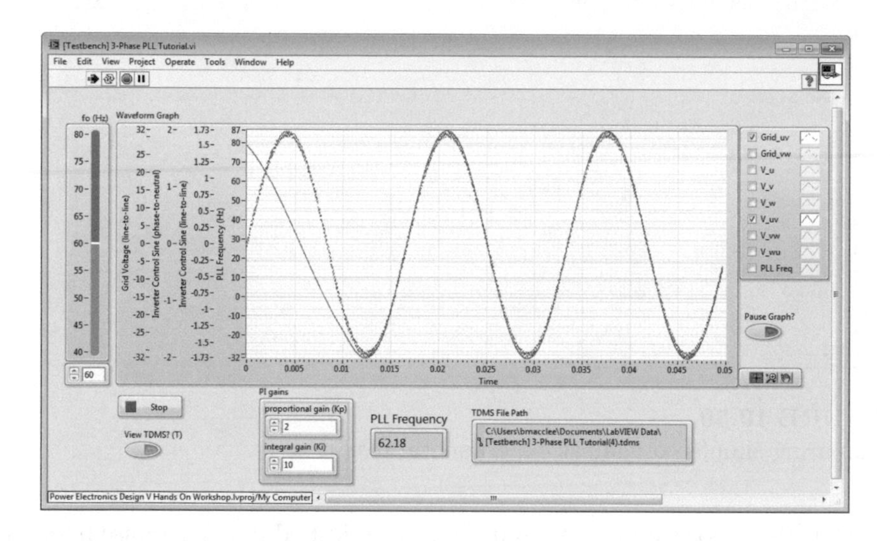

FIGURE 10.51
Frontal panel of a three-phase PLL

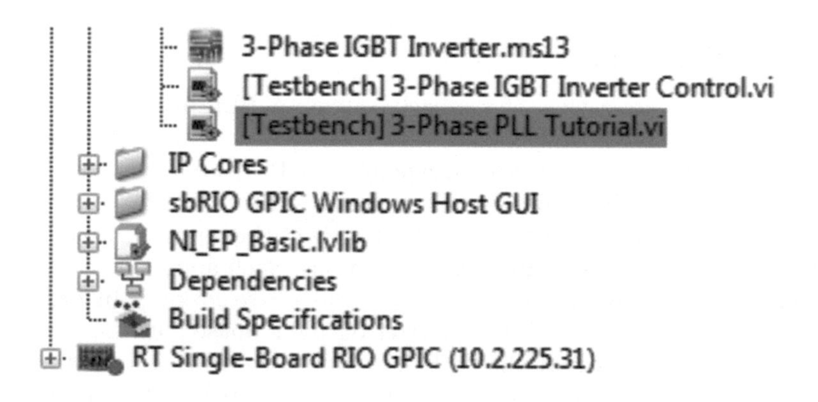

FIGURE 10.52
PLL files in the project

slider on the left to control frequency of the simulated power grid waveforms is added, as shown in Figure 10.53.

The **Desktop Execution Node** to open up the FPGA application **[FPGA] PLL Testbench.vi** is presented in Figure 10.54. For this application, the **Clock Ticks** setting is set to **325** ticks. Thus, for every 325 ticks of the 40 MHz FPGA clock, every 8.125 microseconds of simulated time, the Desktop Execution Node will return the register values for the items chosen in the **Selected Resources** dialogue (see Figure 10.55). Note that the FPGA resources you can choose from include both front panel controls and indicators as well as I/O resources. In this case, no changes to the configuration are necessary, so click **Cancel** to exit.

When the TestBench application is run, the PLL frequency is initially 30 Hz and takes a half cycle to begin tracking the noisy simulated grid waveforms. You can see the frequency of PLL in the indicator labeled **PLL Frequency**. Due to the noisy nature of the 3-phase grid waveforms and the aggressive tuning of the PLL proportional-integral (PI) control system, the PLL Frequency oscillates around the correct grid frequency (see Figure 10.56).

As it was presented, the cos-simulation is a powerful tool for designing power electronics in power systems which has to be considered during the creating process in power electronics. As a result, extremely complex systems can be created, such as the micro-grid experimental laboratory shown in Figure 10.57.

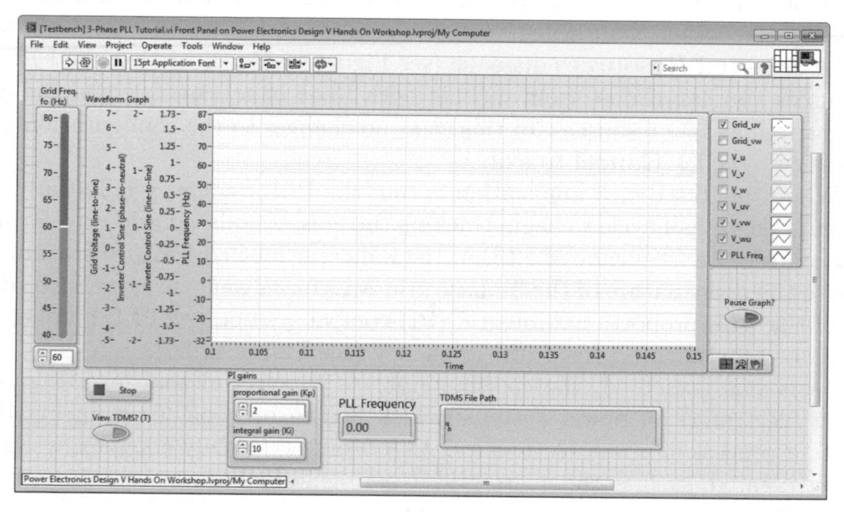

(a)

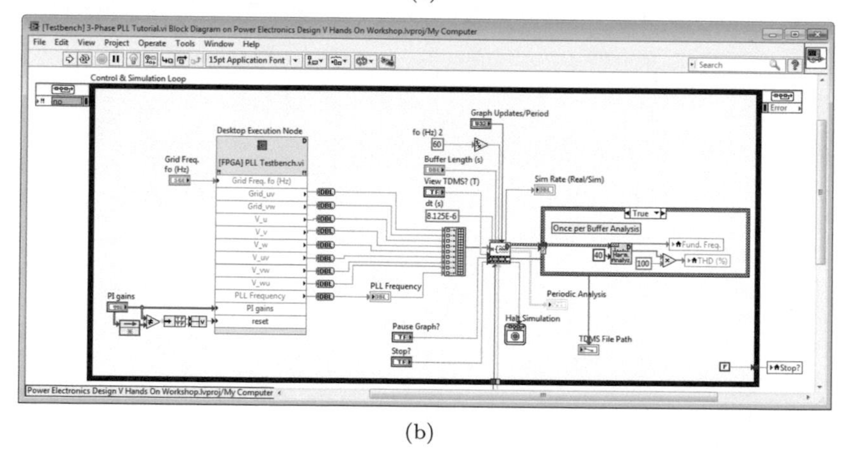

(b)

FIGURE 10.53
Front panel (a) and block diagram (b) for the PLL co-simulation

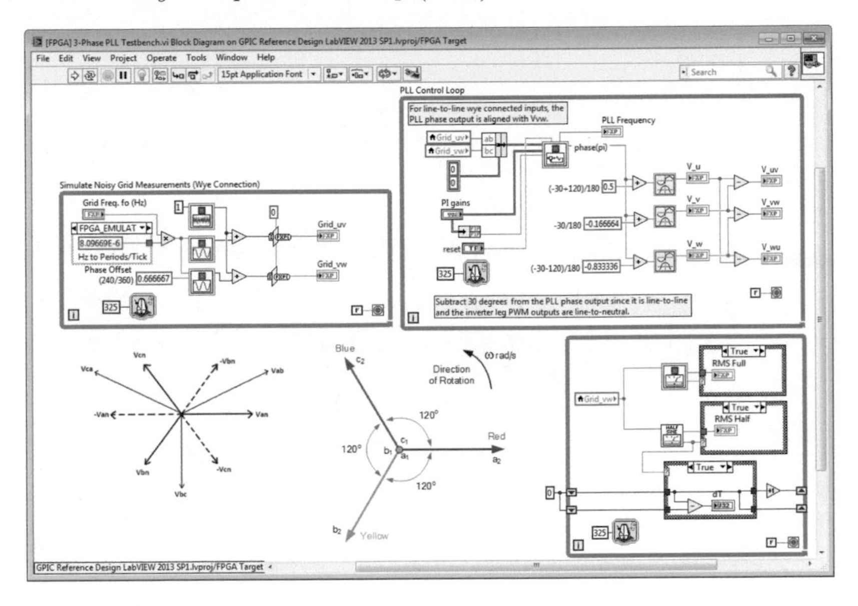

FIGURE 10.54
PLL control loop

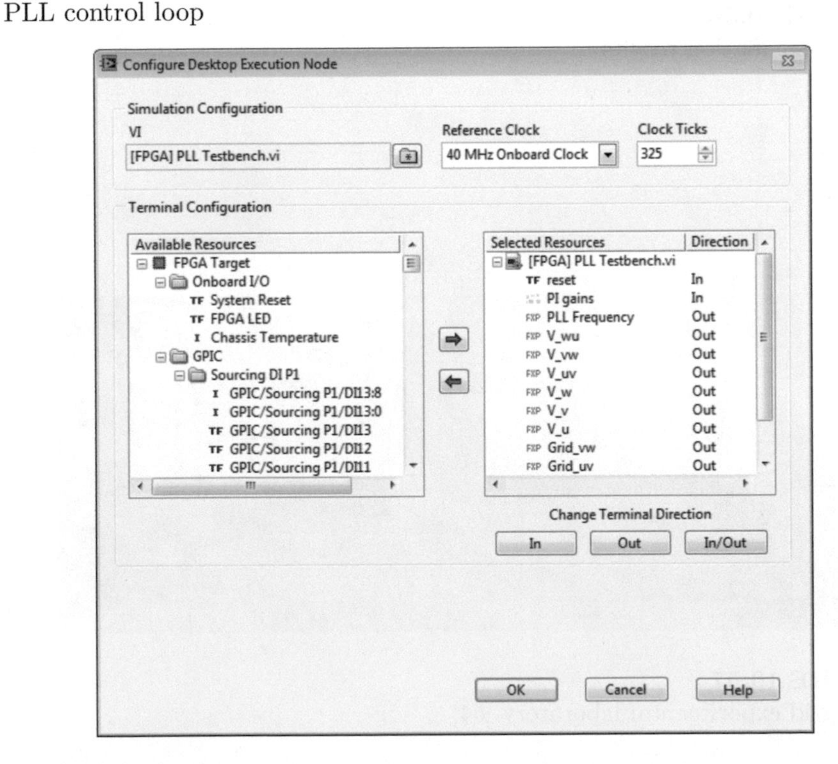

FIGURE 10.55
Configure desktop execution node

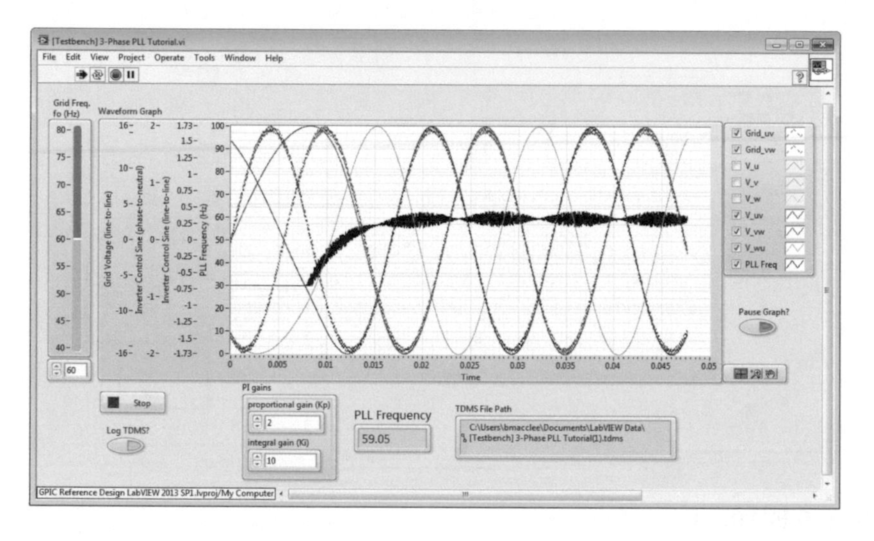

FIGURE 10.56
PLL Co-simulation results

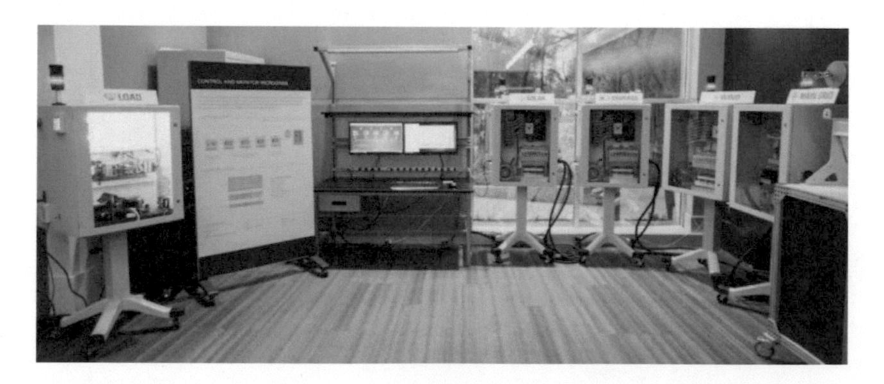

FIGURE 10.57
Micro-grid experimental laboratory [54]

Bibliography

[1] GE: GE wind turbine technology selected for new wind power station in Japan. http://site.ge-energy.com/about/press/en/2003_press/100103.htm. (Accessed 05/19/2017).

[2] NI multisim user manual — National Instruments. http://www.ni.com/pdf/manuals/374483d.pdf. (Accessed 05/19/2017).

[3] Simulation fundamentals: Cosimulation in NI multisim — National Instruments. http://www.ni.com/white-paper/6369/en/. (Accessed 05/19/2017).

[4] Stardom and fa-m3 ensure smooth supply of power to grid by wind farm equipped with large-capacity NAS batteries — Yokogawa Electric Corporation. https://www.yokogawa.com/library/resources/references/stardom-and-fa-m3-ensure-smooth-supply-of-power-to-grid-by-wind-farm-equipped-with-large-capacity-nas-batteries/. (Accessed 05/19/2017).

[5] Storage boosts the power of renewable energy — Renewable Energy World. http://www.renewableenergyworld.com/articles/2008/06/storage-boosts-the-power-of-renewable-energy-52716.html. (Accessed 05/19/2017).

[6] Using the LabVIEW FPGA desktop execution node — National Instruments. http://www.ni.com/white-paper/51859/en/. (Accessed 05/19/2017).

[7] www.cir-strategy.com/uploads/kuhrt.pdf. http://www.cir-strategy.com/uploads/Kuhrt.pdf. (Accessed 05/19/2017).

[8] Smart grid system report. Technical report, U.S. Department of Security, July 2009. Accessed: 2017-05-18.

[9] IEEE draft guide for smart grid interoperability of energy technology and information technology operation with the electric power system (EPS), and end-use applications and loads. P2030/D6.0, July 2011. Inactive-Draft.

[10] M.H. Baker L. Bohmann K. Clark K. Habashi L. Gyugyi J. Lemay A.S. Mehraban A.K. Myers J. Reeve F. Sener D.R. Torgerson R.R. Wood.

A-A. Edris, R. Adapa. Proposed terms and definitions for flexible ac transmission system FACTS. *IEEE transacitions on Power Delivery*, 12(4):1848 – 1853, 1997.

[11] Thomas Ackermann. *Wind power in power systems.* John Wiley & Sons, 2005.

[12] SA Ahmed and HO Mahammed. A statistical analysis of wind power density based on the Weibull and Ralyeigh models of "Penjwen Region" Sulaimani/Iraq. *Jordan Journal of Mechanical and Industrial Engineering*, 6(2):135–140, 2012.

[13] Vladislav Akhmatov. *Induction generators for wind power.* Multi-Science Pub., 2005.

[14] C. K. Alexander and M. Sadiku. *Fundamentals of Electric Circuits.* McGraw-Hill, 4th edition, 2009.

[15] P. M Anderson. *Analysis of faulted power systems.* IEEE Press, New York, 1995.

[16] European Wind Energy Association. *Pure power-wind energy targets for 2020 and 2030.* EWEA, 2011.

[17] Fernando D Bianchi, Ricardo J Mantz, and Hernán De Battista. *The Wind and Wind Turbines.* Springer, 2007.

[18] B.K. Bose. *Modern Power Electronics and AC Drives.* Eastern Economy Edition. Prentice Hall PTR, 2002.

[19] Tony Burton, Nick Jenkins, David Sharpe, and Ervin Bossanyi. Wind Energy Handbook. 2nd edition, 2011.

[20] Natural Resources Canada. Wind energy. `http://www.nrcan.gc.ca/energy/renewables/wind/7299`, October 2016. (Accessed 2017-10-03).

[21] Anders Carlsson. The back-to-back converter. *Lund: Department of Industrial Electrical Engineering and Automation Lund Institute of Technology*, 7, 1998.

[22] Juan Manuel Carrasco, Leopoldo Garcia Franquelo, Jan T Bialasiewicz, Eduardo Galván, Ramón Carlos PortilloGuisado, MA Martin Prats, José Ignacio León, and Narciso Moreno-Alfonso. Power-electronic systems for the grid integration of renewable energy sources: A survey. *IEEE Transactions on industrial electronics*, 53(4):1002–1016, 2006.

[23] J. Casazza, J. Casazza, and F. Delea. *Understanding electric power systems: an overview of the technology and the marketplace*, volume 13. John Wiley & Sons, 2003.

[24] J. J. Cathey. *Electric machines: analysis and design applying MATLAB.* McGraw-Hill Higher Education, 2000.

[25] Joe C.L. Chan and Duncan S. Wong. Cyber attacks and the smart grid. *IEEE Smart Grid Newsletter.*

[26] Stephen J Chapman. *Electric machinery and power system fundamentals.* McGraw-Hill, 2002.

[27] Alexander C.K. and Sadiku M.N.O. *Fundamentals of electric circuits.* McGraw-Hill, Boston, USA, 2009.

[28] Leon Clarke, James Edmonds, Henry Jacoby, Hugh Pitcher, John Reilly, and Richard Richels. Scenarios of greenhouse gas emissions and atmospheric concentrations. *US Department of Energy Publications*, page 6, 2007.

[29] Global Wind Energy Council and Global Wind Statistics. Electronic resource. *European Commission. Mode of access: <http://ec.europa.eu/energy/international/russia/dialogue/doc/2012_eu_russian_members_council.pdf>*. Date of access, 25, 2013.

[30] Kothari D. *Modern power system analysis.* Tata McGraw-Hill Education, 2003.

[31] US DoE. Annual energy outlook 2014 with projections to 2040. *Washington, DC: US Energy Information Administration, US Department of Energy*, 2014.

[32] M. A. El-Sharkawi. *Electric Energy: An introduction.* CRC Press, Boca Raton, FL, 3rd edition, 2013.

[33] H. Enríquez. *Tecnologías de generación de energía eléctrica.* Limusa, Mexico, 1st edition, 2009.

[34] Wagner C. F. and Evans R. D. *Symmetrical components.* McGraw-Hill Education, 1931.

[35] J. R. Fanchi. *Energy in the 21st century.* World Scientific Publishing Co., Singapore, 1st edition, 2006.

[36] Hewitson L. G., Brown M., and Balakrishnan R. *Practical power systems protection.* Oxford, 2005.

[37] Richard Gagnon, Gilbert Sybille, Serge Bernard, Daniel Paré, Silvano Casoria, and Christian Larose. Modeling and real-time simulation of a doubly-fed induction generator driven by a wind turbine. In *Intl. Conference on Power Systems Transients, Canada*, 2005.

[38] J. D. Glover, M. S. Sarma, and T. J. Overbye. *Power System Analysis & Design.* Cengage Learning, 2011.

[39] J. D. Glover, M. S. Sarma, and T. J. Overbye. *Power System Analysis & Design, SI Version.* Cengage Learning, 2011.

[40] Duane C Hanselman. *Brushless permanent magnet motor design.* The Writers' Collective, 2003.

[41] Lars Henrik Hansen, Peter Hauge Madsen, F Blaabjerg, HC Christensen, U Lindhard, and K Eskildsen. Generators and power electronics technology for wind turbines. In *Industrial Electronics Society, 2001. IECON'01. The 27th Annual Conference of the IEEE*, volume 3, pages 2000–2005. IEEE, 2001.

[42] L Holdsworth, XG Wu, Janaka Bandara Ekanayake, and Nick Jenkins. Comparison of fixed speed and doubly-fed induction wind turbines during power system disturbances. *IEE Proceedings-Generation, Transmission and Distribution*, 150(3):343–352, 2003.

[43] J Hossain. World wind resource assessment report. Technical report, World Wind Energy Association, 2014.

[44] Ekanayake J.B. *Smart grid: technology and applications.* Wiley, West Sussex, UK, 2012.

[45] Martin Kaltschmitt, Wolfgang Streicher, and Andreas Wiese. *Renewable energy: technology, economics and environment.* Springer Science & Business Media, 2007.

[46] E. W. Kimbark. *Power System Stability.* IEEE Press, 1995.

[47] D. P. Kothari and I. J. Nagrath. *Modern power system analysis.* Tata McGraw-Hill Education, 2003.

[48] Paul C Krause, Oleg Wasynczuk, Scott D Sudhoff, and Steven Pekarek. *Analysis of electric machinery and drive systems*, volume 75. John Wiley & Sons, 2013.

[49] Blackburn J. L. *Protective relaying: principles and applications.* M. Dekker, 1987.

[50] De la Rosa F.C. *Harmonics and power systems.* Taylor & Francis, Boca Raton, FL, 2006.

[51] E. Lorenzo and E.L. Pigueiras. *Electricidad solar: ingeniería de los sistemas fotovoltaicos.* Universidad Politécnica de Madrid, Instituto de Eneregía Solar, Spain, 1994.

[52] Joachim Luther, Michael Nast, M. Norbert Fisch, Dirk Christoffers, Fritz Pfisterer, Dieter Meissner, and Joachim Nitsch. Solar Technology. In *Ullmann's Encyclopedia of Industrial Chemistry.* Wiley-VCH Verlag GmbH & Co. KGaA, 2000.

[53] Weedy B. M., Cory B. J., N. Jenkins, Ekanayake J. B., and G. Strbac. *Electric power systems.* John Wiley & Sons, 2012.

[54] Brian MacCleery. NI power electronics control design v training workshop co-simulation — National Instruments. `http://www.ni.com/white-paper/6369/en/`, June 2014. (Accessed 05/19/2017).

[55] Patricia Ross McCubbin. EPA's endangerment finding for greenhouse cases and the potential duty to adopt national ambient air quality standards to address global climate change. *S. Ill. ULJ*, 33:437, 2008.

[56] Iulian Munteanu, Antoneta Iuliana Bratcu, N Cutululis, and E Ceanga. Optimal control of wind energy systems: towards a global approach. *IEEE Control Systems Magazine*, 2009.

[57] (U. S.) National Research Council and (U. S.) National Academy of Engineering. *Electricity from renewable resources: Status, prospects, and impediments.* National Academic Press, 1st edition, 2010.

[58] Vaughn C Nelson and Kenneth L Starcher. *Introduction to renewable energy.* CRC Press, Boca Raton, FL, 2015.

[59] Office of Energy Efficiency & Renewable Energy. Power america. `https://energy.gov/eere/amo/power-america`. (Accessed 2017-10-03.)

[60] John Paschal. *Practical guide to power factor correction and harmonics and your electric bill.* Ec and M Books, Overland Park, KS, 2001.

[61] G. L. Pollack and D. R. Stump. *Electromagnetism.* Addison-Wesley, 2002.

[62] Pedro Ponce, Arturo Molina, and Brian MacCleery. Integrated intelligent control and fault system for wind generators. *Intelligent Automation & Soft Computing*, 19(3):373–389, 2013.

[63] Pedro Ponce-Cruz and Fernando D Ramírez-Figueroa. *Intelligent Control Systems with LabVIEW™*. Springer Science & Business Media, 2009.

[64] Neumann R. *Symmetrical Component Analysis of Unsymmetrical Polyphase Systems.* Sir I. Pitman & Sons, 1939.

[65] Power Regen. Solar photovoltaic power system handbook. Technical report, REGEN power, Western Australia, 2011.

[66] Andre Richter, Erwin van der Laan, Wolfgang Ketter, and Konstantina Valogianni. Transitioning from the traditional to the smart grid: Lessons learned from closed-loop supply chains. In *2012 International Conference on Smart Grid Technology, Economics and Policies (SG-TEP)*. IEEE, December 2012.

[67] Maurice George Say. *Alternating current machines.* Halsted Press, 1976.

[68] P. C. Sen. *Principles of electric machines and power electronics.* John Wiley & Sons, New York, 2nd edition, 1996.

[69] Marcelo Godoy Simoes, Robin Roche, Elias Kyriakides, Sid Suryanarayanan, Benjamin Blunier, Kerry D. McBee, Phuong H. Nguyen, Paulo F. Ribeiro, and Abdellatif Miraoui. A comparison of smart grid technologies and progresses in Europe and the U.S. *IEEE transacitions on industrial electronics*, 48(4):1154 – 1162, 2012.

[70] Susan Solomon. *Climate change 2007—the physical science basis: Working group I contribution to the fourth assessment report of the IPCC*, volume 4. Cambridge University Press, 2007.

[71] Siddharth Sridhar, Adam Hahn, and Manimaran Govindarasu. Cyber–physical system security for the electric power grid. In *Proceedings of the IEEE*, volume 100. IEEE, 2012.

[72] R. Strzelecki and G. Benysek. *Power Electronics in Smart Electrical Energy Networks*, volume 1 of *Power Systems.* Springer, London, 1st edition, 2008.

[73] Aman Abdulla Tanvir, Adel Merabet, and Rachid Beguenane. Real-time control of active and reactive power for doubly fed induction generator (DFIG)-based wind energy conversion system. *Energies*, 8(9):10389–10408, 2015.

[74] S Tegen, E Lantz, M Hand, B Maples, A Smith, and P Schwabe. Cost of wind energy review, National Renewable Energy Laboratory. Technical report, Technical Report, 2011.

[75] DM Triezenberg. Electric power systems. Perdue University: West Lafayette, IN, 1978.

[76] Juan Carlos Vega de Kuyper and Santiago Ramírez Morales. Fuentes de energía, renovables y no renovables. Aplicaciones. 2014.

[77] Blume S. W. *Electric power system basics: for the nonelectrical professional.* Wiley-Interscience; IEEE Press, 2007.

[78] C. L. Wadhwa. *Electrical power systems.* New Age International, 2005.

[79] G.J. Wakileh. *Power systems harmonics: fundamentals, analysis, and filter design.* Springer, New York, 2001.

[80] Hayt W.H. *Engineering circuit analysis.* McGraw-Hill Higher Education, Boston, 2006.

[81] David C White and Herbert H Woodson. *Electromechanical energy conversion.* Wiley, 1959.

Index